Coal Mining: Advanced Technologies

Coal Mining: Advanced Technologies

Editor: Tommy McCarthy

Callisto Reference

www.callistoreference.com

Callisto Reference,
118-35 Queens Blvd., Suite 400,
Forest Hills, NY 11375, USA

Visit us on the World Wide Web at:
www.callistoreference.com

ISBN: 978-1-64116-271-5 (Hardback)

Printed in China.

Cataloging-in-Publication Data

Coal mining : advanced technologies / edited by Tommy McCarthy.
p. cm.
Includes bibliographical references and index.
ISBN 978-1-64116-271-5
1. Coal mines and mining. 2. Clean coal technologies. 3. Mines and mineral resources.
4. Energy industries--Technological innovations. I. McCarthy, Tommy.

TN802 .C63 2020
622.334--dc23

Table of Contents

Preface

Coal is a natural resource that is highly valued for its energy content. The steel and cement industries depend on coal for the extraction of iron from iron ore and for the manufacture of cement. The process of extracting coal from a coal pit is known as coal mining. The technical and economic feasibility of a coal extraction process depend on coal seam continuity, thickness, quality, depth, overburden characteristics, topography, surface drainage patterns, etc. Mining can be done on the surface or deep underground. Strip mining, contour mining and mountaintop removal mining are the three primary types of surface mining. There have been several strides in technological advancements in coal mining in recent times. The use of sophisticated sensing equipment for monitoring air quality, hazardous gas monitoring, gas drainage, improved mining techniques, and the employment of highly skilled task force adept at the use of complex state-of-the-art equipment has made coal mining productive and safe. This book provides comprehensive insights into coal mining. It discusses the fundamentals as well as modern approaches of coal mining. Those in search of information to further their knowledge will be greatly assisted by this book.

After months of intensive research and writing, this book is the end result of all who devoted their time and efforts in the initiation and progress of this book. It will surely be a source of reference in enhancing the required knowledge of the new developments in the area. During the course of developing this book, certain measures such as accuracy, authenticity and research focused analytical studies were given preference in order to produce a comprehensive book in the area of study.

This book would not have been possible without the efforts of the authors and the publisher. I extend my sincere thanks to them. Secondly, I express my gratitude to my family and well-wishers. And most importantly, I thank my students for constantly expressing their willingness and curiosity in enhancing their knowledge in the field, which encourages me to take up further research projects for the advancement of the area.

Editor

1

Coal mining in northeast India: an overview of environmental issues and treatment approaches

Mayuri Chabukdhara[1] · **O. P. Singh**[2]

Abstract Northeast India has a good deposit of sub-bituminous tertiary coal. The northeast Indian coals have unusual physico-chemical characteristics such as high sulfur, volatile matter and vitrinite content, and low ash content. In addition, many environmental sensitive organic and mineral bound elements such as Fe, Mg, Bi, Al, V, Cu, Cd, Ni, Pb, and Mn etc. remain enriched in these coals. Such characteristics are associated with more severe environmental impacts due to mining and its utilization in coal based industries. Environmental challenges include large scale landscape damage, soil erosion, loss of forest ecosystem and wildlife habitat, air, water and soil pollution. Several physical and chemical methods are reported in literature for the removal of mineral matter, total sulfur and different forms of sulfur from high sulfur coal in northeast India. This paper may help different researchers and stakeholders to understand current state of research in the field. Initiatives may be taken towards sustainable use of coal resources by adopting innovative clean technologies and by implementing effective control measures and regulatory policies.

Keywords Northeast India · Sub-bituminous coal · Environmental issues · Innovative technologies · Management and regulatory policies

1 Introduction

Coal is the most important and abundant fossil fuel in India. With increased population, growing economy and a quest for improved quality of life, energy demand in India is rising. Mining is not only fulfilling the increasing energy demand of industry, but also plays an important role in the economic development of the country (Chaulya and Chakraborty 1995). Power sector is the largest consumers of coal followed by iron, steel and cement segments in the last four decades (Fig. 1). Other smaller consumers include fertilizer, textile (including jute and jute products), paper and the brick industry. Coal mining and its utilization is associated with substantial environmental challenges as it creates significant and often irreversible impacts upon the terrestrial and aquatic environment.

Open cast or surface mining is dominant in India and it not only alters the nature of groundwater–surface water interactions but also contributes to major air pollutants to the atmosphere and results in dramatic changes in the landscape. Most coal mining districts in India have been declared as critically polluted areas (CPAs) by MoEF in 2009 (CSE 2012).

In northeast India, coal mining was initiated by Medlicott in 1869 and 1874 (Sarma 2005a, b). The Cenozoic coals in the northeast states of India with its unusual physico-chemical characteristics have been playing an important role in the Indian economy for the last few decades (Baruah 2009; Saikia et al. 2014a). Due to its unique properties and consequent environmental issues,

✉ Mayuri Chabukdhara
mayuri_chabukdhara@yahoo.co.in

[1] Department of Environmental Biology and Wildlife Sciences, Cotton College State University, Guwahati, Assam 781001, India

[2] Department of Environmental Studies, North Eastern Hill University, Shillong, Meghalaya 793022, India

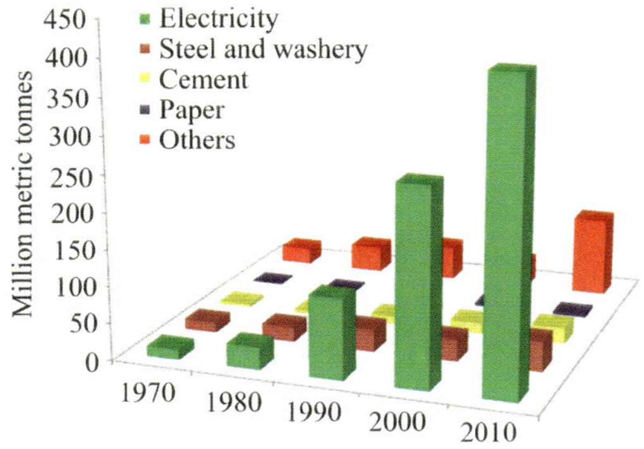

Fig. 1 Consumption of raw coal by different industries in India. *Source*: India Energy Book 2012

coals in northeast India draw special attention (Zamuda and Sharpe 2007; Saikia et al. 2014a).

The main objective of this paper is to summarize coal characteristics and associated environmental issues in northeast India. In addition, this paper also reviews the current state of research in the field of various treatment approaches to reduce environmental impacts of coal.

2 Coal distribution and its characteristics in northeast India

As on April 2014, India's inventory of coal resource was 300 Billion Tons (BT) comprising of: Proven—125 BT; Indicated—142 BT and Inferred—32 BT (Ministry of Coal 2014). Northeast India contributes 105 Million Tons (MT) of the Gondwana coal and 1492 MT of tertiary coal reserves. Meghalaya and Assam in northeast India contain 73 % of the total tertiary coal reserves. Nagaland and Arunachal Pradesh contribute 21 % and 6 % of the total tertiary coal reserves, respectively. Coal inventory of northeast India is given in Table 1.

Sub-bituminous tertiary coal of northeast India was deposited under the influence of marine environment

Table 1 Northeast India's coal inventory of Gondwana and tertiary coal in million tones (Ministry of coal, India as on April 1, 2014)

Coalfield	State	Proved	Indicated	Inferred	Total
Gondwana	Assam	0	4	0	4
	Sikkim	0	58	43	101
	Total	0	62	43	105
Tertiary	Arunachal Pradesh	31	40	19	90
	Assam	465	43	3	511
	Meghalaya	89	17	471	577
	Nagaland	9	0	307	316
Total		594	99	799	1492

Source Ministry of coal (2014)

(Rajarathnam et al. 1996). These coals have high sulfur and low ash content, with high organic sulfur, due to the influence of marine sources during diagenesis (Chandra et al. 1983; Singh and Singh 2000; Ward et al. 2007; Widodo et al. 2010). Coal can be termed as low sulfur (<1 % sulfur content), medium sulfur (1 %–<3 % sulfur content) and high sulfur coals (>3 % sulfur content) based on their sulfur contents (Chou 2012). In addition to high sulfur content, northeastern coals have a high content of volatile matter and vitrinite contents, yielding double the amount of tar in comparison to other Indian coals. Mining of these coals generates a large volume of waste materials. These coals generally contain 2 %–8 % total sulfur, where 75 %–90 % is of organic sulfur, while the rest is in inorganic form viz. sulfate and pyritic sulfur (Baruah and Khare 2007a). Ledo and Baragolai coal of Makum coalfield, Assam, India contains 28.2 % and 21.5 % of inorganic and 71.7 % and 78.5 % of organic sulfur, respectively (Baruah et al. 2006). In a study, proximate and ultimate analysis of coal collected from Makum coalfield, Assam showed 3.31 %, 2.95 % and 2.16 % of ash content, total sulfur and organic sulfur, respectively (Saikia et al. 2014a). The volatile matter and vitrinite content of the Makum coalfield, Assam were 42.3 % and around 93 %, respectively (Saikia et al. 2014a). Total sulfur of Namchik coalfield, Arunachal Pradesh ranged 1.23 %–4.84 %, with organic sulfur constituting ~41 %–74 % of total sulfur and volatile matter ranged 41.8 %–46.6 % (Chandra et al. 1984). Similarly, in Bapung coals of Jaintia Hills, Meghalaya, organic sulfur was more abundant among the different sulfur species constituting an average 62 % of organic sulfur of the total sulfur content of 4.59 % in it (Ahmed and Rahim 1996). Total sulfur sometimes exceeding 7 wt% out of which the organic sulfur content accounts for about 75 % and the rest is inorganic sulfur. Tiru valley coals of Nagaland, India are sub-bituminous to bituminous-D in rank characterized by low to medium moisture (4 %–7 %), moderately high volatile matter content (22 % and 42 %) and high sulfur (5 %–11 %) content (Singh et al. 2012a, b). Ash, volatile matter and total sulfur content of Northern Mongchen and Moulong Kimong coalfields, Nagaland, India ranged 2.01 %–19.5 %, 34.9 %–44.8 % and 3.23 %–5.21 %, respectively (Das et al. 2015).

In addition to high sulfur and volatile matter, and low ash content of northeast coal, many environmental sensitive organic and mineral bound elements remain enriched in these coals that can cause air, water and land pollution. Sub-bituminous coals of Assam obtained from Makum coalfield showed that Fe, Co, Ni, Cu and Zn are significantly mineral bound, Mg, Ca and Mn are organic bound, while Cd is 50 % bound to either organic or mineral matter (Baruah et al. 2003). The aqueous leaching of these coals showed their tendency to atmospheric weathering and highly acidic water formed during the leaching process enhanced the mobilization of associated trace and heavy metals (Fe, Mg, Bi, Al, V, Cu, Cd, Ni, Pb, and Mn) above the regulatory levels (Baruah et al. 2006). Element concentrations such as Cr, Mn, Ni, Cu, Zn, As and Pb in coals obtained from Makum coalfield, Assam were 5, 23, 5, 2, 27, 1 and 4 mg/kg, respectively. Concentrations of these elements in coals from Moulong Kimong coalfield, Nagaland were 4, 2289, 3, 2, 49, 2 and 1 mg/kg, respectively (Saikia et al. 2014a). Study further indicated that many of these elements were associated with hematite, magnetite, and goethite in the coals.

3 Environmental issues associated with coal mining and its utilization in the region

Unscientific mining of minerals poses a serious threat to the environment, resulting in reduction of forest cover and loss of biodiversity, soil erosion and pollution of air, water and land. The primitive and unscientific 'rat-hole' method of mining adopted by private operators and related activities have caused large-scale environmental degradation and severe ecosystem destruction in Meghalaya (Swer and Singh 2003, 2004; Sarma 2005a, b). Large scale denudation of forest cover, scarcity of water, air and water pollution, degradation of soil and agricultural lands, land subsidence, haphazard dumping of coal and overburden are some of the conspicuous environmental implications of coal mining in north eastern coal mines of Meghalaya, India (Swer and Singh 2004). Based on a study in the Nokrek Biosphere Reserve in Meghalaya, India, it is revealed that coal mining has adversely affected the vegetation and the density of trees, shrubs and herbs in mined areas (Sarma and Barik 2011).

The mining and cleaning of coal at local processing sites creates large quantities of ambient particulate matter (Ghose and Banerjee 1995; Ghose and Majee 2000). Opencast mining operations contribute major air pollutants to the atmosphere and are responsible for environmental degradation by deteriorating the air quality in respect to dust, fine coal particles and other gaseous pollutants (Mukhopadhyay et al. 2010). The major sources of air pollution in coal mining area include: drilling and blasting, loading and unloading of coal and overburden, movements of heavy vehicles on haul roads, dragline operations, crushing of coal in feeder breakers, presence of fire, exposed pit faces, wind erosion and exhaust of heavy earthmover machinery (Nair and Sinha 1987; Ghose and Majee 2007; Huertas et al. 2011). According to Ghose and Banerjee (1997), air pollution caused by washeries is more acute than any other coal processing operations.

Based on a report on ambient air quality around northeastern coalmines in Margherita, Assam, the maximum daily average values of SPM (Suspended Particulate Matter), RPM (Respirable Particulate Matter), SO_2 and NO_x were found to be 214, 60, 25 and 52 $\mu g/m^3$, respectively (Envirocon 2010). Except for SPM, all other values were within CPCB guidelines (CPCB 2009). Atmospheric concentration of gaseous NH_3, SO_2 and NO_2 released from the mining activities in open cast mine area of Tirap colliery, Margherita (Assam), ranged between 4.7–40.03, 1.47–6.14, and 1.92–2.40 $\mu g/m^3$, respectively, and particulate NH_4^+ in PM_{10} and $PM_{2.5}$ ranged between 0.02–0.07 and 0.008–0.03 μ g/m^3, respectively (Sarmah et al. 2012). The study further suggested that low emission and deposition of NO_x and SO_x prevents the greater formation of acidic species due to neutralization with NH_4^+. Source apportionment of $PM_{2.5}$ levels at the suburban site of northeast India (Khare and Baruah 2010) showed that largest contribution to aerosol mass in $PM_{2.5}$ is from crustal sources (38 %) followed by coal combustion (26 %), industrial and vehicular emissions (19 %), wood burning (9 %) and secondary aerosol formation (8 %). Among different elements, emissons of Te, Fe, Mn, Cd, Sn and Sb were related to coal burning (Khare and Baruah 2010).

Coke industry is one of the major coal utilization industries in northeast India. To assess the impact of coke oven burning high sulfur and volatile matter containing coal on ambient air quality, levels of SO_2, $PM_{2.5}$ and trace metals were investigated (Khare and Baruah 2011). The study showed that total emissions of $PM_{2.5}$, total carbon (TC), black carbon (BC) and organic carbon (OC) ranged between 72–306, 49–217, 0.71–2.9 and 48–214 t/year, respectively and the concentration of trace metals was in the decreasing order as: $Te > Mn \sim V > Cr > Co > Mo > Cu > Zn \sim Sb > Sn > Cd \sim Ni > As > Se > Hg$. The study further indicated that emission rates of metals were dependent on the volatility of the metals, condition of coke ovens and rank of coal (Khare and Baruah 2011).

In addition to air pollution, problems of AMD (Acid Mine Drainage) are intensely localized in the coalfields of northeast India, where ecology of the surrounding area is badly disrupted. The rejects and coals dumped near the pit entrance are exposed to the environment. Being highly enriched with sulfur, pyrite present in these materials is oxidized and hydrolysed and therefore is well known for the generation of AMD (Tiwary 2001; Baruah et al. 2005, 2006; Baruah 2009; Baruah and Khare 2007a). Metals concentrations in mine water in India and the world is shown in Table 2. As it is visible, metals such as Fe, Cu, Mn, Zn, Ni and Pb in mine water of northeast India (Jaintia and Makum) showed higher concentrations as compared to other mining sites in India. Zn and Pb showed the

Table 2 Metal contents in mine water (μg/L) in few coal mines in northeast India, India and the world

Parameter	Fe	Cu	Mn	As	Zn	Ni	Pb	Cr	Cd	References
Jaintia coalfield (Meghalaya, India)	118,400	320	4070	–	4220	1080	430	60	30	Sahoo et al. (2012)
Jharia coalfield (Jharkhand, India)	423	32.3	136	3.4	106.1	17.6	14.9	8.1		Singh et al. (2009)
Raniganj (West Bengal, India)	329	18.8	39.4	10.06	60	45.6	22.6	44.6		Singh et al. (2009)
West Bokaro coalfields (Jharkhand, India)	652	46	1431	7.21	194	154	34.3	81.2		Singh et al. (2009)
Makum (Assam, India)	105,300	310	10,200		1530	3120	270	56	35	Equeenuddin et al. (2010)
Karnen (Iran)	192,500	350	30,900		2070	1060	180	850	18	Shahabpour et al. (2005)
Dogye coalmine (Korea)	176,300	430	8360		2120					Chon and Hwang (2000)

maximum concentrations in Jaintia coalfield of Meghalaya, northeast India and Ni showed the maximum level in Makum coalfield of Assam. Such high concentrations of metals in these sites can be attributed to higher leaching under acidic conditions in these coalfields. However, elemental contents in leachate water are controlled by three factors: the oxidation rate of pyrite, the acidity of the leachate water and the mineralogy of the rejects (Baruah and Khare 2010). Further, it depends on the element content in the coal. Concentration of toxic elements present in northeastern and other coals in India is shown in Table 3. Toxic metal such as Cd showed the maximum level in Jaintia coal of Meghalaya, northeast India. In a study on elemental leaching of Meghalaya coals, elements such as Al, P, S, K, Ti, Cr, Co, Zn showed negative correlations with pH (Baruah and Khare 2010).The release of Al, Si, P, Cl, K, Ti, Mn, Co and Ni concentrations in the leachates depends on pyrite oxidation and dissolution (Yue and Zhao 2008), whereas Cd, Sn, Sb and Te contents in the leachates are mainly controlled by adsorption on Fe hydroxides, which is indirectly influenced by pH. The concentrations of trace and potentially harmful elements (Sb, As, Cd, Cr, Co, Cu, Pb, Mn, Ni, V, and Zn) in the Meghalaya coals mine rejects ranged (mg/kg): 11.1–12.6, 1.3–25.9, 5–5.1, 259–361, 20.9–22, 23.6–32.9, 98–149, 87–104, 36.4–58, 50–55 and 35.8–55, respectively, and among these Sb and Cd showed high enrichment factor showing build up in the environment (Baruah and Khare 2010).

Impact of AMD in the streams and groundwater at the vicinity of collieries is a growing problem in northeast India. The Meghalaya State Pollution Control Board, Shillong (MSPCB 2007) reported a case of massive fish death in Lukha River on the eastern border of Jaintia Hills district, which was attributed to AMD contaminating the stream water and sediments. Swer and Singh (2003, 2004) have reported the lack of commonly found aquatic life forms such as fish, frogs and benthic macroinvertibrate such as Plecoptera, Ephemeroptera and Tricoptera in water bodies of coal mining areas in Jaintia Hills, Meghalaya. Overall, socio-economic and ecological impacts in the area includes: severe scarcity of freshwater resources for domestic use and drinking purposes by the local community causing breach of basic human right; lack of aquatic life in many rivers and streams and reduced vegetation diversity; decreased agricultural productivity etc. (Swer and Singh 2004). Swer and Singh (2004) further reported that water quality in the Jaintia Hills of Meghalaya is highly affected as evidenced by low pH (in the range of

Table 3 Concentration of some toxic elements present in northeast India and other Indian coals/lignite (mg/kg)

Parameters	As	Cu	Mn	Zn	Ni	Pb	Cr	Cd	References
Jaintia coalfield (Meghalaya, India)	1–3	2.8–40	36.6–81.5	8.5–36.6	2–9.8	2.4–13.7	17.9–55.5	5	Baruah and Khare (2010)
Makum coalfield (Assam, India)	0.04–0.24	9.86–30.35	15.27–63.81	–	–	5.06–24.13	–	–	Mukherjee and Srivastava (2005)
Jammu and Kashmir, India	9.5	16.7	39	17.3	42.5	13.5	31.5	1.8	Banerjee et al. (2000)
Damodar Koel Valley coal	8.2	21.4	57.7	33.3	28	17.9	47.5	2.2	Banerjee et al. (2000)
Wardha Godavari Valley coal	2.1	29.5	58.6	29.2	25	4.5	54.5	2.8	Banerjee et al. (2000)
Pench Kanhan Tawa Valley coal	5.8	24.3	85	26	22.7	10.2	33.7	2.1	Banerjee et al. (2000)

3–5), high conductivity, high concentration of sulphates, iron and other toxic metals, low dissolved oxygen (DO) and high biological oxygen demand (BOD). Such low pH, low DO, higher sulphate content and turbidity in water of coal mining areas are affecting the aquatic life.

Singh and Sinha (1992) reported variation of pH in northeastern coalfields, pH 2.8–4.1 in Churcha, pH 4.2–5.0 in West Chirimir, pH 5.2–5.6, pH 5.3–6.0 in Rakhikhol and pH 4.0–4.6 in Gorbi coalfields. Highly acidic mine water with high sulphate (up to 1500 mg/L) and Fe (40 mg/L) were reported in Margherita group of mines in Assam (Rawat and Singh 1982). Bhole (1994) reported pH of 3.9, 3.10 and 4.3 in Ledo, Tirap and Bargolia mines of Assam. Based on a similar study carried out in Makum coalfields in Assam by Equeenuddin et al. (2010), it was found that the mine discharges were highly acidic (up to pH 2.3) to alkaline (up to pH 7.6) in nature with high concentration of SO_4^{2-} and mine water was highly enriched with Fe, Al, Mn, Ni, Pb and Cd. In addition, ground water close to the collieries and AMD affected creeks were highly contaminated by Mn, Fe and Pb but major rivers were not much impacted by AMD due to their large volume of water. Different physico-chemical parameters of surface and groundwater near coalfields in northeast and other parts of India are shown in Table 4. As can be seen in Table 4, pH of surface water near Jaintia coalfield, Meghalaya, India is highly acidic as compared to surface water in other sites in India. The maximum concentrations of metals detected in groundwater near Makum coalfield, Assam, India was (mg/L): 0.018 for Cr, 0.2 for Ni, 0.108 for Zn, 2.18 for Mn, 3.9 for Fe, 1.1 for Al, 0.061 for Pb, and 0.009 for Cu, in river water, the maximum concentrations were (mg/L): 0.06 for Ni, 0.016 for Zn, 0.94 for Mn, 2.47 for Fe, 0.42 for Al, 0.017 for Cd, 0.056 for Pb and 0.021 for Cu (Equeenuddin et al. 2010). In a study by Abhishek et al. (2006), water quality parameters in groundwater in Jharia coalfield ranged: pH (6.72–7.94), TDS (213–530 mg/L), SO_4^{2-} (8.8–41.2 mg/L), Cl^- (19.8–96 mg/L), NO_3^- (3–77.7 mg/L), Fe (0.13–2.18 mg/L), Zn (0.02–0.04 mg/L), Pb (0.01–0.04 mg/L). The maximum TDS, NO_3^- and Fe concentrations exceeded the Bureau of Indian Standards (BIS) limit for drinking water quality. In surface water, water quality parameters varied between (Abhishek et al. 2006): pH (7.15–7.76), EC (250.6–470.6 μS/cm), TDS (237–616 mg/L), DO (2.5–5.8 mg/L), BOD (3.8–13.7 mg/L), Pb (0.01–0.03 mg/L), Zn (0.03–0.09 mg/L) and Fe (0.15–1.91).

Metals concentrations in stream sediments around Makum coal field of Assam ranged (mg/kg): 5.5–71.7, 100–386, 3.1–21.1, 0.48–2.1, 23.1–231, 101–9163 and 17.8–264 for Cu, Cr, Pb, Cd, Zn, Mn and Ni, respectively (Equeenuddin et al. 2013). The study further indicated that higher concentrations of all metals were available in exchangeable fraction under strongly acidic environment. Based on their mobility and potential bioavailability, metals were in the order of Cd > Pb > Mn > Ni ≥ Zn > Cu > Cr.

4 Management and treatment strategies to reduce environmental impacts of coals

Since coal mining and its utilization in coal based industries is associated with environmental issues, it is necessary to manage or mitigate its impact on environment or clean coal prior to its utilization. An attempt was made by Dowarah et al. (2009) to achieve eco-restoration of a high-sulfur containing coal mine overburden dumping site through primary and secondary ecological succession of native plant species in Tirap Collieries, Assam, India. The study revealed that planting of herbaceous monocots with fibrous root systems such as citronella, lemon grass, *Saccharum spontaneum*, lianes and shrub species accelerates the ecological processes in an adverse mine overburden environment of Tirap colliery and a secondary sere ecological succession was observed in the restored mine site. In addition, 80 %–100 % vegetation coverage was observed, the plant species density was more than 80 %, and soil organic matter increased from 0.001 %–0.005 % to 0.5 %–1.3 %. Restoration refers to reinstatement of the pre-mining ecosystem in all its structural and functional aspects (Bradshaw 2000). Re-vegetation plays a crucial role in enhancing the soil fertility status in mine spoil and in the stabilization of dump slopes by creating mechanical reinforcement of dump material and enhancing shear strength of dump material (Singh 2011; Singh et al. 2012a, b). Soil structure development, nutrient cycling, and soil chemical and physical limitations to plant growth are mediated and mitigated by microorganisms and they play a very important role in eco-restoration (Singh and Singh 2006).

Mineral matter and sulfur exhibit harmful effects on utilization of coal. De-sulfurization and de-ashing are essential for sustainable utilization of low rank high sulfur coals used in different industries (Baruah et al. 2006; Baruah and Khare 2007b; Saikia et al. 2013). Sequential solvent extraction was found to be an effective method of desulfurization of high sulfur containing Assam coal, especially for organic sulfur, which could be removed up to 89 % (Das and Sharma 2001). Investigation on desulfurization of coal samples from Boragolai and Ledo collieries of Makum coal field, Assam, India using alkali treatment leads to over 70 % removal of inorganic sulfur, and removal of sulfur increased with increase in alkali concentration and treatment time (Mukherjee and Borthakur 2003). In another study, for the same coal, solvent

Table 4 Physico-chemical characteristics of surface water around coalmines in northeast and other parts of India (in mg/L except pH)

Item	Parameter	pH	TDS	Ca	Mg	Fe (µg/L)	Cu (µg/L)	Mn (µg/L)	Zn	Ni	Pb	Cr	Cd	References
Surface water	Jaintia coalfield (Meghalaya, India)	2.6–5.6	174–2078	0.33–108.2	0.72–26.27	1.5–66.1	bdl-0.09	0.01–3.25	0.011–2.05		bdl-0.46	bdl-0.05	bdl-0.06	Sahoo et al. (2012)
	Jharia coalfield(Jharkhand, India)	6.83–9.71	5–885	6.9–100.1	3.8–136.9	0.01–0.82		0.01–3.29						Sarkar et al. (2007)
	Singrauli, Madhya Pradesh	7.6–8.6	315–1425	17.6–38.5	1.95–12.7	0.072–2.27	0.004–0.013	0.005–0.153	0.015–0.058	0.012–.034		bdl-0.200		Khan et al. (2013)
	Makum (Assam, India)	6.1–7.4		6.2–31.4	2.5–16.5	0.41–2.47	bdl-0.021	0–0.94	bdl-0.016	bdl-0.06	0.008–0.056		bdl-0.017	Equeenuddin et al. (2010)
Groundwater	Jaintia coalfield (Meghalaya, India)	4.8–6.8	93–234	2.6–23.5	2.2–11.7	0.024–2.3	bdl-0.01		bdl-0.06		bdl-0.03	bdl-0.01		Sahoo et al. (2012)
	Jharia coalfield (Jharkhand, India)	4.6–7.68	5–320	16.4–175.6	4.4–174.5	0.001–11.94	0.002–0.21		0.01–3.95					Sarkar et al. (2007)
	Jharia coalfield (Jharkhand, India)					0.693	0.0282	0.74	0.153	0.0223	0.0125			Chandra and Jain (2013)
	Singrauli coal field (Madhya Pradesh, India)	7.83–8.7	176–1845	9.62–41.68	bdl-26.3	bdl-1.45	0.003–0.068	0.002–0.222	0.014–0.797	0.009–0.03		0.004–0.266		Khan et al. (2013)
	Makum (Assam, India)	4.2–7.8		4.8–27.52	2.3–12.5	0.15–3.76	bdl-0.009	0.01–2.18	bdl-0.108	bdl-0.2	0.0–0.061	bdl-0.018		Equeenuddin et al. (2010)

extraction and alkali treatment showed successful removal of organic and inorganic sulfur. Solvent extraction using dimethyl formamide (DMF) increased desulfurization of the oxidized Baragolai and Ledo coals up to 95 % and 93 % for inorganic sulfur and 31 % and 23 % organic sulfur, respectively, while the alkali treatment showed complete removal of inorganic sulfur and a maximum of 33 % and 26.4 % organic sulfur for these coals, respectively (Baruah and Khare 2007b). Alkali treatment of high sulfur Assam coal using mixtures (1:1) of 16 % sodium hydroxide and potassium hydroxide solution followed by 10 % hydrochloric acid could remove 50 %–54 % of the ash, total inorganic sulfur, and around 25 % organic sulfur (Mukherjee 2003). 9.4 % of the total organic sulfur was removed by electron transfer process (Borah and Baruah 1999). In another study, approximately 93 % and 98 % of the pyritic sulfur was removed in the case of the Baragolai and Ledo coal of Makum, Assam, respectively, using 15 % (v/v) hydrogen peroxide + 0.1 N sulfuric acid (Mukherjee and Srivastava 2004). An attempt was made to clean some low rank medium to high sulfur coal samples from northeast India using low ultrasonic energy (20 kHz) in the presence of H_2O_2 solutions and it showed removal of 31 %, 48 %, 51 %, 48 % and 32 % of total sulfur, organic sulfur, pyritic, sulfate sulfur and ash, respectively (Saikia et al. 2014b). In a similar study, treatment using application of ultrasonic energy (20 kHz) in aqueous and mixed alkali media (1:1 KOH and NaOH) on coals collected from Assam and Nagaland, India showed that the maximum removal of ash, pyritic sulfur, sulphate sulfur and total sulfur were 87.52 %, 83.92 %, 12.50 % and 18.80 %, respectively (Saikia et al. 2014c). Ultrasound assisted coal de-sulfurization and de-ashing is partially green approach that has been recently studied by other researchers (Hoffmann et al. 1996; Ze et al. 2007; Wang and Yang, 2007; Mello et al. 2009; Shen et al. 2012).

In addition to several physico-chemical desulfurization methods, biodesulfurization using *Thiobacillus ferrooxidans* (ATCC 13984) was attempted for Assam coal (Dastidar et al. 2000). Results showed that the rate of pyritic sulfur removal was retarded at higher concentrations of ferrous and ferric ions that need to be controlled to maintain high rate of removal (Dastidar et al. 2000). In general, AMD can be remediated by two generic approaches i.e. active or passive treatment (Skousen et al. 1998; Wolkersdorfer 2008). Active treatment requires the use of alkaline materials (lime, limestone, hydrated lime, caustic soda, soda ash, etc.) or aeration to reduce acidity and precipitate metals, while passive (abiotic and biological) treatment allows chemical and biological processes to take place naturally in a controlled environment (Costello 2003; Johnson and Hallberg 2005; Sheoran and Sheoran 2006; Rios et al. 2008; Sheoran et al. 2010). A pilot plant

Table 5 Pollution standards for air quality in India

Pollutant	Time-weighted averages	Concentration in ambient air (mg/L)		
		New coal mines (after December 1998)	Existing coalfields/ mines	Old coal mines (Jharia, Raniganj, Bokaro)
SPM	Annual average	360	430	500
	24 h	500	600	700
RPM	Annual average	180	215	250
	24 h	250	300	300
SO_2	Annual average	80	80	80
	24 h	120	120	120
NO_x	Annual average	80	80	80
	24 h	120	120	120

consisting of sequential alkalinity producing (SAP) system coupled with biological processes was designed for treatment of AMD from coalmines of Meghalaya, northeast India (Baruah et al. 2010). The treatment system was found to be effective in reducing TDS, conductivity, sulphate and toxic elements.

In India, the Ministry of Environment and Forests (MoEF) plays a key role in regulating the environmental impacts of mining and in providing clearances for mining in forest lands. Some environmental protection measures include: prevention of pollution at source; ensuring polluters pay principle; protection of heavily polluted areas and river stretches; encouragement of development and application of best available technological solutions; and involving the public in decision making (Mehta 2002). Under Mineral Concession Rules, 1960, it is required to specify the area indicating impact of mining activity on forest, land and environment, scheme for restoration of the area by afforestation, adoption of pollution control devices. According to Article 23 of the Mineral Conservation and Development Rules (1988), conditions for the abandonment of any mine need to be laid down by the mining company and provision of a plan for dealing with the environment, and is liable to protect and control pollution during the mining and post mining operations. The law further lays guidelines to restore or protect the flora of the area under the mining lease and nearby areas, technically, economically and environmentally.

The main environmental acts that impact the mining industry in India are: The Wildlife (Protection) Act, 1972 (amended in 1991); The Water (Prevention and Control of Pollution) Act, 1974 (amended in 1988); The Forest (Conservation) Act, 1980 (amended in 1988); The Air (Prevention and Control of Pollution) Act, 1981 (amended in 1988); and The Environment (Protection) Act, 1986 (with rules 1986 and 1987). Separate pollution standards for air quality and coal mine effluents has been laid down by Central Pollution Control Boards for coal mining in India (Tables 5, 6).

Table 6 Pollution standards for coal mine effluents

Parameter	Level
pH	5.5–9.0
TSS (mg/L)	100
Oil and grease (mg/L)	10
COD (mg/L)	250
BOD (mg/L)	30
Phenolics (mg/L)	1.0

In order to achieve sustainable utilization of coal resources integrated approach considering various aspects to reduce its environmental impacts is necessary. Proper implementation of regulatory rules and policies is as important as other management strategies to deal with environmental issues.

5 Conclusions

Demand for coal in India is projected to increase dramatically in short to medium term. This would result in increased coal mining in different parts of India including northeast region. Since, coals in northeast India is characterised by high sulfur and volatile matter contents that exhibits more potential harmful impacts, extra efforts are required to manage these coals to reduce its environmental impacts in the region. More studies need to be done in the field to assess the impact of coal mining on biodiversity, soil, air, surface and ground water in northeast India. Although several researches on desulfurization, de-ashing and demineralization techniques have been made, effort should be made to do further research on developing effective, low cost and environmental friendly technologies to clean coal and to use these techniques in the field. Further, it is essential to encourage and emphasize on alternative clean sources of energy to meet future energy demands.

References

Abhishek R, Tiwary K, Sinha SK (2006) Status of surface and groundwater quality in coal mining and industrial areas of Jharia coalfield. Ind J Environ Prot 26:905–910

Ahmed M, Rahim A (1996) Abundance of sulfur in Eocene coal beds from Bapung, Northeast India. Int J Coal Geolgy 30:315–318

Banerjee NN, Ghosh B, Das A (2000) Trace metals in Indian Coals; CFRI golden jubilee monograph. Allied Publishers Ltd, New Delhi, p 13

Baruah BP (2009) Environmental studies around makum coalfields. LAP Lambert Academic Publishing, India

Baruah BP, Khare P (2007a) Pyrolysis of high sulfur Indian coals. Energy Fuels 21:3346–3352

Baruah BP, Khare P (2007b) Desulfurization of oxidized Indian coals with solvent extraction and alkali treatment. Energy Fuels 21:2156–2164

Baruah BP, Khare P (2010) Mobility of trace and potentially harmful elements in the environment from high sulfur Indian coal mines. Appl Geochem 25:1621–1631

Baruah MK, Kotoky P, Borah GC (2003) Distribution and nature of organic/mineral bound elements in Assam coals, India. Fuel 82:1783–1791

Baruah BP, Kotoky P, Rao PG (2005) Genesis of acid mine drainage from coalfields of Assam, India. In: Proceedings of international seminar on coal science and technology—emerging global dimensions: global coal. Allied Publishers. ISBN:81-7764-818-7

Baruah BP, Saikia BK, Kotoky P, Rao PG (2006) Aqueous leaching of high sulfur sub-bituminous coals in Assam, India. Energy Fuels 20:1550–1555

Baruah BP, Khare P, Rao PG (2010) Management of acid mine drainage (AMD) in Indian coal mines In: Proceedings of international seminar on mineral processing technology (MPT-2010), pp 1163–1170

Bhole AG (1994) Acid mine drainage and its treatment. Proceedings of International Symposium on the impact of mining on the envirnment. Paithankar et al. (ed) Oxford & IBH Pub., Nagpur, Jan 11–16, pp 131–141

Borah D, Baruah MK (1999) Electron transfer process 1. Removal of organic sulphur from high sulphur Indian coal. Fuel 78:1083–1088

Bradshaw AD (2000) The use of natural processes in reclamation–advantages and difficulties. Landsc Urban Plan 51:89–100

Centre for Science and Environment, CSE (2012) Coal mining, pp 1–5. http://www.cseindia.org/userfiles/fsheet2.pdf. Accessed 5 Apr 2015

Chandra A, Jain MK (2013) Evaluation of heavy metals contamination due to overburden leachate in groundwater of coal mining area. J Chem Biol Phys Sci 3:2317–2322

Chandra D, Mazumdar K, Basumallick S (1983) Distribution of sulfur in the tertiary coals of Meghalaya, India. Int J Coal Geol 3:63–75

Chandra D, Ghose S, Chaudhuri SG (1984) Abnormalities in the chemical properties of tertiary coals of Upper Assam and Arunachal Pradesh. Fuel 63:1318–1323

Chaulya SK, Chakraborty MK (1995) Perspective of new national mineral policy and environmental control for mineral sector. In: Proceedings of national seminar on status of mineral exploitation in India, New Delhi, India, pp 114–123

Chon HT, Hwang JH (2000) Geochemical characteristics of the acid mine drainage in the water system in the vicinity of the Dogye coalmine in Korea. Environ Geochem Health 22:155–172

Chou CL (2012) Sulfur in coals: a review of geochemistry and origins. Int J Coal Geol 100:1–13

Costello C (2003) Acid mine drainage: innovative treatment technologies. US Environmental Protection Agency, Office of Solid Waste and Emergency Response. Technology Innovation Office, Washington, DC. http://www.cluin.org/download/studentpapers/costello_amd.pdf

CPCB (Central Pollution Control Board) (2009) National Ambient Air Quality Standards (NAAQS), under Gazette Notification B-29016/20/90/PCI-1, New Delhi

Das A, Sharma DK (2001) Organic desulfurization of assam coal and its sulfur-rich lithotypes by sequential solvent extraction to obtain cleaner fuel. Energy Source 23:687–697

Das T, Saikia BK, Baruah BP, Das D (2015) Characterizations of humic acid isolated from coals of two Nagaland coalfields of India in relation to their origin. J Geol Soc India 86:468–474

Dastidar MG, Malik A, Roychoudhury PK (2000) Biodesulfurization of Indian (Assam) coal using *Thiobacillus ferrooxidans* (ATCC 13984). Energy Conver Manag 41:375–388

Dowarah J, DekaBoruah HP, Gogoi J, Pathak N, Saikia N, Handique AK (2009) Eco-restoration of a high-sulphur coal mine overburden dumping site in northeast India: a case study. J Earth Syst Sci 118:597–608

Envirocon (2010) A report on ambient air quality, water/waste water analysis & noise level measurement at north eastern coalfields. Coal India Ltd, Margherita, pp 1–20

Equeenuddin SM, Tripathy S, Sahoo PK, Panigrahi MK (2010) Hydrogeochemical characteristics of acid mine drainage and water pollution at Makum Coalfield, India. J Geochem Explor 105:75–82

Equeenuddin SM, Tripathy S, Sahoo PK, Panigrahi MK (2013) Metal behavior in sediment associated with acid mine drainage stream: role of pH. J Geochem Explor 124:230–237

Ghose MK, Banerjee SK (1995) Status of air pollution caused by coal washery projects in India. Environ Monit Assess 38:97–105

Ghose MK, Banerjee SK (1997) Physico-chemical characteristics of air-borne dust emitted by coal washery in India. Energy Environ Monit 13:11–17

Ghose MK, Majee SR (2000) Assessment of dust generation due to opencast coal mining—an Indian case study. Environ Monit Assess 61:257–265

Ghose MK, Majee SR (2007) Characteristics of hazardous airborne dust around an Indian surface coal mining area. Environ Monit Assess 130:17–25

Hoffmann MR, Hua I, Hochemer R (1996) Application of ultrasonic irradiation for the degradation of chemical contaminants in water. Ultrason Sonochem 3:163–172

Huertas J, Camacho D, Huertas M (2011) Standardized emissions inventory methodology for open pit mining areas. Environ Sci Pollut Res 19:2784–2794

Johnson DB, Hallberg KB (2005) Acid mine drainage remediation options: a review. Sci Tot Environ 338:3–14

Khare P, Baruah BP (2010) Elemental characterization and source identification of $PM_{2.5}$ using multivariate analysis at the suburban site of North-East India. Atmos Res 98:148–162

Khare P, Baruah BP (2011) Estimation of emissions of SO_2, $PM_{2.5}$, and metals released from coke ovens using high sulfur coals. Environ Progr Sustain Energy 30:123–129

Khan I, Javed A, Khurshid S (2013) Physico-chemical analysis of surface and groundwater around Singrauli Coal Field, District Singrauli, Madhya Pradesh, India. Environ Earth Sci

68:1849–1861
Mehta PS (2002) The Indian mining sector: effect on the environment and FDI inflows. In: Conference on foreign direct investment and the environment, 7–8 February, Paris, France
Mello PA, Duarte FA, Nunez MAG, Alencar MS, Moreira EM, Korn M et al (2009) Ultrasound-assisted oxidative process for sulfur removal from petroleum product feedstock. Ultrason Sonochem 16:732–736
Ministry of Coal (2014) Government of India, provisional coal statistics 2013–14, coal reserves. http://coal.nic.in. Accessed 12 Aug 2015
Mukherjee S (2003) Demineralization and desulfurization of high-sulfur Assam coal with alkali treatment. Energy Fuels 17:559–564
Mukherjee S, Borthakur PC (2003) Effects of alkali treatment on ash and sulphur removal from Assam coal. Fuel Process Technol 85:93–101
Mukherjee S, Srivastava SK (2004) Kinetics and energetics of high-sulfur northeastern India coal desulfurization using acidic hydrogen peroxide. Energy Fuels 18:1764–1769
Mukherjee S, Srivastava SK (2005) Trace elements in high-sulfur Assam coals from the makum coalfield in the northeastern region of India. Energy Fuels 19:882–891
Mukhopadhyay S, Pal S, Mukherjee AK, Ghosh AR (2010) Ambient air quality in opencast coal mining areas of Bankola area (under Eastern coal field ltd.) of Asansol–Raniganj regions. Ecoscan 4:19–24
Nair PK, Sinha JK (1987) Dust control at deep hole drilling for open peak mines and development of a arrester. J Mines Met Fuel 35(8):360–364
Rajarathnam S, Chandra D, Handique GK (1996) An overview of chemical properties of marine-influenced Oligocene coal from the northeastern part of the Assam-Arakan basin, India. Intern J Coal Geol 29:337–361
Rawat NS, Singh G (1982) The role of micro-organisms in the formation of acid mine drainage in the north eastern coal field of India. Int J Mine Wat 2:29–36
Rios CA, Williams CD, Roberts CL (2008) Removal of heavy metals from acid mine drainage (AMD) using coal fly ash, natural clinker and synthetic zeolites. J Hazard Mater 156:23–35
Sahoo PK, Tripathy S, Equeenuddin SM, Panigrahi MK (2012) Geochemical characteristics of coal mine discharge vis-à-vis behavior of rare earth elements at Jaintia Hills coalfield, northeastern India. J Geochem Explor 112:235–243
Saikia BK, Kakati N, Khound K, Baruah BP (2013) Chemical kinetics of oxidative desulfurization of Indian coals. Int J Oil Gas Coal Tech 6:720–727
Saikia BK, Ward CR, Oliveira MLS, Hower JC, Baruah BP, Braga M, Silva LF (2014a) Geochemistry and nano-mineralogy of two medium-sulfur Northeast Indian coals. Intern J Coal Geol 121:26–34
Saikia BK, Dutta AM, Baruah BP (2014b) Feasibility studies of de-sulfurization and de-ashing of low grade medium to high sulfur coals by low energy ultrasonication. Fuel 123:12–18
Saikia BK, Dutta AM, Saikia L, Ahmed S, Baruah BP (2014c) Ultrasonic assisted cleaning of high sulphur Indian coals in water and mixed alkali. Fuel Process Technol 123:107–113
Sarkar BC, Mahanta BN, Saikia K, Paul PR, Singh G (2007) Geo-environmental quality assessment in Jharia coalfield, India using multivariate statistics and GIS. Environ Geol 51:1177–1196
Sarma K (2005a) Impact of coal mining on vegetation: a case study in Jaintia Hills district of Meghalaya, India. M.Sc. diss., International Institute for Geoinformation Science and Earth Observation (ITC), Enschede, the Netherlands
Sarma K (2005b) Impact of Coal Mining on Vegetation: A Case Study in Jaintia Hills District of Meghalaya, India, M.Sc Thesis, International Institute for Geo-information Science and Earth Observation, the Netherlands and Indian Institute of Remote Sensing (IIRS), India
Sarma K, Barik SK (2011) Coal mining impact on vegetation of the Nokrek Biosphere Reserve, Meghalaya, India. Biodiversity 12:154–164
Sarmah M, Khare P, Baruah BP (2012) Gaseous emissions during the coal mining activity and neutralizing capacity of ammonium. Water Air Soil Pollut 223:4795–4800
Shahabpour J, Doorandish M, Abbasnejad A (2005) Mine-drainage water from coal mines of Kerman region, Iran. Environ Geol 47:915–925
Shen Y, Sun T, Liub X, Jiaa J (2012) Rapid desulfurisation of CWS via ultrasonic enhanced metal boron hydrides reduction under ambient conditions. RSC Adv 2:4189–4197
Sheoran AS, Sheoran V (2006) Heavy metal removal mechanism of acid mine drainage in wetlands: a critical review. Miner Eng 19:105–116
Sheoran AS, Sheoran V, Choudhary RP (2010) Bioremediation of acid-rock drainage by sulphate-reducing prokariotes: a review. Miner Eng 23:1073–1100
Singh TN (2011) Assessment of coal mine waste dump behavior using numerical modeling. In: Fuenkajorn K, Phien-wej N (eds) Rock mechanics. Proceedings of the third Thailand symposium. ISBN 978 974 533 636 0, pp 25–36
Singh MP, Singh AK (2000) Petrographic characteristics and depositional conditions of Eocene coals of platform basins, Meghalaya, India. Int J Coal Geol 42:315–356
Singh AN, Singh JS (2006) Experiments on ecological resto-ration of coalmine spoil using native trees in a dry tropical environment, India: a synthesis. New For 31:25–39
Singh G, Sinha DK (1992) The problem of acid mine drainage its occurrence and effects. In: Proceedings of Environment Management of Mining Operations, Department of Environment & Forest Government of India, pp 156–167
Singh AK, Mondal GC, Tewary BK, Sinha A (2009) Major ion chemistry, solute acquisition processes and quality assessment of mine water in Damodar valley coalfields, India. In: International mine water conference proceedings ISBN Number: 978-0-9802623-5-3, 19th–23rd October, Pretoria, South Africa, pp 267–276
Singh PK, Singh MP, Singh AK, Naik AS (2012a) Petrographic and geochemical characterization of coals from Tiru valley Nagaland, NE India. Energy Explor Exploit 30:171–192
Singh RS, Tripathi N, Chaulya SK (2012b) Ecological study of revegetated coal mine spoil of an Indian dry tropical ecosystem along an age gradient. Biodegradation 23:837–849
Skousen J, Rose A, Geidel G, Foreman J, Evans R, Hellier W (1998) A handbook of technologies for avoidance and reclamation of acid mine drainage. National Mine Land Reclamation Center, West Virginia University, Morgantown, p 131
Swer S, Singh OP (2003) Coal mining impacting water quality and aquatic biodiversity inJ aintia Hills District of Meghalaya. ENVIS Bull Himal Ecol 11:26–33
Swer S, Singh OP (2004) Status of water quality in coal mining areas of Meghalaya, India. In: National seminar on environmental engineering with special emphasis on mining environment. NSEEME, 19–20 March, Dhanbad, India. pp 26–33
Tiwary RK (2001) Environmental impact of coal mining on water regime and its management. Water Air Soil Pollut 132:185–199
Wang Y, Yang RT (2007) Desulfurization of liquid fuels by adsorption on carbon-based sorbents and ultrasound-assisted sorbent regeneration. Langmuir 23:3825–3831
Ward CR, Li Z, Gurba LW (2007) Variations in elemental composition of macerals with vitrinite reflectance and organic sulphur in the Greta coal measures, New South Wales, Australia. Intern J Coal Geol 69:205–219

Widodo S, Oschmann W, Bechtel A, Sachsenhofer RF, Anggayana K, Puettmann W (2010) Distribution of sulfur and pyrite in coal seams from Kutai basin (east Kalimantan, Indonesia): implications for paleoenvironmental conditions. Intern J Coal Geol 81:151–162

Wolkersdorfer Ch (2008) Water management at abandoned flooded underground mines. Fundamentals, tracer tests, modelling, water treatment. Springer, New York, pp 235–275

Yue M, Zhao F (2008) Leaching experiments to study the release of trace elements from mineral separates from Chinese coals. Int J Coal Geol 73:43–51

Zamuda CD, Sharpe MA (2007) A case for enhanced use of clean coal in India: an essential step towards energy security and environmental protection. In: Workshop on Coal Beneficiation and Utilization of Rejects. Ranchi, India

Ze KW, Xin XH, Tao CJ (2007) Study of enhanced fine coal desulfurization and de-ashing by ultrasonic floatation. J Chin Univ Min Technol 17:358–362

2

Green infrastructure: a planning concept for the urban transformation of former coal-mining cities

Peter Wirth[1] · **Jiang Chang**[2] · **Ralf-Uwe Syrbe**[1] · **Wolfgang Wende**[1] · **Tinghao Hu**[2]

Abstract Green infrastructure is a concept aimed at realizing a strategically planned network of valuable natural and semi-natural areas, designed and managed to deliver a wide range of ecosystem services and to protect biodiversity in rural and urban settings. In the general post-mining context of China and Germany, this paper suggests and outlines an approach that combines green infrastructure with specific concepts of post-mining landscapes. While Germany has a long tradition of post-mining restoration, concepts of green infrastructure are still poorly developed. China, on the other hand, has taken its first steps in the restoration of coal regions, and could profit from the new concepts while drawing on Germany's experience. The potentials of relevant strategies are investigated here through two case studies from Germany and China. Although there are significant differences in the post-mining development strategies of these two countries, it is clear that the green infrastructure concept can contribute significantly to a growth in ecosystem services. Four advantages of the green infrastructure concept are seen: First, it enables a systematic and highly objective assessment of mining impacts on the ecosystem. Second, the concept assists in the implementation of urban sustainability goals. Third, it is rooted in the local characteristics of landscapes. And finally, the green infrastructure concept seems to be highly compatible with tools of landscape and regional planning.

Keywords Post-mining development · Ecosystem services · Urban planning · Green space · China · Germany

1 Introduction

Mining is a highly dynamic process of intervention in landscapes, commonly accompanied by environmental damage and fundamental social and economic impacts. Such impacts not only persist during the period of active mining but also after reserves are exhausted and mines closed. Almost all mining regions around the world face a similar set of challenges in post-mining landscapes. Long-term mining leaves a legacy of considerable environmental degradation (Wende 2015; Chang and Yang 2015). The environmental impacts are usually accompanied by serious problems of urban and regional development such as economic decline, out-migration and demographic shrinkage (Wirth and Lintz 2007).

Over the last three decades, transformation processes in former mining regions have been investigated by researchers from many fields such as geography, sociology, engineering and planning (for an overview see Wirth et al. 2012). Descriptive and explanatory studies of post-mining areas have dealt with political, cultural, social and economic change. National programs have been set up in order to rehabilitate and economically transform affected areas, such as the restoration of former lignite, copper and uranium mining areas in the east of Germany (Lintz and Wirth 2009; Wirth et al. 2012; BMWi 2015). Considering the

✉ Peter Wirth
p.wirth@ioer.de

1 Leibniz Institute of Ecological Urban and Regional Development (IOER), Weberplatz 1, 01217 Dresden, Germany

2 China University of Mining and Technology (CUMT), Daxue Road 1, Xuzhou 221116, China

dynamic nature of such restoration processes, it is unsurprising that ecological and environmental planning tools have also been developed (Wende 2011, 2015).

The concept of green infrastructure (GI) is comparatively new. Arising in the USA in the 1990s (Firehock 2010), GI became popular in many countries at the beginning of the new millennium (Wright 2011). There is a wide consensus that GI can deliver significant environmental, social and economic benefits (Mell 2008). The GI concept encompasses all spatial levels. It focuses on connectivity of urban green space, aims at multi-functionality, forges a link to the spatial provision of ecosystem services and highlights the need for close collaboration between scientific, political and administrative actors (cf. Haase et al. 2014; Pauleit et al. 2011; Rouse and Bunster-Ossa 2013).

In 2013 the European Commission highlighted a new policy concept entitled 'Green Infrastructure (GI)' to close the gap between conventional land use and urban development on the one hand and sustainable urban development on the other. This GI concept tries to achieve a strategically planned network of valuable natural and semi-natural areas, designed and managed to deliver a wide range of ecosystem services (ES) and to preserve biodiversity in rural and urban settings (European Union 2013). Thus, the concept follows the idea or *Leitbild* of improving ES and human well-being as well as promoting sustainable development in rural areas and cities.

In this article we suggest and show that the GI concept is also applicable to post-mining cities and regions. Clearly, mining activities have significant long-term impacts on the environment, on ecosystem services as well as biodiversity (Wirth and Lintz 2007; Cao 2007). All too often such impacts are tremendously damaging to ecosystems. Consequently, our leading research questions are: How can the idea of GI be applied to meet the specific requirements of mining regions and post-mining landscapes? And what are the strengths of the GI concept in relation to landscape restoration in coal-mining areas? We propose that the GI approach has a remarkable potential for the planning and management of former coal-mining areas burdened with a legacy of subsidence, coal gangue dumps, old industrial facilities and brownfields.

In the general post-mining context of China and Germany, this paper aims to develop ways of combining GI with specific concepts of post-mining landscapes by examining two different regional settings. Coal is one of the most important sources of energy in China, and in Germany will remain an important fuel until the completion of the country's *Energiewende* towards renewables. Germany has already gained some experience in both ecological restoration and green city reconstruction using GI strategies (see WBGU 2016; BMUB 2017). While this application of GI is still at an early stage, these initial experiences could be of significant benefit for China and other regions around the world. In the following we investigate coal-mining cities in China and Germany, capturing temporal-spatial differences in scales and landscape structures with the aim of revealing GI strategies and laws to upgrade urban ecosystems. A particular focus is on the protection and enhancement of ecosystem services using the GI concept.

The article is subdivided into five sections. Following this introduction, Sect. 2 considers the state of research on the concepts of post-mining and GI. In Sect. 3, two case studies from China and Germany are presented to illustrate possible applications of GI in post-mining areas. In Sect. 4, these two case studies are appraised, compared and evaluated against the background of the conceptual framework of Sect. 2 in order to reveal opportunities and limits of the different practical approaches. In the final section we derive some conclusions from the findings, which should be applicable to other regional and international contexts.

2 General considerations of post-mining development, GI and ES

2.1 Connections between post-mining regions and the GI concept

One legacy of mining operations in China is a large number of post-mining landscapes, generally situated in urban areas that thrived during the country's period of industrialization. Clearly, such landscapes are associated with a number of negative impacts on the environment as well as living conditions (Chen et al. 2010; Luo and Chen 2011; Chang and Yang 2015; Ma and Hu 2015). Comprehensive land reclamation planning for mining areas and mine rehabilitation are still at an exploratory level and lack a theoretical foundation, while research on ecological restoration and rehabilitation remains at a qualitative, descriptive stage. In particular, the country requires a stable, socio-economically and environmentally coordinated regulatory framework for the coal sector, one that encompasses the question of mine-closure (Cao 2017). The process of implementing land restoration has been hindered by problems such as inadequate management, a lack of feedback and monitoring, and conflict with the various aims of urban development (Chang et al. 2011).

It is widely agreed that the successful transformation of mining cities and regions can only be achieved through comprehensive environmental and the socio-economic change. Damage to the landscape through mining activities should be remedied and compensated through a series of measures which are embedded in an overall strategy

(Hennek and Unselt 2002; Wende et al. 2009; Wende 2015; Chang and Yang 2015). Ideally, these need to be developed and conceptualized in detail before the start of mining activities, including an assessment of ES (Syrbe 2015). In Germany there are several examples of successful mining transformations and the re-use of post-mining landscapes, realized through a combination of environmental restoration and socio-economic redevelopment (Ache 2000; Wissen 2001; Wirth et al. 2012; Harfst and Wirth 2011; Lockie et al. 2009). Against this background, the GI concept has a number of useful features (see Hansen and Pauleit 2014, p. 517), which could assist cities and regions in the transition process from mining to post-mining:

- Integration: urban green is viewed as a form of infrastructure, which can be integrated with other urban infrastructures in terms of physical and functional relations;
- Multifunctionality: GI planning encompasses the ecological, social, economic, abiotic/biotic and cultural functions of green spaces;
- Connectivity: GI planning considers physical and functional connections between green spaces at different scales;
- Multi-scale approach: GI can be applied to initiatives at different scales, from individual parcels to the community, regional, and state level;
- Multi-object approach: GI includes all kinds of green and blue space, natural and semi-natural areas, water bodies, parks and gardens, etc.

This experience can be of great assistance to China, which is currently confronted with problems similar to those of Europe some 30–40 years ago. The reprocessing and transfer of such experience could also generate ideas and opportunities for the development of corresponding geographical theory, ecological and environmental planning methodology and technology for wider use. Hence, Germany's experience should not be merely transferred to China but used to enhance methodology and to improve implementation in both countries (Müller et al. 2005).

Examination of the overall state of the art reveals a significant research gap on how to integrate the concepts of post-mining and GI to improve citizens' quality of life in affected cities.

2.2 Potentials of ES against the background of post-mining development

Ecosystems as well as all kinds of vegetated areas (here called *green spaces*) provide a wide range of ecosystem services (ES) to cities. The benefits for urban residents largely depend on the quality of these urban green spaces. The concept of ES is now commonly employed in the environmental debate on the development of urban nature conservation strategies. The upgrading of green spaces highlights urban ES, particularly regarding their benefits for health and recreation. Urban green spaces are found in a broad variety of types, ranging from high maintenance urban parks to natural areas and buffer spaces between facilities of technical infrastructure and other land uses. Diverse benefits are generated by urban green spaces, reflecting the heterogeneity of such spaces (Panduro and Veie 2013). The location, vertical structure, composition and configuration of urban green spaces all influence the ecological qualities and services (Grunewald et al. 2017).

Whereas the ES concept is commonly employed in urban and regional landscape planning (e.g. Grunewald and Bastian 2015; Maraja et al. 2016; Panduro and Veie 2013), there is less experience of ES in relation to mining cities and post-mining development. In general, post-mining regions are characterized by multiple forms of environmental damage (such as landscape degradation, land instability, landslides, soil erosion), by pollution of the air, soil and water as well as by a general lack of green spaces (Wirth et al. 2012). In view of such problems, long-term mining activities (in many cases 100 years and more) clearly hamper urban development. Yet after the cessation of mining, the damaged areas can be interpreted as a potential for urban development, for the re-cultivation of new green spaces easily accessible by local residents and which can serve as attractors for the urban future. One attempt in Europe to adapt the ES concept to post-mining areas is the study by Larondelle and Haase (2012). The authors quantify several ecosystem services in a large lignite mining area in Germany in order to assess the impacts of mining on the landscape and to construct future development perspectives.

In the following, we discuss the physical legacy of mining and its ES potentials, specifically in the case of cities affected by the repercussions of hard coal-mining, a topic of considerable relevance in China today. The Common International Classification of Ecosystem Services (Haines-Young 2013) defines three basic classes of ES, which shall be used here to structure the current section. These are: *regulation and maintenance*, i.e. the mediation of flows and of climate; *cultural*, i.e. physical, intellectual and spiritual services as well as the interaction with nature; and *provisioning*, e.g. the supply of food or water.

Firstly, *regulation and maintenance* services contribute significantly to environmental quality in cities. Depending on the environmental situation, the most important regulation services are climate and water regulation, carbon sequestration, as well as pest and erosion control.

Maintenance services support the lifecycles of biota as well as protecting the local and regional gene pools.

The main functional element of GI is to ensure a well-connected habitat network, permeable to several organisms. Brownfields, former mining sites as well as abandoned landscape elements connected to mining activities such as dumps, mine railways, conveyor and pipeline systems are hugely important as potential sites for the establishment of future green networks. Though often small in size, such sites can provide important pathways or at least stepping-stones for wildlife and gene exchange if secured for this purpose and maintained accordingly. Therefore, if a basic design for GI is drawn up for each mining city, this can help to guide post-mining planning (Grunewald et al. 2017).

Urban areas often suffer from so-called urban heat islands (UHIs), characterized by aridity, heat and lower wind strengths compared to the rural surroundings (Arnfield 2003). Typical physical legacies of mining activities such as wet areas affected by subsidence or dry, wind-exposed dumps can help to mitigate the impact of UHIs if such sites are made safe, free of pollution and planted with specific forms of vegetation to create valuable habitats. Climate-regulating ecosystem services are becoming ever more important to help counter the impact of climate change. In particular, water-logged ground as well as water bodies, which frequently emerge in former mining areas, can mitigate strong temperature differences or serve as pathways for the inflow of cold, fresh air from rural to urban areas. Also, the shift in airflows caused by dumps and hollows could boost air movement and thus balance the UHI.

Green spaces help to regulate the urban water cycle. The main elements of this cycle, namely evapotranspiration, surface runoff, infiltration, groundwater-recharge, subsurface flow, spring discharge and stream flow, provide the ecosystem services of water regeneration and purification without the need for large-scale technical investments. Yet building activities and an expanding transport system increase the extent of sealed surfaces, reducing possible infiltration and space for vegetation. This hinders groundwater recharge, affects the supply of freshwater and increases the amount and speed of surface runoff, thereby raising the likelihood of flooding and hence necessitating additional physical safeguards (Weller et al. 2012). One possible solution is the so-called 'sponge city' concept, developed in China, whereby rainwater is absorbed, stored and released in the vegetation, soil and substrate of a city much like a sponge (Wang et al. 2015). A large problem is that mining cities show an unstable flow system that occasionally may even reverse direction, so that drainage systems can break down. Therefore, well-adjusted groundwater and runoff models must be applied when designing the surface water system in such areas.

Secondly, cultural values are primarily connected with the accessibility of green spaces. In their study of the situation in Germany, Maraja et al. (2016) have identified recreation and tourism, nature experience, aesthetic values, the sense of place and local identity, places for social relations as well as educational, spiritual and religious values as the most important cultural ES. Abandoned mines and areas damaged by mining activities are ideally suited for upgrading as green spaces precisely because their specific problems preclude the development of housing or new industries. Hence, the general aim is to make these areas safe and accessible to the local population.

When maintained in a good state, urban green spaces enable a close experience of nature as well as providing places for relaxation or aesthetic enjoyment, recreation and sporting activities. In so doing, they contribute to citizens' psychological and physical regeneration. In a study of 30 Chinese cities, Smyth et al. (2008) found that residents with greater access to urban parks experienced higher levels of well-being. Similar results have been found for Germany (Krekel et al. 2016). In particular, urban grassland, forest and water bodies have a positive impact on mental and physical health.

Thirdly, provisioning services are often connected with the production of food such as urban agriculture, fishing, forestry and gardening. Such activities, however, make up only a small part of the supply of resources required by local residents. The more valuable part of urban provisioning services consists of resources supplied by ecosystems such as water for drinking and production, raw materials (fibers, timber) as well as renewable energy (hydropower, biogas, firewood, etc.). This latter resource is of increasing importance, especially in Germany. Familiar ecosystem products that are becoming more and more popular in cities include honey and fruits from meadows with scattered trees. These examples show how the sustainable use of ES can be combined with the conservation of valuable green spaces if such utilization is not oriented towards the optimization of only one kind of product. Mining cities often encompass areas that are not suited to the development of housing or industrial activities; alternatively, they have the potential to generate renewable forms of energy. In cases where former mining areas cannot be used for food production, they can be redeveloped with short rotation coppices or other kinds of plantation to provide renewable raw materials or energy crops. Biofuel crops can also be planted on polluted soils if the problem of residues and emissions can be resolved. The introduction of such biofuel crops can serve to establish a new industrial profile, and can be combined with other forms of renewable sources of energy such as solar power, wind power as well as hydropower or storage. Another development option particularly applicable to areas afflicted by post-mining subsidence is to establish so-called *paludicultures* (Wichtmann et al. 2016). This is the cultivation of wetlands and shallow lakes with adapted plants such as reeds and willows.

In summary, the examples above show how the physical legacy of mining activities can be upgraded, redeveloped and used in a way that simultaneously generates and maintains various ecosystem services such as cultural experience, climate regulation and energy production. In this way local residents are able to draw multiple benefits from a single site, thereby improving the quality of life in former mining cities.

3 Case studies

In order to analyze the supporting role which GI could play in the transformation of coal-mining areas, two contrasting case study regions were chosen: Zwickau–Lugau–Oelsnitz (Germany) and Xuzhou (China). These are suitable for analysis because: (1) Several research projects have already been conducted in these regions, producing a great deal of information; (2) the current authors were involved in research and planning tasks in these regions; and (3) the regions have already made some progress in restoration, allowing for a thorough assessment of the application of GI.

The two research projects underpinning this article were carried out independently of one another in Germany and China. Research by the German team started in 2009 with the INTERREG IV-B project Re-Source (2009–2012), dealing with ecological restoration and post-industrial strategy development (Harfst and Wirth 2012). Further data collection in the Zwickau–Lugau–Oelsnitz region was undertaken in 2015 by one of the authors as an external expert to the Saxon Horticultural Show in Oelsnitz. Basic data for the German case study was drawn from the analysis of relevant planning documents regarding the post-mining period from 1998 to 2017, statistics on the socio-economic development of the Zwickau–Lugau–Oelsnitz area, semi-structured interviews as well as three group discussions with regional stakeholders (2010, 2012, 2015). The basic dataset was updated in 2017 to allow comparison with the data from China.

Initial research into the Chinese region was undertaken in the project "Ecological Restoration Planning of Mining Subsidence Areas in Xuzhou", supported by the Government Xuzhou (2008–2010). Further data was gathered by a "Study on reutilization of industrial wasteland in mining towns from the perspective of urban and rural overall planning", which ran from 2012 to 2015 (Feng et al. 2016). In 2016, the committee of Lake Pan'an launched a monitoring project "Protection and Restoration of Pan'an-lake wetland park", compiled by the Jiangsu Academy of Forestry and the company Nanjing Puhou Ecological Technology Ltd. The Chinese authors of the current paper were involved in the project, which intends to establish a daily monitoring system for the wetland park which includes specific measures such as a wetland ecological monitoring project, publicity and education, wetland drainage, water quality restoration, and wetland management.

The German and China research teams cooperate since 2012, beginning with the project "Partners for Sustainability: Post-carbon Energy Concepts for Mining Areas", funded by the Robert Bosch Foundation. When it became clear that both countries were facing similar developmental problems, the idea was born to pursue a comparative approach. The joint efforts gained momentum in the framework of the conference "Land Reclamation in Ecological Fragile Areas" held in October 2017 in Xian/China. By harmonizing the research methods, updating data and consolidating information regarding the areas under investigation, the two case studies presented here evolved to follow approximately the same design. In order to describe changes to ES in the process of mine land restoration, we use the ES impact method originally developed by Bastian et al. (2013) and Lupp et al. (2015) for the comparison of different scenarios. Under this method, we select the most variable and important ecosystem services impacted by the given landscape changes and make a qualitative comparison of the differences in ES supply between the two case studies areas (rather than considering different scenarios in one area).

3.1 German case: the Zwickau–Lugau–Oelsnitz coal-mining district

3.1.1 Geographical features and mining history

Located in the state of Saxony (Germany), the former mining district of Zwickau–Lugau–Oelsnitz covers an area of 273 km^2. In 2015 the population was 153,699. The district encompasses the city of Zwickau, serving as a regional hub, as well as eight surrounding suburban or rural communities, namely Reinsdorf, Mülsen, Lichtenstein, Hohndorf, Gersdorf, Lugau, Oelsnitz/Erzgebirge and Hartenstein (Fig. 1). Situated in the Chemnitz-Zwickau agglomeration, the local population density is rather high at 563 inhabitants per km^2 (the national figure is 230 per km^2). Nevertheless, the population has declined in all municipalities over recent decades. Specifically, the area has lost around 30% of its inhabitants since 1990.

Zwickau is the biggest city of the former mining district with a population of 91123 (2015). It is the region's administrative, industrial and cultural center, whose reach extends far beyond the Zwickau–Lugau–Oelsnitz area. The

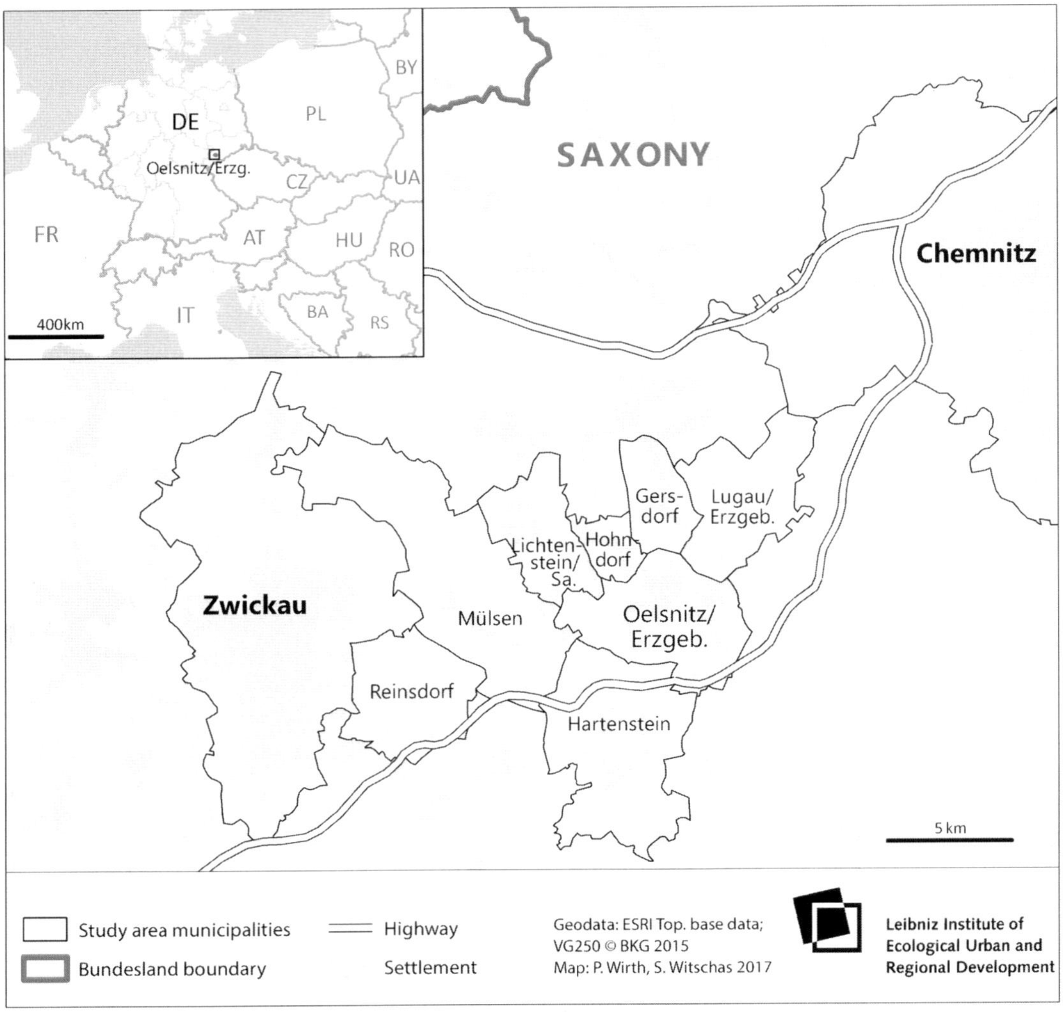

Fig. 1 Study area of Zwickau–Lugau–Oelsnitz

city is the capital of the administrative district (*Landkreis*) of Zwickau and accommodates various regionally important cultural facilities such as museums and theatres as well as a university of applied sciences (for more information see Harfst and Wirth 2012).

The entire Zwickau–Lugau–Oelsnitz area is part of the old industrial heartland of southern Saxony, stretching from Zwickau in the west to Chemnitz in the east. This area underwent rapid and intensive industrialization from the 1850s onwards, making it a central European hub of various industries such as textile production, manufacturing and coal-mining. While the region retained its industrial character under the socialist economic system of the German Democratic Republic (GDR) until 1989, coal production was already phased out in the 1970s due to its unprofitability.

3.1.2 Impacts of mining

The first evidence of hard coal mining in the Zwickau region dates back to 1348 (Hoth et al. 2009). During the main period of industrial mining from ca. 1850 to 1978, some 207 million tons of coal was extracted from the Zwickau coalfield, while the Lugau–Oelsnitz field produced 142 million tons between 1844 and 1971. Although mining activities were entirely underground, this work had significant environmental impacts on the surface. Post-mining development is still dealing with three main legacies from mining activities: firstly, the dumps of gangue (i. e. worthless rock); secondly, a number of subsidence areas; and thirdly, the numerous brownfields (Harfst and Wirth 2012).

One of today's greatest challenges for the region is the incomplete rehabilitation of former mining sites after their closure in the 1970s. For example, dumps were only partly afforested or simply abandoned to spontaneous vegetation. Today the region contains 80 dumps, the smallest less than

1 hectare in size and the biggest extending 26 hectares. In total, these cover 336 hectares of land (Felix et al. 2009). After the cessation of mining, most shafts were decommissioned and sealed. After many years of physical neglect, areas of subsidence can now be found in all places where underground mining was conducted. Clearly, these are associated with various long-term problems. Surface subsidence caused by the remaining unstable mining galleries has reached a maximum of 16 meters in Zwickau (Hoth et al. 2009) and 17 m in Oelsnitz (Berrios and Kolitsch 2001). Furthermore, groundwater levels are rising and there is a danger of toxic washout. In addition, subsidence causes persistent and recurring damage to buildings and infrastructure, and is a major obstacle to urban and infrastructure planning and development in the region.

3.1.3 Ecosystem potentials and services of post-mining areas

The potentials of coal-mining sites in the Zwickau–Lugau–Oelsnitz region are mainly determined by the existence of areas of subsidence, coal gangue heaps as well as brownfields. In previous years, the mining heaps, which are spread over the whole area, have been viewed as offering the greatest potential. Despite their incomplete rehabilitation (see above), such heaps are used for recreation and sporting activities such as strolling, dog-walking, jogging and mountain biking ① (*NB*: encircled numbers refer to Table 1). In Lugau a motocross track has been built on top of a mining heap ①. In addition to these cultural services, biomass production (firewood) is likely to become more important over the next few years due to Germany's nationwide transition to renewable energies ②. In the city of Zwickau, mining heaps also help to regulate the urban climate ③ by fostering an inflow of fresh air and buffering the summer heat. Furthermore, many of these heaps provide dry habitats seldom found in the surrounding cultural landscape for plants and animals ④.

In contrast, subsidence is predominantly seen as a problem for urban development in the region. In Zwickau, subsidence affects the districts of Bockwa and Schedewitz, where built-up areas are at permanent risk of flooding by groundwater. However, subsidence can also bring new potentials: The steepening of the course of the river Mulde enabled its use for canoeing ⑤ (Hoth et al. 2009). As a precaution, the groundwater level has been lowered by pumping stations. Consequently, the subsidence areas are generally viewed as a burden rather than a potential for future development ⑥. Regarding brownfields, which encompass large areas of shaft towers and other surface mining facilities, disused railway tracks and terminals, coking plants, power stations, etc., there are two contrasting trends: Many areas have been reused for reindustrialization projects in the 40 years since the end of mining, while (as will be shown in the next section) others form a basis for GI. These new green spaces include former infrastructural elements such as mining railways, roads,

Table 1 Changes to ecosystem services when transforming mining legacies into green spaces (case of Zwickau–Lugau–Oelsnitz)

Mining legacies	Areas of subsidence	Coal gangue heaps	Brownfields
Restructuring measures	Deconstruction, wetland use, lake areas	Stabilization, covering, planting	Deconstruction, surface unsealing, decontamination
Regulating ES			
Local climate regulation	⇧ (s. Sect. 2.2)	⇧ ③	–
Carbon sequestration	–	⇧ ⑧	–
Water regulation	⇩ ⑥ (s. Sect. 2.2)	–	⇧ (s. Sect. 2.2)
Cultural ES			
Opportunities for sports	⇧ ⑤ (aquatics)	⇧ ①	⇧ (new paths)
Habitat and species	⇧ (s. Sect. 2.2)	⇧ ④	⇧ ⑦
Heritage/sense of place	–	⇧ ⑧	⇧ ⑨
Recreation	⇧ ⑧	⇧ ⑧	⇧ ⑧ ⑨
Provisioning ES			
Renewable energy	⇧ (s. Sect. 2.2)	⇧ ②	–
Food: (fruits/honey)	⇧ (s. Sect. 2.2)	⇧ (s. Sect. 2.2)	⇧ ⑨

Symbols: ① encircled numbers refer to examples in the text, ⇧ service increasing, ⇩ service decreasing, – Service unimportant or non-uniform change

Fig. 2 Hiking trail and viewing tower on the former coal-mining dump 'Deutschlandschachthalde' in Oelsnitz (photo: Bieberstein/IOER)

pipelines and conveyor systems. Such elongated areas are preferred links for biotope networking ⑦.

Amongst the manifold activities, some projects can be highlighted. For example, Oelsnitz municipality developed a local recreation area at the region's biggest dump, namely the 'Deutschlandschachthalde' (Fig. 2). A viewing tower was erected on the mainly afforested heap in 2000, serving as the main attraction on a hiking trail that connects several key sites of former mining activities in the town and its surroundings ⑧.

The most ambitious post-mining project in the investigated region has been the Saxon Horticultural Show of 2015, located around the former coal loading station in Oelsnitz. The project encompassed the design of a modern exhibition area covering 15 hectares. The concept included green spaces for trees and flowers as well as recreation sites such as a skating course and a graduation tower (for the production of salt through evaporation). Over a period of 6 months, more than 400000 people visited the exhibition. Subsequently, the area has been used as a municipal park ⑨.

3.2 Chinese case: Xuzhou

3.2.1 Urban development and coal-mining activities

Xuzhou City is located in the northwest of Jiangsu province, China (Fig. 3). With five urban districts and five counties,[1] it covers a total area of 11258 km^2 and has 10.3 million residents, among them 3.3 million in the urban center (2016). Xuzhou, the first city to be founded in Jiangsu province, is one of the nine ancient cities in China and enjoys a rich history stretching back nearly 2600 years. Coal-mining has been carried out for 900 years in Xuzhou. The setting up of the Liguo Iron and Coal Mining Bureau in 1882 marked the beginning of the large-scale exploitation and utilization of coal, as well as launching modern industrial processes in Xuzhou (Chang et al. 2011).

Following the foundation of the People's Republic of China in 1949, more than one thousand state-owned enterprises in the fields of coal-mining, machine manufacture, chemicals, building materials and metallurgy were established in Xuzhou. All of these were strongly dependent on local coal resources. Enjoying the only fuel resource in Jiangsu province, the output of heavy industry in Xuzhou accounted for a quarter of the entire output of Jiangsu province at the peak period. According to the Master Plan of Xuzhou City for 1979, priority was to be given to the development of coal resources and electric power as well as the creation of a transport hub and regional business center (Shen et al. 2012). At the end of the twentieth century, the old industrial system was greatly impacted by shifting structures of power generation, accompanied by the rapid development of other sectors such as manufacturing, commerce and services. In line with central governmental policy, the provincial government began to support the process of urban transformation. A central document on the "Revitalization of the Old Industrial Base of Xuzhou" stressed the development of manufacturing, tourism and services as new pillars of the economy (People's Government of Jiangsu Province 2008). Exploiting the historic, cultural and ecological background, ecological tourism became a development goal in the 2007 Master Plan of Xuzhou City.

3.2.2 Impacts of mining

Centuries of coal-mining activities have brought considerable economic benefits to Xuzhou. At the same time the physical legacy of mining has generated a number of obstacles to urban development. The most important of these are areas of subsidence that negatively impact the urban ecological environment (Chang and Koetter 2005; Chang and Feng 2008). According to data from 2012, 16133 ha of land susceptible to subsidence from mining activities could be located in the five urban districts of Xuzhou; 72% of these areas had not been upgraded and remained subject to subsidence. Most sites are located in fringe areas where the ecological and social environments are highly sensitive. The areas affected by subsidence form a semicircle to the north of the main urban area, restricting the city's development in a swathe of land from the north-west to the north-east. At other sites, ecosystems have been disturbed, resulting in the isolation of valuable green spaces. The areas of subsidence are mainly located in

[1] The administrative level of "county" is equal to "urban district" in prefecture-level cities in China.

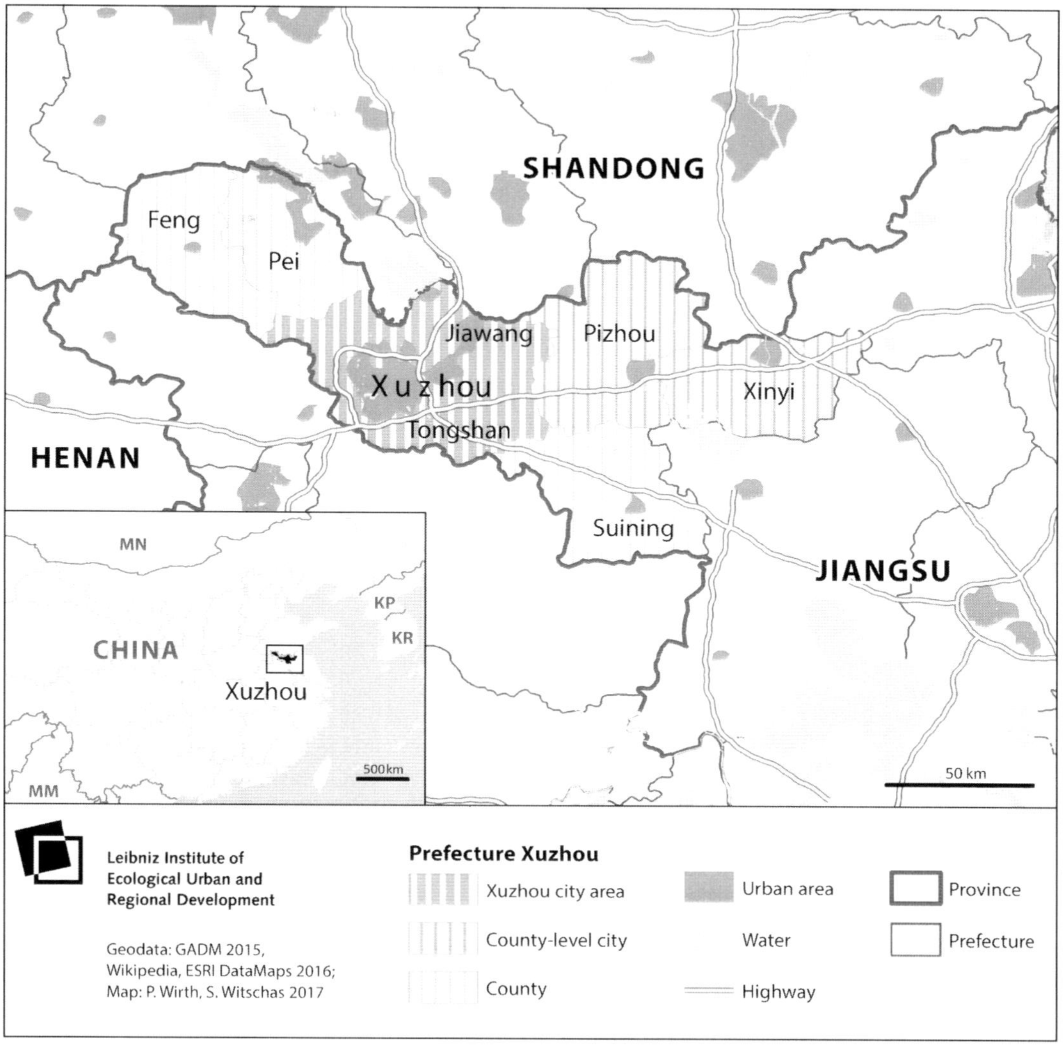

Fig. 3 Study area of Xuzhou

Tongshan district and Jiawang district in Xuzhou urban area (Feng et al. 2016).

3.2.3 Ecosystem potentials and services of post-mining areas

Although subsidence hampers urban expansion, affected areas constitute an important resource for the ecological development of Xuzhou (see also the encircled numbers, which refer to Table 2). Due to the long-term abandonment of such sites and lack of human disturbance, local ecosystems are in a process of self-recovery and natural succession, leading to the formation of areas with rich ecological potential and providing space for endangered species (Chang et al. 2011). In fact, these lands have never been "abandoned", but rather help to form an integrated ecological network (Feng et al. 2016). Areas of subsidence, in particular the surface water system ①, have a special significance for the maintenance and control of the ecosystem network when restoration and reconstruction have been implemented in some key patches and core areas. After restoration, post-mining lands can provide important ecosystem services, namely:

(1) Agricultural use, which is one of the most typical recovery goals for post-mining lands, provides the majority of provisioning services. The transformation of post-mining lands into agricultural and forested land can provide food, wood and other natural resources for residents.
(2) The establishment of wildlife areas and nature reserves in post-mining landscapes helps to foster biodiversity ②.
(3) After restoration, post-mining landscapes can have a positive impact on the regional ecological environment through improved climate and air quality ③, carbon sequestration and storage ④, the prevention

of soil erosion, improved soil fertility by natural methods, higher levels of pollination ⑤, etc.

(4) Abandoned mines close to urban areas can be used as parks, stadiums, golf courses and fishing zones, thereby boosting the range of community and cultural services.

(5) Due to the high sulphur content, vegetation will generally not grow on the remains of coal gangue heaps. Instead, the heaps are used to provide underground backfill to prevent subsidence in mining areas, as substitute material for making bricks or concrete and as fuel for power plants. Renewable energy can be generated by new-constructed solar power plants within mining lands, particularly in areas of shallow subsidence ⑥.

A typical example of the implementation of this program is the construction of Pan'an Wetland Park (Fig. 4). Created on vast areas affected by subsidence from coal-mining activities, the park has a total area of 16 km^2. In order to implement the restoration work in the Pan'an lake area, the city established a construction model including basic farmland management ⑦, the reclamation of subsidence areas, ecological restoration and landscape development in wetlands. An afforestation and restoration project was launched in 2011. After 2 years of restoration work, an area of 4.3 km^2 was reclaimed to form a continuous lake and water catchment area. More than 100 kinds of trees, 200 varieties of shrubs and 300 varieties of aquatic plants ⑧ were planted (Zhang 2015). A huge wetland landscape was created to meet the demands of tourism. By the end of 2015, Pan'an Wetland Park had attracted more than 3.6 million visitors ⑨. Its construction demonstrated how a barrier to development can be transformed into a regional attraction. A series of large public events such as an international music festival, a cross-country run ⑩ as well as film and television activities has greatly enriched citizens' lives ⑪ (Wang et al. 2016).

From a scientific perspective, this ecological restoration plan for areas afflicted by post-mining subsidence in Xuzhou is realized through a combination of environmental restoration, urban sprawl control, village relocation and new rural development (Lin et al. 2009; Chang et al. 2011). In the meantime, a cooperative platform has been established for various interest groups such as local residents, scholars, policymakers, investors as well as operators in order to increase their awareness of social, economic and ecological factors.

4 Discussion

Here we return to the research questions: How can the idea of GI be applied to meet the specific requirements of mining regions and post-mining landscapes? And what are the strengths of the GI concept in relation to landscape restoration in coal mining areas? In formulating answers, we first have to remember that it was not the explicit goal of actors in the investigated cases in Germany and China to create what we call here 'green infrastructure'. First and foremost, the motivation of local actors was to overcome the negative environmental impacts of coal-mining, to raise the living conditions of inhabitants, and to remove obstacles to urban development. The realized measures were based on concepts that combined clean-up activities, greening and improved wellbeing. Hence, we have to consider the effects of the tested strategies and determine how they are related to the planning concept of GI.

4.1 Comparing the two cases

Firstly, we can point to a range of *similarities,* which are mainly related to the nature of hard coal-mining and its impacts the landscapes and ecosystems, as well as societal needs in the mining regions. Clearly, the character of landscapes has been substantially transformed by mining activities, whether through the creation of dumps or widespread subsidence. Furthermore, ecosystems have been negatively impacted through soil and water pollution as well as damaged wildlife habitats. Without rehabilitation measures, many affected areas are unable to play a role in urban and regional development concepts and programs. Even after rehabilitation, many planning restrictions remain for urban development, for example due to large underground cavities, toxic contamination of soils and soil erosion. Yet such restrictions do not entirely prevent the utilization of ES in such areas or their integration into urban and regional concepts. As explained in Sect. 3, post-mining sites can help regulate the local climate, they can be used for the production of raw materials and bio-energy, for recreation or to foster biodiversity. Clearly, all classes of ES as described in Sect. 2 (provisioning, regulation and maintenance, cultural) are relevant to the case study areas. Comparing Tables 1 and 2, we note in particular the similarity in ES changes regarding water regulation and food production, i.e. although water regulation remains a problem due to the occurrence of subsidence in both countries, it can be enhanced elsewhere; further, food provision can generally be enhanced in post-mining landscapes.

From a societal perspective, we can observe a paradigm shift. During the period of industrialization, the coal-mining areas were hotbeds of innovation, technological

Table 2 Changes to ecosystem services when transforming mining legacies into green spaces (case of Xuzhou)

Mining legacies	Areas of subsidence	Coal gangue heaps	Brownfields
Restructuring measures	Deconstruction, wetland use, lake areas	Backfill, industrial reuse of the material	Deconstruction, surface unsealing, decontamination
Regulating ES			
Local climate regulation	⇧ ③	–	⇧ ③
Carbon sequestration	⇧ ④⑧	–	⇧ ④
Water regulation	⇩ ①	–	⇧ (s. Sect. 2.2)
Pollination	⇧ ⑤	–	⇧ ⑤
Cultural ES			
Opportunities for sports	⇧ ⑩	–	⇧ ⑩
Habitat and species	⇩ ② ⇧ ⑧	–	⇧ ⑧
Heritage/ identification	⇧ ⑪	–	⇧ (s. Sect. 2.2)
Recreation	⇧ ⑨⑩⑪	–	⇧ ⑨
Provisioning ES			
Food (crops, fish)	⇧ ⑦	–	⇧ ⑦
Raw materials	–	⇧ ⑥	–
Energy	⇧ ⑥	⇧ ⑥	–

Symbols: ① encircled numbers refer to examples in the text, ⇧ service increasing, ⇩ service decreasing, – service unimportant or non-uniform change

Fig. 4 Pan'an Wetland Park in Xuzhou/China, created through the restoration of post-mining areas (photo: Syrbe/IOER)

progress and economic vitality, irrespective of the political and economic system. Economically, the cities benefited greatly from the extracting industries, and their populations were mostly increasing. Ecological damage was ignored for a long period or dismissed as a minor concern. In the post-industrial period, there is great appreciation of the problems left by mining, specifically the environmental damage and the impacts on the future development of cities. The question arises of how best to deal with this physical legacy, often seen as a burden on urban development. In both case studies we can observe a change in urban and regional development strategies, most obviously in the case of Xuzhou City, whose primary function has shifted from a 'resource city' to a place of historic, cultural and ecological importance. This is an excellent condition for applying the GI approach.

The *differences* in handling the physical legacy of mining in the presented cases have a number of causes. Specifically, we can point to the disparate dimensions and technologies of mining, as well as local characteristics such as the location of mining facilities in relation to built-up areas. In the Chinese case, areas of subsidence play a decisive role in discussions around mine closure and restoration. Such areas form a semicircle on the outskirts of the settlement area. This is the reason why wetland areas have come to dominate post-mining planning in Xuzhou. In contrast, subsidence areas from mining are a considerable burden in Zwickau–Lugau–Oelsnitz, where some built-up areas are at permanent risk of flooding by groundwater. Instead, the main focus of urban planning in the German case is on the manifold mining dumps. Spread over the whole of the former mining district in an irregular pattern, they can potentially be transformed into sites for recreation, sports as well as new habitats.

In spite of these differences in the physical character of the former mining sites, their ecological potentials have been identified by urban and regional planners for some years. Today, we note that many of the implemented projects in the presented case studies fit very well with the new perspective of GI. The advantage of GI seems to be in the systematic conceptualization of ES to include potentials for urban and regional development. Of course, the cases of China and German present contrasting political strategies and social demands (cf. Table 1 and 2). For instance, the demand for renewable energy and aquatic sports is higher in Germany, while the demand for some provisioning services (fish, honey, wood) and, in particular, space for specific cultural activities is more pronounced in China.

4.2 Opportunities of the GI concept

Under quite different initial conditions, our two case studies in China and Germany show the value of the physical legacy of mining if sites are developed into functional green spaces to supply basic services to residents. Therefore, the GI concept opens up the possibility of systematically developing not only the technical but also positive ecological factors in a city, particularly under consideration of spatial connectivity. Due to its integrated perspective, the GI approach encourages the assessment and consideration of a multitude of ecosystem services as benchmarks for livability and sustainability, whose improvement will foster the successful development of mining cities. Although the character of the affected cities and regions in our examples were negatively impacted by mining activities, post-mining landscapes could subsequently be upgraded and living conditions improved. In the observed cases, however, the planning and implementation of projects were in part intuitive, depending at each step on the priorities of local and state policies. Since a more comprehensive planning approach is demanded politically (if not yet legally-anchored in all cases), the way to this from an ecological point of view can be the GI concept.

Adoption of the GI concept will encourage more systematic exploitation of ecosystem potentials in post-mining areas. Instead of focusing on just a single function (such as land development) or ecosystem service, a wider consideration of the potentials of the diverse pattern of green spaces typical of post-mining landscapes can ensure delivery of a greater range of services to the local population. Both Tables 1 and 2 show predominant improvements in the ES of the common types of mining sites and ecosystems. In order to develop such a range of services, it is first important to ask the question: Which ES have been degraded by mining and which can be most improved in the post-mining landscape? Then it is necessary to pinpoint existing demand for such services. Subsequently, various scenarios can be discussed on the preferred kind of ES, whether related to biodiversity, climate regulation, freshwater provision, fiber or food production (cp. Larondelle and Haase 2012). Following this approach, results can be systematically evaluated and the effects of measures appraised. In summary, we can say that the GI concept is a means to optimize urban, regional and landscape planning.

5 Conclusions

The findings of this paper illustrate the potential of the GI concept for the development of post-mining landscapes. The case study regions help to reveal the strengths and versatility of the GI concept. Four basic advantages can be highlighted: *Firstly*, the concept allows a systematic and highly objective assessment of mining impacts on the ecosystem, free of commercial and other individual interests. *Secondly*, the GI concept facilitates the incorporation of sustainable urban development goals, for instance those of the United Nations Sustainable Development Goals (UN 2015). *Thirdly*, the concept is rooted in the local characteristics of landscapes, allowing a tailor-made GI approach for regions in different parts of the world. *Lastly*, the concept seems to be highly compatible with tools of landscape and regional planning, fitting well with the application of Strategic Environmental Assessment (SEA). Clearly, further basic research has to be undertaken on how best to integrate the concept of GI in post-mining development, including a consideration of institutional framework conditions.

Restoration plans for former coal-mining areas should focus on upgrading areas of subsidence, mining heaps and brownfields to new (secondary) habitats and green areas, both as compensation for the lost (primary) pre-mining habitats and to create new potential for the successful post-mining development of the region. At the same time it is vital to develop methods of GI assessment for innovative post-mining concepts. Individual case studies such as Zwickau–Lugau–Oelsnitz, Germany, and Xuzhou, China, could serve as role models for this research, not just in these two countries but also at an international level. The GI concept can help foster the sustainable development of post-mining areas, providing a good quality of life and new perspectives for local people.

Acknowledgements The basic research for the Chinese case study was supported by the National Natural Science Foundation of China (NSFC), which funded the project "Research on Space–Time Evolution Laws and Optimization Model of green infrastructure in Coal Resource Based Cities"(No. 41671524). In the German case study, the basic research was supported by the European Union's Interreg programme CENTRAL EUROPE (ReSource project). The authors would like to thank these two organizations.

References

Ache P (2000) Cities in old industrial regions between local innovative Milieu and Urban Governance—reflections on City Region Governance. Eur Plan Stud 8(6):693–709. https://doi.org/10.1080/713666434

Arnfield AJ (2003) Two decades of urban climate research: a review of turbulence, exchanges of energy and water, and the urban heat island. Int J Climatol 23:1–26. https://doi.org/10.1002/joc.859

Bastian O, Lupp G, Syrbe RU, Steinhäußer R (2013) Ecosystem services and energy crops—spatial differentiation of risks. Ecológia 32(1):13–29

Berrios D, Kolitsch S (2001) Untersuchung von Sickerwässern der Steinkohlenhalden im Raum Oelsnitz (Investigation of leakage water in coal mining dumps in the Oelsnitz area). TU Bergakademie Freiberg

BMUB (2017) Weißbuch Stadtgrün. Grün in der Stadt—Für eine lebenswerte Zukunft. Bundesministerium für Umwelt, Naturschutz, Bau und Reaktorsicherheit (Federal Ministry for Environment, Nature Conservation, Building and Nuclear Safety). http://www.bmub.bund.de/fileadmin/Daten_BMU/Pools/Broschueren/weissbuch_stadtgruen_bf.pdf. Accessed 15 Aug 2017

BMWi (2015) Wismut—landscapes designed and preserved. German Federal Ministry for Economic Affairs and Energy http://www.wismut.de/www/webroot/de/download.php?download=3509. Accessed 28 March 2017

Cao X (2007) Regulating mine land reclamation in developing countries: the case of China. Land Use Policy 24:472–483. https://doi.org/10.1016/j.landusepol.2006.07.002

Cao X (2017) Policy and regulatory responses to coalmine closure and coal resources consolidation for sustainability in Shanxi, China. J Clean Prod 145:199–208. https://doi.org/10.1016/j.jclepro.2017.01.050

Chang J, Feng S (2008) Strategies on redevelopment of mining city industrial wasteland. Urban Dev Stud 2:54–57

Chang J, Koetter T (2005) From abandoned mine land to landscape park. J China Coal Soc 30:399–402

Chang J, Yang Y (2015) Suitability evaluation of abandoned mine lands supported by GIS: a case study of Yangzhuang mining area in Huaibei. In: Hu ZQ (ed) Legislation technology and practice of mine land reclamation. CRC Press, London, pp 117–125

Chang J, Wende W, Luo P, Deng Y (2011) Re-use of the mining wasteland. Tongji University Press, Shanghai

Chen N, Peng W, Xiao D, Lei Q (2010) Selective solar tunnel-concept and experiment. Acta Energ Sol Sin 31(4):442–446

European Union (2013) Green infrastructure (GI)—enhancing Europe's Natural Capital. COM (2013)249 communication green infrastructure. https://biobs.jrc.ec.europa.eu/sites/default/files/generated/files/policy/COM%282013%29249%20Communication%20Green%20Infrastructure.pdf. Accessed 31 Mar 2017

Felix M, Sohr A, Riedel P, Assmann L (2009) Gefährdungspotenzial Steinkohlenhalden Zwickau/Oelsnitz. Kurzbericht (Risk potential of hard coal mining dumps in Zwickau/Oelsnitz. Summary report). http://www.umwelt.sachsen.de/umwelt/download/luft/42_Kurzbericht_SteinkohlenHalden_SN.pdf. Accessed 9 Aug 2017

Feng S, Chang J, Hou W (2016) A framework for setting restoration priorities for coal subsidence areas based on green infrastructure (GI). Acta Ecol Sin 36(9):2724–2731

Firehock K (2010) A short history of the term green infrastructure and selected literature [WWW Document]. http://www.gicinc.org/PDFs/GI%20History.pdf. Accessed 22 Feb 2018

Grunewald K, Bastian O (2015) Ecosystem services—concept, methods and case studies. Springer, Berlin

Grunewald K, Li J, Xie G, Kümper-Schlake L (2017) Towards Green Cities Urban biodiversity and ecosystem services in China and Germany. Springer, Berlin

Haase D, Larondelle N, Andersson E, Artmann M, Borgström S, Breuste J, Gomez-Baggethun E, Gren Å, Hamstead Z, Hansen R, Kabisch N, Kremer P, Langemeyer J, Rall EL, McPhearson T, Pauleit S, Qureshi S, Schwarz N, Voigt A, Wurster D, Elmqvist T (2014) A Quantitative review of urban ecosystem service assessments: concepts, models, and implementation. Ambio 43:413–433. https://doi.org/10.1007/s13280-014-0504-0

Haines-Young R (2013) Common international classification of ecosystem services (CICES): Consultation on version 4 August–December 2012. Report to the European Environmental Agency https://cices.eu/content/uploads/sites/8/2012/07/CICES-V43_Revised-Final_Report_29012013.pdf. Accessed 11 Aug 2017

Hansen R, Pauleit S (2014) From multifunctionality to multiple ecosystem services? A conceptual framework for multifunctionality in green infrastructure planning for urban areas. Ambio 43:516–529. https://doi.org/10.1007/s13280-014-0510-2

Harfst J, Wirth P (2011) Structural change in former mining regions: problems, potentials and capacities in multi-level-governance systems. Procedia Soc Behav Sci 14:167–176. https://doi.org/10.1016/j.sbspro.2011.03.033

Harfst J, Wirth P (2012) Zwickau–Lugau–Oelsnitz (Germany)—the long shadow of mining. In: Wirth P, Černič Mali B, Fischer W (eds) Post-mining regions in central Europe. problems, potentials, possibilities. Oekom, Munich, pp 118–128

Hennek F, Unselt C (2002) Sicherung von Naturschutzflächen in Bergbaufolgelandschaften. Landwirtschaftsverlag Münster

Hoth K, Brause H, Döring H et al (2009) Die Steinkohlenlagerstätte Zwickau (The hard coal deposit Zwickau). Bergbau in Sachsen (Mining in Saxony) https://publikationen.sachsen.de/bdb/artikel/12184/documents/15040. Accessed 16 Aug 2017

Krekel C, Kolbe J, Wüstemann H (2016) The greener, the happier? The effect of urban land use on residential well-being. Ecol Econ 121:117–127. https://doi.org/10.1016/j.ecolecon.2015.11.005

Larondelle N, Haase D (2012) Valuing post-mining landscapes using an ecosystem services approach—an example from Germany. Ecol Ind 18:567–574. https://doi.org/10.1016/j.ecolind.2012.01.008

Lin ZR, Chang J, Wang W (2009) Research on planning of subsidence land's ecological restoration in Xuzhou Mining Area under the Framework of urban-rural coordination. Mod Urban Res 10:91–95

Lintz G, Wirth P (2009) The importance of Leitbilder for structural change in small towns. In: Strubelt W (ed) Guiding principles for spatial development in Germany, German annual of spatial research and policy. Springer, Berlin, pp 75–95

Lockie S, Franettovich M, Petkova-Timmer V, Rolfe J, Ivanova G (2009) Coal mining and the resource community cycle: a longitudinal assessment of the social impacts of the Coppabella coal mine. Environ Impact Assess Rev 29:330–339. https://doi.org/10.1016/j.eiar.2009.01.008

Luo PJ, Chen N (2011) Abandoned coal mine tunnels: future heating/power supply centers. Min Sci Technol V21(5):637–640. https://doi.org/10.1016/j.mstc.2011.10.011

Lupp G, Steinhäußer R, Bastian O, Syrbe RU (2015) Impacts of increasing bioenergy use on ecosystem services on nature and society exemplified in the German district of Goerlitz. Biomass Bioenergy 83:131–140

Ma BG, Hu ZQ (2015) Bioremediation of acid-mine drainage contaminated with acid and heavy metals in coal mine by sulfate-reducing bacteria. In: Hu ZQ (ed) Legislation, technology and practice of mine land reclamation. CRC Press, London, pp 423–427

Maraja R, Barkmann J, Tscharntke T (2016) Perceptions of cultural ecosystem services from urban green. Ecosyst Serv 17:33–39. https://doi.org/10.1016/j.ecoser.2015.11.007

Mell I (2008) Green infrastructure: concepts and planning. FORUM Ej 8:69–80

Müller B, Finka M, Lintz G (2005) Rise and decline of industry in central and eastern Europe. Springer, Berlin

Panduro TE, Veie KL (2013) Classification and valuation of urban green spaces—a hedonic house price valuation. Landsc Urban Plann 120:119–128. https://doi.org/10.1016/j.landurbplan.2013.08.009

Pauleit S, Liu L, Ahern J, Kazmierczak A (2011) Multifunctional green infrastructure planning to promote ecological services in the city. In: Niemelä J (ed) Handbook of urban ecology. Oxford University Press, New York, pp 272–285

People's Government of Jiangsu Province (2008) Opinions about acceleration of the revitalization of the old industrial base of Xuzhou (in Chinese language). http://www.jiangsu.gov.cn/jsgov/tj/bgt/201311/t20131104405504.html. Accessed 16 Aug 2017

Rouse DC, Bunster-Ossa IF (2013) Green infrastructure: a landscape approach. American Planning Association, New York

Shen S, Lin L, Jiang G (2012) Theoretical and empirical of urban planning assessment. Southeast University Press, Nanjing

Smyth R, Mishra V, Qian X (2008) The environment and well-being in urban China. Ecol Econ 68(1):547–555. https://doi.org/10.1016/j.ecolecon.2008.05.017

Syrbe RU (2015) Recultivation and sustainable development of post-mining landscapes. In: Hu ZQ (ed) Legislation, technology and practice of mine land reclamation. CRC Press, London, pp 489–492

UN (2015) United Nations Sustainable Development Goals [WWW Document]. http://www.un.org/sustainabledevelopment/sustainable-development-goals/. Accessed 16 Feb 2018

Wang W, Li J, Wang E, Zhang L, Cao Y, Xu H (2015) Introduction of Sponge City construction. Constr Sci Technol 01:19–21 **(in Chinese)**

Wang WL, Zhang M, Shi Q (2016) Research on replication mechanism of "ecological restoration for coal mining area Xuzhou model". Coal Econ Res 36:27–31

WBGU (2016) Humanity on the move: unlocking the transformative power of cities. Wissenschaftlicher Beirat der Bundesregierung Globale Umweltveränderungen (German advisory council on global change), Berlin

Weller B, Naumann T, Jakubetz S (2012) Gebäude unter den Einwirkungen des Klimawandels. REGKLAM-Publikationsreihe, Heft 3, Rhombos, Berlin

Wende W (2011) Ecological potential and environmental planning of post-mining landscapes—Germany as a role model for China. In: Chang J, Wende W, Luo P, Deng Y (eds) Re-use of the mining wasteland. Tongji University Press, Shanghai, pp 22–39

Wende W (2015) Environmental planning of post-mining landscapes. In: Hu ZQ (ed) Legislation, technology and practice of mine land reclamation. Proceedings of the Beijing international symposium land reclamation and ecological restoration, CRC Press, pp 471–474

Wende W, Marschall I, Heiland S, Lipp T, Reinke M, Schaal P, Schmidt C (2009) Umsetzung von Maßnahmenvorschlägen örtlicher Landschaftspläne. Naturschutz und Landschaftsplanung 41(5):145–149

Wichtmann W, Schröder C, Joosten H (2016) Paludiculture-productive use of wet peatlands. Climate protection-biodiversity-regional economic benefits. Schweizerbart Science Publishers, Stuttgart

Wirth P, Lintz G (2007) Strategies of rehabilitation and development in European mining regions. In: Good (Best) practice cases in regional development after mining and industry. Symposium, May 17–18, 2006, Institute of Geography and Regional Science, University of Graz, Austria. Graz: Institut f. Geographie u. Raumforschung d. Universität, (Grazer Schriften für Geographie und Raumforschung; 42):75–85

Wirth P, Černič Mali B, Fischer W (2012) Post-mining regions in central Europe-problems, potentials, possibilities. oekom Verlag, München

Wissen M (2001) Strukturpolitische Intervention und ungleiche Entwicklung. Zur Rolle des Staates im Strukturwandel. Geographische Revue 3(1):3–22

Wright H (2011) Understanding green infrastructure: the development of a contested concept in England. Local Environ 16:1003–1019. https://doi.org/10.1080/13549839.2011.631993

Zhang XH (2015) Application of ground cover plants in main island of Pananhu Wetland Park. J Anhui Agric Sci 43:4–10

3

Surface movement and deformation characteristics due to high-intensive coal mining in the windy and sandy region

Zhenqi Hu[1] · Chao Chen[1] · Wu Xiao[1] · Xinjing Wang[2] · Mingjie Gao[1]

Abstract As China's energy strategy moving westward, the surface movement and deformation characteristics due to high-intensive coal mining in the windy and sandy region become a research hotspot. Surface movement observation stations were established to monitor movement and deformation in one super-large working face. Based on field measurements, the surface movement and deformation characteristics were obtained, including angle parameters, subsidence prediction parameters, etc. Besides, the angle and subsidence prediction parameters in similar mining areas are summarized; the mechanism of surface movement and deformation was analyzed with the combination of key stratum theory, mining and geological conditions. The research also indicates that compared with conventional working faces, uniform subsidence area of the subsidence trough in the windy and sandy region is larger, the trough margins are relative steep and deformation values present convergence at the margins, the extent of the trough shrink towards the goaf and the influence time of mining activities lasts shorter; the overlying rock movement and breaking characteristics presents regional particularity in the study area, while the single key stratum, thin bedrock and thick sand that can rapidly propagate movement and deformation are the deep factors, contributing to it.

Keywords Windy and sandy region · High-intensive mining · Surface movement and deformation · Key stratum

1 Introduction

Underground coal mining resulted in the overburden collapse and deformation, furthermore, caused the movement and deformation of the land surface (Gray 1990; Hu and Xiao 2013, 2014a, b, c), which are the main forms of the mining impact to the eco-environment and is crucial for ecological restoration (Hu et al. 2014a, b, c; Xiao et al. 2014a, b). There are abundant coal resources in the windy and sandy region of northwestern China, and its occurrence characteristics, including shallow cover depth, thin bedrock, overlying hard strata, near-flat seam with thick sand on the surface, which are suitable for high-intensive coal mining characterized with large advancing speed, high recovery rate and big width/depth ratio, are significantly different from those in eastern China (Du 2010; Zhang et al. 2011; Liu 2014). The coal mines in the eastern China has long been the major mining areas (Xiao and Hu 2014), as China's energy strategy moving westward, high-intensive coal mining is and will be in operation in the Shendong coalfield where the ecological environment is fragile, thus, the surface movement and deformation due to coal mining will be a prominent ecological problem, which poses a great threat to the ecological safety, however, few studies regarding the windy and sandy region can be referred (Wang et al. 2015). With the continually innovation of fully high-intensive mechanized technology, the high-intensive coal mining with super-large working (SLW) face has been widely applied in large coal bases due

✉ Zhenqi Hu
huzq@cumtb.edu.cn

1 Institute of Land Reclamation and Ecological Restoration, China University of Mining and Technology (Beijing), Beijing, China

2 College of Resource and Environment, North China University of Water and Electric Power, Zhengzhou, China

to its advantages as diverse as large panel dimensions, large advancing speed (10 m/d at least), high recovery rate, etc. (Wang 2014). Wu et al. (2010) considered that the large working face reduced the disturbance magnitude compared with the conventional working face in Yanzhou coalfield. Tan et al. (2008) had revealed the surface movement and deformation characteristics of a large working face in eastern China based on field measurements. The research of Teng and Liu (2002) in Wuyang coal mine with high-intensive mining shows similar results to Wu. Li (2010) applied the 3D laser scanning technology in the research on the surface movement characteristics in Yanzhou coalfield, found the convergence of deformation values. Besides, the numerical simulation of mining subsidence cracks is also a research focus (Peng et al. 2009, 2010; Zhang et al. 2011; Liu et al. 2015). The methods of simulation and calculation have been recently applied in the research on overlying strata movement characteristics, ground pressure behavior, subsidence prediction associated with high-intensive mining in the windy and sandy region (Huang 2002; Wang et al. 2007; Xuan 2008; Fan et al. 2011; Ju et al. 2015). Sun (2008) compared the surface movement and deformation characteristics in the windy and sandy region with that in the loess area by using numerical simulation, Tang et al. (2007) proposed a subsidence prediction model and a parameter confirmation method in the windy and sandy region based on the stochastic medium theory. Gu et al. (2014) has obtained the surface movement and deformation characteristics under the condition of subcritical extraction in similar coal mines. The researches concerning the surface movement and deformation characteristics under the condition of high-intensive mining increase annually, however, there are still some unsatisfactory phenomena and need to conduct further researches, those are: (1) many achievements in the similar mining and geological conditions are mostly in the eastern China, theories and methods are more mature, and less researches concerned about the surface movement and deformation characteristics under the condition of critical extraction in the windy and sandy region; (2) most of the existing researches in the windy and sandy region are based on the numerical simulation and functional model, lacking comparison with measured data; (3) It is still unclear whether there is generality of surface movement and deformation characteristics under the condition of high-intensive mining, with certain mining and geological conditions in the windy and sandy region.

Based on above, surface movement observation stations were established in the 12406 super-large working (SLW) face in Bulianta coal mine, to get a few phases of measured data based on the long-term dynamic monitoring, and the surface movement and deformation characteristics in the typical windy and sandy region were revealed. The angle parameters have been acquired after the analysis of the data, and the parameters of probability integral method have also been acquired via non-liner fitting of Origin software, the generalities of the surface movement and deformation characteristics are concluded after comparison and analysis of the surface movement and deformation characteristics and parameters under similar conditions, which would provide theoretical reference for land reclamation and ecological restoration under the condition of high-intensive mining in the windy and sandy region.

2 Study area

2.1 Physical geography situation

The Bulianta coal mine of Shendong coalfield, located in Ordos, Inner Mongolia (Fig. 1). The geographical position is for the north latitude 39°18′31″–39°22′43″, east longitude 110°4′16″–110°10′18″, and the coal mine covers an area of 36.70 km^2 with an altitude from 1130 to 1260 m above sea level.

Shendong coal field is located in the transition zone of Ordos Plateau and Muus Desert, and the surface is covered by quaternary loose windy bed, with hilly topography, terrain is relatively flat, with high elevation in the west and lower in the east. The study area has a semi-arid and desert plateau continental climate, with mean annual precipitation of 357 mm, maximum annual precipitation of 531 mm, and minimum annual precipitation of 195 mm. More than half of the precipitation occurs during the summer (from July to September). The annual average temperature is 6.2 °C, extreme minimum temperature of −31.4 °C, and the extreme maximum temperature of 36.6 °C. The average frost-free period is between 130 and 140 days. The annual sunshine hours is from 2740 to 3100 h. Potential yearly evaporation is around 2163 mm, about 6 times of the annual evaporation. In absence of earthquake date, according to the *China earthquake intensity zoning map*, the earthquake intensity degree of the study area is below 6.

2.2 Mining and geological conditions

This paper takes the 12406 SLW face of Bulianta coal mine as the study object, which has commenced to mine the 1^{-2} coal seam on Apirl and completed on December, 2011.The virgin zone lied on uphill, and the adjacent old goaf lied on the downhill. The SLW face mined 1^{-2} coal seam, with a dip angle from 1° to 3°, which is stable and belongs to flat seams. The thickness of the coal seam ranges from 4.19 m to 5.56 m with an average thickness of 4.81 m. The depth of the coal seam was from 190 m to 220 m. The

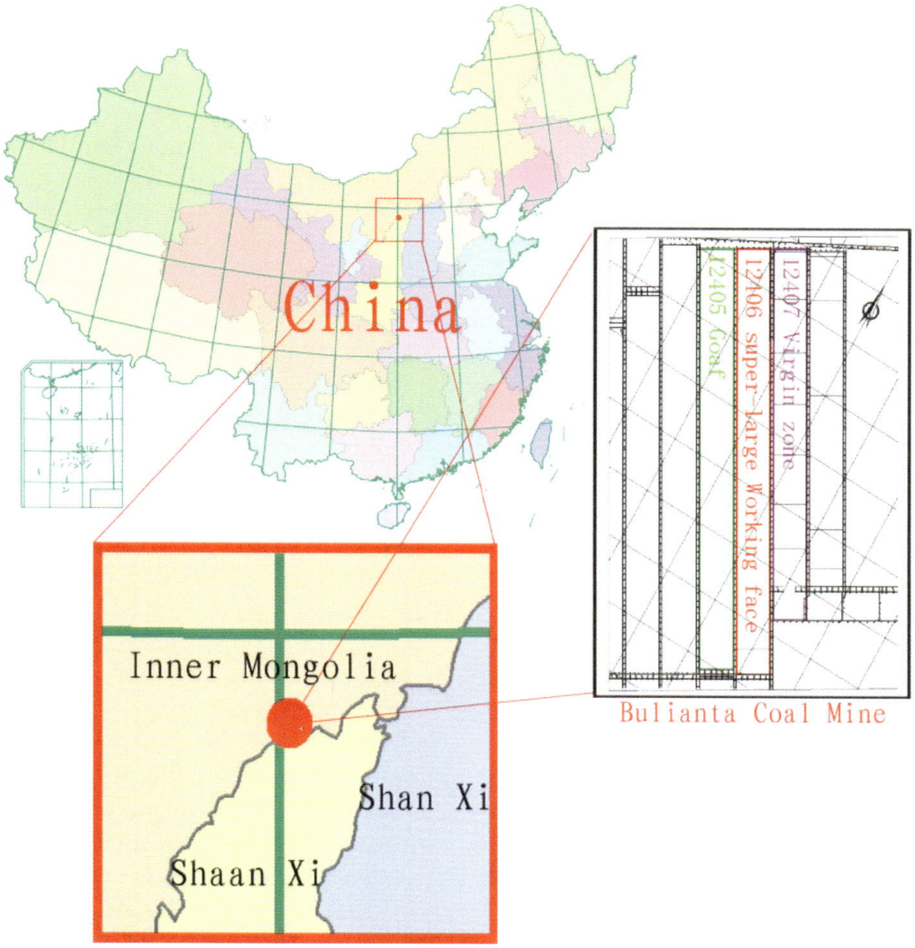

Fig. 1 Location of the study area

overburden was composed of 180–200 m medium-hard rock which is partial to hard and 8–27 m quaternary loose bed, a typical geological section was shown in Fig. 2. Surface is covered by loose sandy soil with strong flow characteristic and thin desert soil crusts. The SLW face adopted longwall fully-mechanized and full-seam mining method, the roof of which is managed through the all fall method, the longwall panel was 3592 m long and 300.5 m wide, average mining thickness (*M*, meter) is 4.5 m, with advancing speed of 12 m/d. At present, all robbing work has been finished.

3 Materials and methods

In order to monitor the surface movement and deformation characteristics, surface movement observation stations were established by adopt the layout method of profile striation observation station according to the situation of landform and terrain and the mining geological conditions in the SLW face (Zou et al. 2003). In consideration of critical extraction that can be achieved in the longitudinal direction, the observation points were just needed to be deployed in the critical extraction region, therefore half loaf of observation line was layout along the longitudinal main section while a loaf of observation line was layout along the transverse main section, two observation line are perpendicular to each other. According to mining subsidence theory and the data of similar mine areas (Zou et al. 2003; Wang 2014), the area that 300–400 m away to the stopping line can achieve critical extraction when the mining depth is approximately 200 m. Based on the above research, observation points were began to be layout 400 m away from the open-off cut outside the SLW face, and it ended until 500 m away from the open-off cut above the longitudinal main section, the distance between each points is 20 m and thus the total length of the longitudinal observation line is 900 m. In order to guarantee the deformation observation results of the transverse direction, the starting point should be layout 400 m away to the boundary of the SLW face, thus the total length of the transverse direction is 1100 m. Considering the effect of

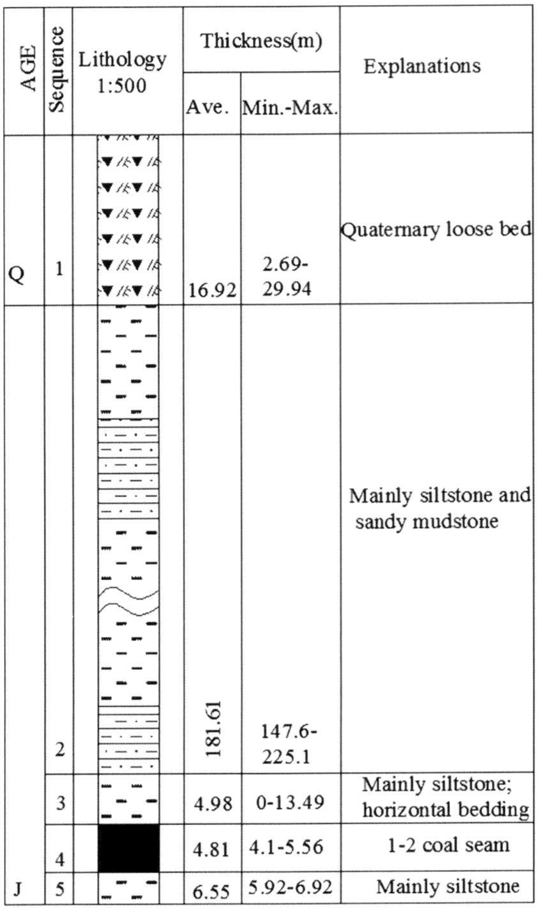

Fig. 2 The lithological column of No. 12406 SLW face

the old adjacent goaf, more specifically, the distance between each points is 30 m on the old adjacent goaf side, and 20 m on the side without mining. The observation points are cement column with a depth of 1 m, which were made in the field, of which the port is 10 cm × 10 cm while the bottom is 30 cm × 30 cm, in spite of this, the part of the cement column on the ground ranges from 5 cm to 10 cm. According to the requirement of the *Coal mine measurement specification*, comprehensive measurement and subsidence measurement had been conducted six times and eleven times, respectively. The layout situation of the surface movement observations is shown as Fig. 3.

4 Results and discussion

4.1 Data analysis

(1) The analysis of subsidence data

According to the observation data, the subsidence curve of longitudinal direction and transverse direction can be obtained respectively (Gao 2012; Wang 2014; Wang et al. 2015), as shown in Fig. 4.

As shown in Fig. 4a, the subsidence areas are becoming larger with the continuous advancing. The range and values of subsidence gradually increased, and then subjected to be stable without any abnormal phenomenon, which followed the general laws of mining subsidence. However, subsidence curve has its specific characteristics, when comparing with those in the mid-eastern China mining areas, such as Yanzhou mining area (Wu et al. 2010), which are: (1) the uniform subsidence areas were relatively larger in subsidence basin and the trough margins was more sharp. There was an obvious difference of the subsidence curve, which showed more convergence at the two ends of open-off cut, leading to the extent of the trough towards the goaf. The subsidence value of 10 mm was taken as the boundary of the affected mining areas, and the boundary was approximately 1.0*H* (*H*: the average mining depth, meter) away to the open-off cut. (2) when the working face was approximately 1.5*H* away to the observation points, the subsidence value reached the peak, and then maintained stable. During this period, the affected time lasted 27 days where around the observation points. The reason is that with the increasing advancing speed, the time of surface deformation continuously decreased, which speed up the recovery process of surface ground. As Fig. 4b shows, the process of surface deformation and the shape of the distribution in the transverse direction followed those in the longitudinal direction basically. Because the subsidence was affected by the adjacent old goaf, the subsidence values and range on the side of old goaf was larger, it is safe to indicate that multi-disturbance can result in the increase of subsidence values and range. In additional, inclination deformation values concentrated at the two sides of the mining boundary due to high-intensive mining, while inclination deformation outside the mining boundary decreased sharply, nevertheless, curvature deformation followed the general laws of mining subsidence.

(2) The analysis of horizontal movement data

As shown in Fig. 5a, the positive horizontal movement was larger than negative horizontal movement, the horizontal movement towards goaf was larger than that towards coal pillars, and the horizontal movement basically approached to goaf, and the pillar which is 1.5*H* away to the open-off cut occurred horizontal deformation. The maximum horizontal movement occurred in the observation point which is 0.2*H* away to the open-off cut, and the maximum negative horizontal movement occurred in the observation point which is 0.5*H* away to the open-off cut. The horizontal movement towards the mining direction was larger than that towards open-off cut overall, and the horizontal movement was near the open-off cut (Fig. 5a). *X* direction of the maximum movement point of transverse line was towards the mining direction, and *Y* direction was

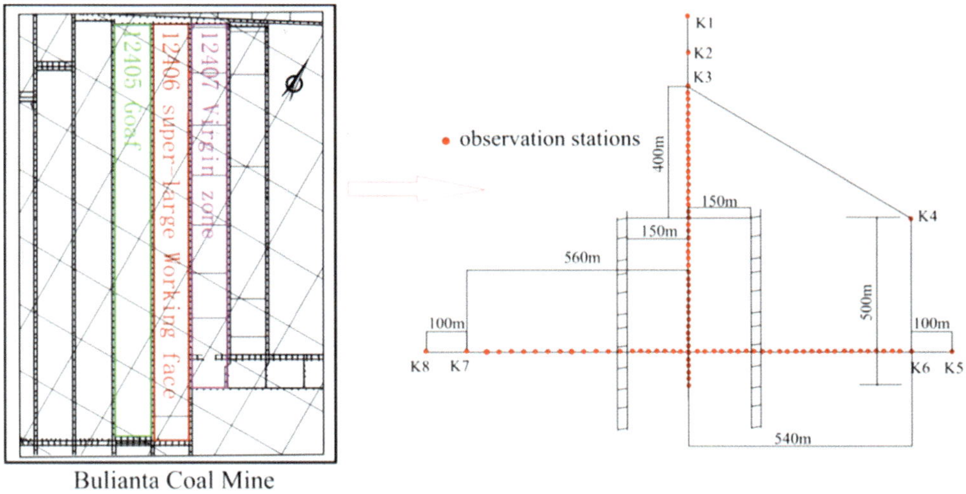

Fig. 3 Layout plan of some SLW faces and the observation stations in 12406 SLW face

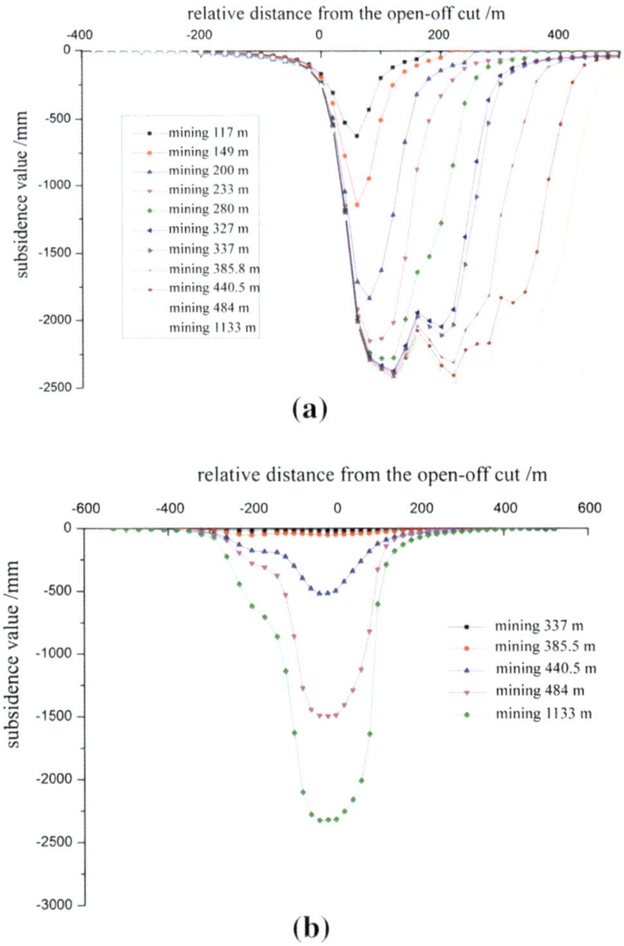

Fig. 4 **a** Subsidence curve of longitudinal direction, **b** Subsidence curve of transverse direction

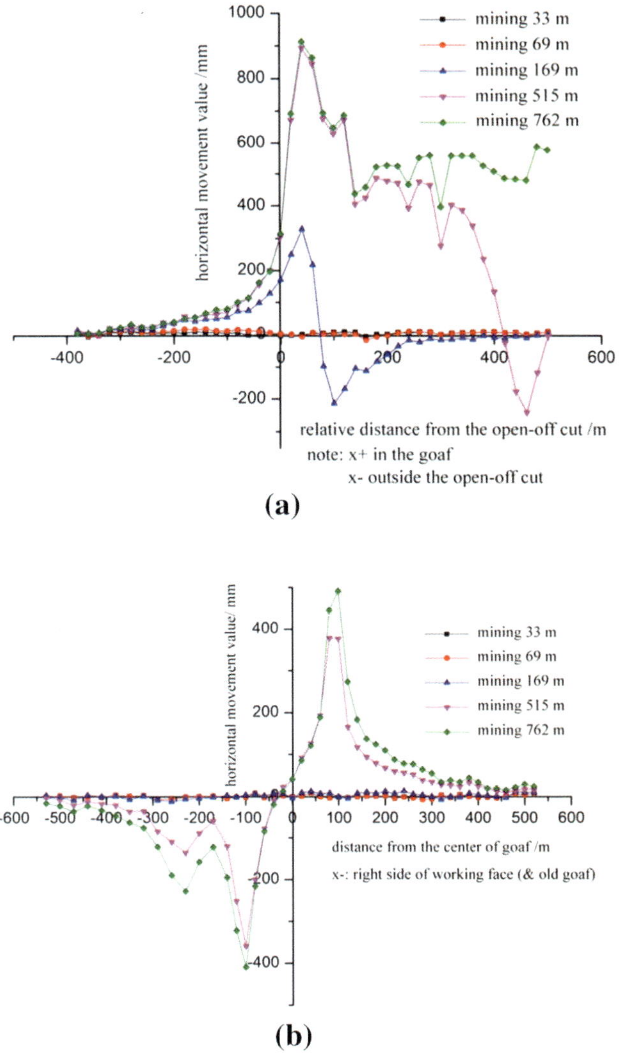

Fig. 5 **a** Horizontal movement curve of longitudinal direction, **b** Horizontal movement curve of transverse

towards the old goaf, which was resulted from that the surface near the old goaf has experienced primary subsidence. When mining began in the SLW face, subsidence values were relatively small, while the general subsidence values was still larger than that in the virgin zone, and the horizontal movement was more complex comparing with that in the new mining areas. The horizontal movement changes were relatively obvious in which the range is approximately 1.2*H* away to the open-off cut, with the horizontal displacement of uphill and downhill pointing to the center of the SLW face. The two sides of surface in the transverse direction exhibited a trend pointing to the center. In addition, the horizontal deformation of longitudinal direction concentrated on the open-off cut and the advancing direction of the SLW face, however, the horizontal deformation in virgin zone side of the transverse direction was generally larger than that in the goaf (Fig. 5b). Therefore, the point with the critical deformation value is closer to the mining boundary, and the damage range reduced further (Gao 2012; Wang 2014; Wang et al. 2015).

(3) Calculation of surface angle parameters

According to the theory of mining subsidence (Zou et al. 2003), the boundary of the trough is generally described by the parameters of the angle, such as the angle of critical extraction, angle of greatest subsidence, limit angle, angle of displacement and angle of crack. Because of the insufficiency of observation data in the longitudinal direction, moreover, the coal seam belongs to the flat seam, which is stable, therefore, the angle parameters of transverse direction were mainly analyzed. Based on the analysis of the subsidence curve, horizontal movement curve and field measured data, the angle parameters of the SLW face obtained are shown in Table 1 (Gao 2012; Wang 2014).

According to the above parameters of the angle, the distribution range of the surface movement and cracks exhibited a trend of internal shrinkage to the adjacent goaf.

(4) Probability integral fitting for parameters based on the Origin software

There are some limitations in the method that based on the definition of the predicting parameters in the probability integral method and the field measured data to calculate subsidence parameters, while the Origin software

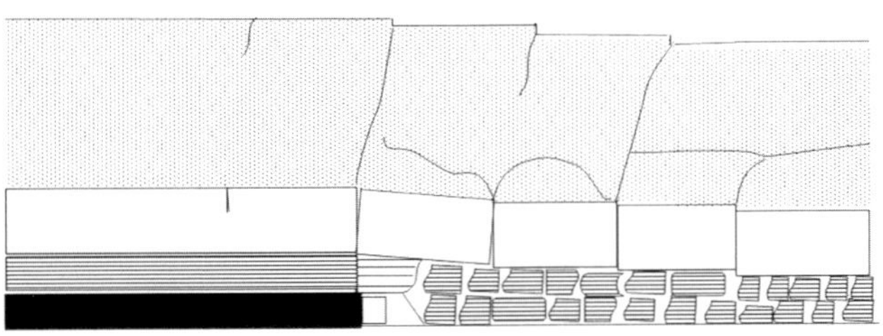

Fig. 6 Overall cut and fall of the overburden and step convergence

has a strong user-defined function, the predicting parameters are obtained via combining theoretical formula of the probability integral method. The fitting progress of the Origin software with the application of probability integral method is described as follows (Gao 2012).

subsidence:

$$W(x) = W_0 \sum \int_0^D \frac{1}{r} \exp\left(-\pi \frac{(x-s)^2}{r^2}\right) ds \tag{1}$$

tilting:

$$i(x) = \frac{W_0}{r} = e^{-\pi \frac{x^2}{r^2}} \tag{2}$$

curvature:

$$k(x) = -\frac{2\pi W_0}{r^3} x e^{-\pi \frac{x^2}{r^2}} \tag{3}$$

horizontal movement:

$$U(x) = bri(x) = bW_0 e^{-\pi \frac{x^2}{r^2}} \tag{4}$$

where, W_0 is the surface maximum subsidence value(critical extraction); D is the mining length;s is the mining unit; r is the main influence radius; b is the horizontal movement coefficient.

Whether the underground ore body reach critical extraction or not can be determined by the following formula.

$$n_1 = K_1 \frac{D_1}{H_0}, \quad n_3 = K_3 \frac{D_3}{H_0} \tag{5}$$

In the formula, K_1 and K_3 are the coefficients which are smaller than 1, D_1 and D_3 respectively, are the length of transverse and longitudinal in the actual goaf, when n_1 and n_3 are larger than 1, it reached critical extraction, on the contrary, it is subcritical extraction, the value of K_1 and K_3 are equal to 0.7.

On the condition of SLW face, $D_1 = 300.5$ m, $D_3 = 3592$ m, $H_0 = 188$ m, the calculated n_1 and n_3 via the formula were larger than 1, namely the SLW surface reached critical extraction.

The parameters are calculated via multiple iterations, the fitting results are shown in Table 2 (Gao 2012; Wang 2014).

4.2 The generality of surface movement and deformation

After referring to the published papers (Xu et al. 2005; Yi 2008; Du 2010; Gao 2012; Wang 2014), the mining geological conditions, parameters of surface movement angle and probability integral method of the selected similar

Table 1 Angle parameters

Angle of greatest subsidence θ (°)	Limit angle (°)			Displacement angle (°)			Crack angle (°)			Angle of critical extraction (°)	
	Longitudinal direction δ_0	Uphill γ_0	Downhill β_0	Longitudinal direction δ	Uphill γ	Downhill β	Longitudinal direction δ''	Uphill γ''	Downhill β''	Uphill ψ_1	Downhill ψ_2
83.9	55	52	57	76	83	84	88	88	89	51	50

SLW face are summarized as shown in the following Tables 3, 4, and 5.

(1) Mining and geological conditions of similar SLW faces

According to the Table 3, the better occurrence condition of coal resource is suitable for high-intensive coal mining in the windy and sandy region.

(2) The generality of static angle parameters

The angle parameters of similar mining areas are summarized in the table, the angle parameters showed some generality despite the differences among them.

(i) The limit angles of each subsidence basin are basically identical, the range of the boundary is approximately equal to the *H*. For the same SLW face, the limit angle of longitudinal direction is approximately equal to those of uphill (downhill), and exhibits a trend that the larger the mining depth the greater the limit angle, the larger mining thickness the smaller the limit angle, the thicker the topsoil the smaller the limit angle. On the condition that the mining depth of summarized SLW faces are approximately equal to each other, the limit angle decreases with the reduction of the ratio (*H*/*M*), thus the subsidence basin is relatively larger.

(ii) On the condition that the ratio is larger than 30, the displacement angle is generally larger than 75°, with the concentrated movement deformation, moreover, the displacement boundary become smaller, pointing to the goaf; If the ratio is smaller than 30, the displacement angle would decrease obviously, which can be attributed to that the ratio is small, the fracture height of overlying strata increased and the fractured zone is easy to propagate to surface, severe destruction of overlying strata above the coal pillar, and the expansion of the displacement boundary. As to the same mining unit, displacement angles of uphill (downhill) are slight larger than that of longitudinal direction.

(3) The generality of the parameters of probability integral method

Parameters of probability integral method of similar SLW faces in windy and sandy region are summarized in Table 5, there are some generality as follows.

(1) In the study area, the subsidence coefficients are basically between 0.55 and 0.61, and it is relatively small, which can be attributed to the following factors: (1) The lithology of overlying strata is medium hard and hard; (2) The ratio between bedrock ply and top soil thickness is relatively large, which can reach 10 generally. As shown above, the

Table 2 Parameters of probability integral method

Subsidence coefficients q	Horizontal movement factor b	Tangent of major influence angle tan β	Deviation of inflection point s (m)
0.55	0.26	2.51	40

Table 3 Mining and geological conditions of similar SLW faces in the windy and sandy region

SLW face	Bulianta 12406	Cuncaota 22111	Liuta 12106	Daliuta 52304	Daliuta 1203	Bulianta 31401
Average mining depth H (m)	200	250	150	225	61	255.4
Average mining thickness M (m)	4.5	2.8	6.9	6.9	4.03	4.2
Ratio of H/M	44.4	89.3	21.7	32.3	15.1	60.7
longitudinal length L (m)	3592	2085	633	4547	938	4629
transverse length L (m)	300.5	224	246.8	301	150	265.25
Topsoil thickness h (m)	17	8	30	30	26.5	/
Average coal seam dip angle (°)	1–3	1	1	1–3	0.3	1–3
Advancing speed C (m/d)	12	9.7	5	7.4	2.4	>12

Table 4 Angle parameters of similar SLW faces in the windy and sandy region

SLW face	Bulianta 12406	Cuncaota 22111	Liuta 12106	Daliuta 52304
Measured limit angle of longitudinal direction (°)	55	48	48.3	42.5
Measured limit angle of uphill (downhill) (°)	54.5	51.9	51.2	42.2
Measured displacement angle of longitudinal direction (°)	76	74.3	61.1	/
Measured displacement angle of uphill (downhill) (°)	83.5	75.9	61.2	82

Note Comprehensive limit angle and displacement angle in this table

Table 5 Parameters of probability integral method of similar SLW faces in windy and sandy region

SLW face	Bulianta 12406	Daliuta 1203	Daliuta 52304	Bulianta 31401
Subsidence coefficient	0.55	0.59	0.61	0.55
Horizontal movement factor	0.26	0.29	0.32	0.127
Tangent of major influence angle	2.51	2.65	2.88	3.40
Deviation of inflection point S_0 (m)	40	20.4	37.5	29
Maximal subsidence velocity (mm/d)	268	/	430	540

subsidence coefficient is mainly related to the geological conditions in the study area.

(2) The horizontal movement factors are basically between 0.20 and 0.30, which accorded with the general values of other ordinary fully-mechanized working faces (Wang 2014). It is safe to conclude that a significant increase doesn't appear in the horizontal displacement due to the high-intensive mining.

(3) The tangent of major influence angle are basically between 2.51 and 3.40, which shows that the main surface deformation value is more concentrated, with a steep surface subsidence basin and there is a obvious difference of the subsidence curve at the two sides of the mining boundary pointing to the goaf, moreover, it is consistent with the field data, which can be mainly attributed to such factors as the rapid mining, the acceleration of the overburden

subsidence velocity, the reduction of relative suspended time.

(4) The surface subsidence velocity increases obviously due to rapid mining, and it also becomes larger with the increase of the mining thickness. In the mining process, the dynamic change rate of the surface subsidence velocity is relatively large, moreover,-compared with the published paper, it shortens the cycle of mining impact on the land ecological environment (Qu et al. 2009).

4.3 The analysis of the mechanism of surface movement and deformation

The process of the surface movement and deformation in the study area is in line with the general mining subsidence characteristics, however, on the condition of high-intensive coal mining, the law of overburden rock movement and breaking exhibits regional particularity owing to the special occurrence characteristics of coal resources in the windy and sandy region, thereby leading to the certain particularity of the surface movement and deformation characteristics. Published researches (Qian et al. 2010) indicated that there exists single key stratum structure, the lithology is hard partial to hard, which determined the particularity of the roof bedrock movement of the SLW face.

In some study areas, the fractured zone is likely to propagate to surface, and ground pressure behavior is severe, resulting in stepped subsidence (Fig. 6) (Qian et al. 2010) with larger subsidence velocity of some measured points. The subsidence curve is steep but relatively stable on the side of open-off cut, whereas the subsidence curve advanced in a larger angle on the SLW face advancing side in the mining process. When achieving critical extraction, a flat base appeared in the subsidence basin of which the range was larger than the general ones. In addition, high-intensive mining are responsible for the reduction of the abscission layer and fracture and accelerating the compaction and stability of damaged overburden, thereby reducing the overburden damage effectively (Yi 2008; Du 2010; Zhang and Li 2013) and promoting the self-healing of the damaged topsoil (Hu et al. 2014a, b, c).

5 Conclusions

(1) The uniform subsidence area of the subsidence basin due to high-intensive coal mining in the SLW face is larger, with relative steep trough margins and concentrated deformation value, the range of the subsidence basin shrinks to the goaf with shorter influence time of mining activities, which is consistent with field observations; The angle of greatest subsidence is 83.9°, the limit angle of longitudinal direction and those of uphill (downhill) are 55°, 52°, 57°, respectively, the displacement angle of longitudinal direction and those of uphill (downhill) are 76°, 83°, 84°, respectively, the crack angle of longitudinal direction and those of uphill (downhill) are 88°, 88°, 89°, respectively, moreover, the angle of critical extraction of uphill (downhill) are 51° and 50° respectively; Among the probability integration method parameters, the subsidence coefficient, the horizontal movement factor and the tangent of major influence angle are 0.55, 0.26 and 2.51 respectively, while the deviation of inflection point is 40 m.

(2) In the similar circumstances, the limit angles of each SLW face are basically identical, the range of the boundary is approximately equal to H. For the same face, the limit angle of longitudinal direction is approximately equal to those of uphill (downhill), and exhibit a trend that the larger the mining depth the greater the limit angle, the larger M the smaller the limit angle, the thicker the topsoil the smaller the limit angle. In the condition that the H is approximately equal to each other, the limit angle decreases with the reduction of the ratio of H/M, thus the subsidence basin is relatively larger; In the condition that the ratio is larger than 30, the displacement angle is larger than 75° generally, with a concentrated movement deformation value, thereby the displacement boundary becomes smaller and shrinks to the goaf; If the ratio is smaller than 30, the displacement angle would decrease obviously. For the same SLW face, displacement angles of uphill (downhill) are slight larger than that of longitudinal direction.

(3) In the similar circumstances, the subsidence coefficients are small, mainly relating to the medium-hard overburden rock which is partial to hard; Under the condition of high-intensive coal mining, the horizontal movement factor did not increase, on the contrary, the tangent of major influence angle and the maximum subsidence velocity are relatively larger.

(4) The law of overburden rock movement and breaking exhibits regional particularity because of the condition of high-intensive coal mining and the special occurrence characteristics of coal resources, such as shallow coal seam, thin bedrock and thick sand bed in the study area, while the single key strata, the thin bedrock and thick sand bed which can rapid propagate movement and deformation are the deep factors that contribute to the certain particularity of surface movement and deformation.

Acknowledgments Financial supports for this work, are provided by the National Natural Science Foundation of China (NSFC) & Shenhua Group Corporation Limited key support project of the coal joint fund (U1361203) and NSFC under Grant No. 41501562. Thanks are also due to some participants for rendering assistant cooperation during studies.

References

Du SZ (2010) The rule of the overlying strata movement and their phytoremediation technology after large-scale underground mining operation in Shendong coal mine area. China University of Mining and Technology (Beijing), Beijing

Fan GW, Zhang DS, Ma LQ (2011) Overburden movement and fracture distribution induced by longwall mining of shallow coal seam in the Shendong coalfield. J China U Min Technol 40(2):196–202

Gao MJ (2012) Research the law of ground movement after mining in shallow large-scale in Shendong coal mine area. China University of Mining and Technology (Beijing), Beijing

Gray RE (1990) Mining subsidence—past, present, future. Int J Min Geol Eng 8(4):400–408

Gu SC, Wang EB, Xiong JQ (2014) Field measurement study on movement law of insufficient mining movement surface ground above shallow depth seam. Coal Eng 6:99–102

Hu ZQ, Xiao W (2013) Optimization of concurrent mining and reclamation plans for single coal seam: a case study in northern Anhui, China. Environ Earth Sci 68(5):1247–1254

Hu ZQ, Wang XJ, He AM (2014a) Distribution characteristic and development rules of ground fissures due to coal mining in windy and sandy region. J China Coal Soc 39(1):11–18

Hu ZQ, Yang GH, Xiao W, Li J, Yang YQ, Yu Y (2014b) Farmland damage and its impact on the overlapped areas of cropland and coal resources in the eastern plains of China. Resour Conserv Recy 86:1–8

Hu ZQ, Long JH, Wang XJ (2014c) Self-healing, natural restoration and artificial restoration of ecological environment for coal mining. J China Coal Soc 39(8):1751–1757

Huang QX (2002) Ground pressure behavior and definition of shallow seams. Chin J Rock Mech Eng 21(8):1174–1177

Ju JF, Xu JL, Zhu WB (2015) Longwall chock sudden closure incident below coal pillar of adjacent upper mined coal seam under shallow cover in the Shendong coalfield. Int J Rock Mech Min Sci 77:192–201

Li L (2010) Study on damage mechanics and repair methods of embankment under the conditions of high-strength underground mining. China University of Mining and Technology, Xuzhou

Liu H (2014) The development law and treatment technology of ground fissures due to underground mining in Loess Hilly Area of Western China. China University of Mining and Technology, Xuzhou

Liu H, Deng KZ, Lei SG, Bian ZF (2015) Mechanism of formation of sliding ground fissure in loess hilly areas caused by underground mining. Int J Min Sci Technol 25(4):553–558

Peng YW, Qi QX, Li HY, Deng ZG (2009) Numerical simulation and analysis on influencing factors of fractured zone height of high-strength underground mining. J China Coal Soc 34(2):145–149

Peng YW, Qi QX, Wang YG, Deng ZG, Li HY, Li CY (2010) Study of field measurement of mining-induced coal fracture field and its application. China J Rock Mech Eng 29(s2):4188–4193

Qian MG, Shi PW, Xu JL (2010) Mining pressure and strata control. China University of Mining and Technology Press, Xuzhou

Qu JF, Li G, Zhang SL, Li KX, Zhang P (2009) Study on exploitation and utilization of water and land resources to mining subsidence in the plain area of higher underground water. China Min Mag 18(3):64–66

Sun HX (2008) Rules comparison of surface movement induced by coal mining under aeolian sand and loess. Coal in Technol 1:6–9

Tan ZX, Wang ZS, Li YJ, Sun XM, Deng KZ (2008) Field research on ground subsidence rules of intensive fully-mechanized mining by sublevel caving. J Min Saf Eng 1:59–62

Tang FQ, Yao WQ, Xia YC (2007) Prediction method of ground subsidence for coal mining in seam under thin base rock. Coal Sci Technol 6:103–105,66

Teng YH, Liu KG (2002) Research on surface ground displacement law under high tension coal mining conditions in Wuyang Mine. Coal Sci Technol 4:9–11,15

Wang JA, Zhao ZH, Hou ZY (2007) Study on the catastrophic collapse of surface land induced by mining under a shallow and hard strata. J China Coal Soc 10:1051–1056

Wang XJ, Hu ZQ, Hu QF, Chen C (2015) Evolution and self-healing characteristic of land ecological environment due to super-large coalface mining in windy and sandy region. J China Coal Soc 9:2166–2172

Wu K, Ge JX, Wang LD, Zhou M (2010) Integrated technology of mining subsidence. China University of Mining and Technology Press, Xuzhou

Xiao W, Hu ZQ (2014) GIS-based pre-mining land damage assessment for underground coal mines in high groundwater area. Int J Min Miner Eng 5(3):245–255

Xiao W, Hu ZQ, Chugh YP, Zhao YL (2014a) Dynamic subsidence simulation and topsoil removal strategy in high-groundwater table and underground coal mining area—a case study in Shandong Province. Int J Min Reclam Environ 28(4):250–263

Xiao W, Hu ZQ, Fu YH (2014b) Zoning of land reclamation in Coal mining area and new progresses for the past 10 years. Int J Coal Sci Technol 1(3):177–183

Wang XJ (2014) Monitoring, evolution and self-healing characteristic of land damage due to high tension coal mining in windy and sandy area. China University of Mining and Technology (Beijing), Beijing

Xu YN, He F, Wu ZS, Yuan HC (2005) Prediction of mining cave-in index of Shendong coal-mine area. China Coal 12:37–40,43

Xuan YQ (2008) Research on movement and evolution law of breaking of overlying strata in shallow coal seam with a thin bedrock. Rock Soil Mech 2:512–516

Yi MS (2008) Study and application of key strata theory in shallow seam of Shendong mining area. China University of Mining and Technology, Xuzhou

Zhang YJ, Li FM (2011) Monitoring analysis of fissure development evolution and height of overburden failure of high tension fully-mechanized caving mining. China J Rock Mech Eng 30(s1):2294–2301

Zhang JM, Li P, Gao L (2013) Research on structural damage of mining overburden rock of oversized fully mechanized coal mining face. Shenhua Sci Technol 4:20–23

Zhang DS, Fan GW, Ma LQ, Wang XF (2011) Aquifer protection during longwall mining of shallow coal seams: a case study in the Shendong Coalfield of China. Int J Coal Geol 86(2–3):190–196

Zou YF, Deng KZ, Ma WM (2003) Mining subsidence engineering. China University of Mining and Technology Press, Xuzhou

4

40 years of the Surface Mining Control and Reclamation Act (SMCRA): what have we learned in the State of Wyoming

Brenda K. Schladweiler[1]

Abstract The Surface Mining Control and Reclamation Act (SMCRA) was enacted in 1977 and was the overriding federal regulation governing mining and reclamation of surface coal mines in the United States of America (USA). Many of the newest surface mines in the USA, at that time, were in the western portion of the nation. Wyoming surface coal mines numbered approximately 20 and were located throughout the state in the coal bearing regions, generally in the south/southwestern portion of the state and the northeastern corner. The Office of Surface Mining Reclamation and Enforcement (OSMRE) is the federal agency tasked with oversight of the implementation of SMCRA. Individual states developed statewide programs that met or exceeded the requirements of SMCRA and, thus, obtained primacy over coal mine mining and reclamation within their boundaries as long as those regulatory conditions were maintained. OSMRE retained oversight on the Indigenous Nations programs within the USA. Much information has been learned on the reclamation side of SMCRA, its regulations and state programs since its passage. This paper and presentation will present some of the basic changes made in reclamation programs over the last 40 years in the State of Wyoming.

Keywords Regulatory · Policies · Historical changes

1 Introduction

Many environmental issues, including those related to surface mining in the United States of America (USA), were being addressed by the federal government in the late 1960's and 1970's. However, reclamation was not new to the mining industry. Wyoming reclamation laws pre-dated the federal Surface Mining Control and Reclamation Act (SMCRA) of 1977. The Open Cut Land Reclamation Act of 1969 and the subsequent Environmental Quality Act of 1973 provided a regulatory framework for state oversight of surface mining, and, in particular, coal mining. Some researchers and agency personnel understood the vast coal resources in Wyoming and the need for such a framework (Thilenius and Glass 1974).

✉ Brenda K. Schladweiler
bschladweiler@bksenvironmental.com

[1] BKS Environmental Associates, Inc. and Accord Resource Solutions, LLC, P.O. Box 3467, Gillette, WY 82717, USA

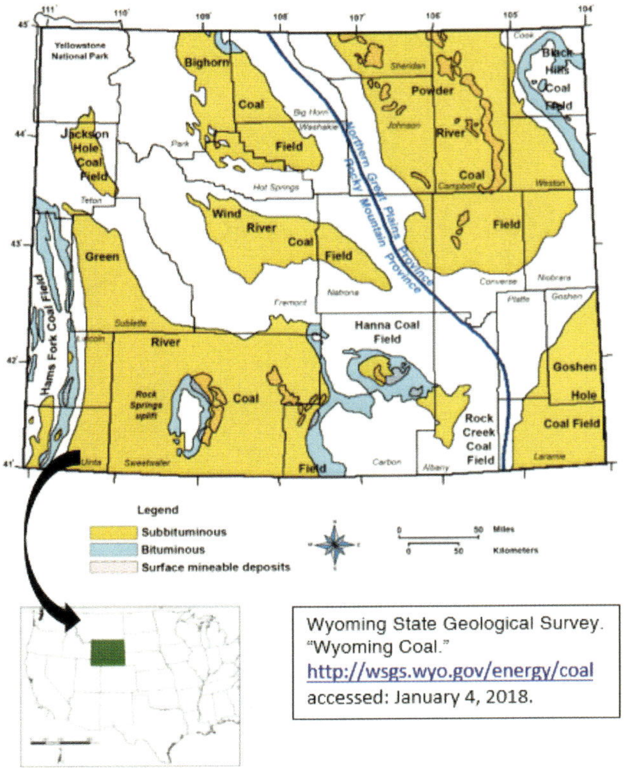

The Office of Surface Mining Reclamation and Enforcement (OSMRE) is the federal agency tasked with oversight of the implementation of SMCRA. The federal government allows those states with surface coal mines to meet or exceed their implementation of the regulations and attain "primacy". States in the Eastern USA have slightly different requirements than the Western United States, e.g., length of the bonding period being 5 years instead of 10.

Learning curves abounded in the late 1970's and 1980's for industry, regulators, researchers and service providers. Much research was carried out during the late 1980's and early 1990's in Wyoming under the Abandoned Coal Mine Land Research Program.

On the 40th anniversary of SMCRA, the author reflects on the knowledge gained over this time. In addition, several people knowledgeable in past or current reclamation efforts were interviewed for insight into the extent of "what we have learned". Much of this information is from the Western USA and, in particular, the State of Wyoming (where mines in 1977 were relatively new). The following general categories of knowledge were discussed in the present paper.

1.1 Influence of agriculture

Many of the early reclamation-based regulations were influenced by agricultural practices in place at the time. Reclamation was considered a phase of mining and not necessarily a field of science in its own right. A simplistic way of looking at reclamation was "just getting something to grow and stabilize a site". In the Eastern United States, that meant forested lands and subsequent land use. In the Western United States, the dominant land uses were grazing lands for domestic livestock and wildlife habitat. Prime farmland was a part of the Midwestern United States, and cropland and hayland were dominant uses.

Eventual knowledge of seeding drastically disturbed lands did not necessarily mirror agricultural systems in the following ways.

1.1.1 Seed mixes

Seed mixes were rudimentary and much less diverse in 1977. The emphasis was on commercially available seed and quick-establishing plants with little understanding or regard to inter- and intraspecific species competition. Ecological diversity, longevity and wildlife habitat value were often secondary considerations. Species such as crested wheatgrass (*Agropyron cristatum*) and smooth brome (*Bromus inermis*) (often used for pasture grass) were commonly used in reclamation but are now considered invasive and thus undesirable. Commonly seeded legumes were also agriculturally-based, i.e., alfalfa (*Medicago sativa*) and yellow sweetclover (*Melilotus officinalis*).

Since 1977, the seed industry has essentially "caught up" to market demands. Natural Resource Conservation Service (NRCS) Plant Material Centers throughout the Western USA developed hardy and prolific varieties of commonly seeded grasses, especially cool-season perennial grasses such as wheatgrasses. Many of these have the ability to outcompete other perennial grasses, forbs and shrubs; therefore, their overall use is now more limited. The seed for many warm season grasses such as blue grama (*Bouteloua gracilis*) was not commercially available and, until wildlife habitat for species of concern became an issue, the use of perennial forbs was limited. Commercial availability of perennial forb seed was also lacking.

In the late 1970's seed sources were limited, as were plant varieties. Markets for seed that could tolerate colder northern climates had not generally been developed prior to SMCRA. Often, seed sources were from more southern climates which, when planted in the north, caused many of the germinating seedlings to winter-kill or die. The push for adapted seed began in the mid-1980's. Hand collection of the less commercially available forb seed began about that time frame; however, such seed was costly and supplies limited but, in many ways, still is.

1.1.2 Seeding equipment

Many of the early seed drills, used for grains and other agricultural crops, were used for seeding reclaimed ground. Much of this ground was rocky with large clods of soil which caused problems for traditional agricultural equipment. Rangeland seeders have now been developed to handle the rockier soil.

In agriculture, a smooth firm seed bed is desirable. While that is still the case for flatter reclaimed areas, the likelihood of flat, smooth, clod-free seed beds in drastically disturbed areas is less likely. Rough is now considered better in some situations, especially steep slopes, especially those exposed to wind or water erosion. Therefore, pitters or imprinters can be helpful equipment in those instances, especially if followed with broadcast seeding/chaining to get good seed/soil contact.

The type of seed also made it difficult for passage through many traditional agricultural drills. Fluffy (trash) perennial grass seed (and some shrubs) containing awns and brackets did not feed adequately through seed chambers and tubing on drills. A new generation of drills was created by manufacturers to handle trash seed.

Broadcasting of shrub and forb seed was later found to improve chances of survival of those life forms. Seeding of shrubs and forbs, if drill seeded with perennial grasses, did not allow long-term survival of these species with the overly competitive cool-season grasses.

1.1.3 Soil amendments

Nitrogen application was once a commonly applied fertilizer to reclaimed systems in the Western USA, especially if used in conjunction with surface hay or straw mulch. However, such applications were found to encourage annual weed growth and, if misapplied, could cause nitrate pollution of nearby waterways. Its use in many reclaimed areas is now discouraged. Phosphorus, on the other hand, can stimulate root growth, and, where limiting in the system, is a beneficial fertilizer.

In a semi-arid/arid environment such as Wyoming, topsoil is limiting. The dark, rich in organic matter soils, such as those found in the Midwestern USA, is limited to the extreme eastern edge of Wyoming. Every effort is made to salvage all available material suitable for plant growth. Organic matter is considered a precious commodity. However, knowledge of how much organic carbon is needed to initiate or maintain nutrient cycling has changed since 1977; a much smaller percentage of organic carbon is necessary (Ingram et al. 2005). This, however, does not negate the continued and judicial salvage of topsoil.

Understanding of the biotic component of soil was limited in 1977, especially as it related to mine land reclamation. The reestablishment of the soil biota and a functioning soil ecosystem is important to the nutrient cycling mentioned above. Organic fertilizers is an acceptable method of reestablishing this important community.

Mulching, in the form of native hay or straw, was once widely used but is now limited. Although native hay can be a desirable seed source, straw seed germination can have a deleterious effect from volunteer individuals during subsequent growing season. The high C:N ratios found in straw material degrade slowly over time, especially in semi-arid/arid environments such as Wyoming.

1.1.4 Weed control

As in most agricultural systems, weeds were often considered a negative aspect of plant growth and measures were taken to control them chemically (through herbicides) or mechanically (through mowing). In semi-arid/arid environments, weeds can play a vital successional role. However, it is important to distinguish the type of weed, i.e., non-noxious annuals or noxious perennials. The former can provide some beneficial temporary aspects to reclamation such as shade for new seedlings with reduced surface temperatures, as well as reduced soil erosion and reduced grazing pressure. In addition, they can provide snow-catch in the winter time and improve soil moisture. Noxious perennials can outcompete native plants, reduce species richness of desired perennials, and provide many more negatives than positives to desirable plant growth (Hejda et al. 2009).

Since 1977, new invasive annual grasses such as cheatgrass (*Bromus tectorum*) have provided extreme challenges to reclamationists (Gasch et al. 2013; Rimer and Evans 2006). More recently Ventenata grass (*Ventenata dubia)* and medusahead (*Taeniatherum caput-medusae)* have caused new concerns. Immediate control of small populations is often the best defense for such aggressive annual grasses. Once established, control can be problematic and often takes a multi-step, multi-year approach to significantly reduce or control invasions of these annual grasses (Kettenring and Adams 2011).

1.2 Wildlife habitat

Much has changed since the days in the 1970's when reestablishment of wildlife habitat was questioned. Doubt existed that large ungulates such as Pronghorn Antelope and Mule Deer would utilize and repopulate reclaimed areas in Wyoming. In addition, no one foresaw that species considered potentially threatened or endangered, such as Sage-grouse, by the U.S. Fish and Wildlife Service would lead much of the research conducted today. Some considerations for wildlife habitat diversity include:

1.2.1 Shrub establishment

As part of reestablishing the land use of wildlife habitat, full shrubs such as Wyoming big sagebrush (*Artemisia tridentata wyomingensis*) had to be reestablished. Such reestablishment was difficult in the northeastern corner of the state which is transitional to the high plains and prairie ecosystems to the east and north. In the early years following the implementation of SMCRA, shrubs were assumed to invade, as was evident on many go-back agricultural hayland systems in Northeastern Wyoming. Early seeding used fourwing saltbush (*Atriplex canescens*) as a substitute for big sagebrush. As mentioned earlier, southern sources of this seed did not allow for long-term establishment. Fall and spring seeding of shrub seed was conducted and often in the same drill rows as other plant lifeforms which often buried small seed and provided competition from perennial grasses. Winter broadcast seeding of big sagebrush and other shrub seeds is now considered a desirable method of shrub establishment.

Early on, many federal agencies were still trying to control big sagebrush through spraying, burning or chaining. It was not until much later that these practices transitioned to protecting big sagebrush stands. Since then, much has been learned about the need for such shrub patches in the landscape and associated research is still ongoing. Not all big sagebrush patches are considered equal in quality for wildlife habitat (Stiver et al. 2010).

In 1996, the Wyoming Department of Environmental Quality-Land Quality Division (WDEQ-LQD), promulgated shrub density standards for shrub reestablishment. This was part of the regulatory performance standards a mine operator is required to meet in order for a reclaimed area to be deemed successful for release of the reclamation bond. The shrub density standard joined other revegetation performance standards for plant cover and plant production that had been in place for some time under Wyoming's primacy.

1.2.2 Landscape diversity and topsoil

In the early implementation of SMCRA, most reclaimed landscapes in Wyoming's surface coal regions were homogenous with less steep slopes as a result of coal removal and backfill sloping, as well as even redistribution of salvaged topsoil. Despite the fact that slope configurations of 5:1 were more desirable in high rainfall climates, these less steep slopes were established in semi-arid/arid areas. Remnant high walls or scarps could not generally be left, even if they were suitable habitat for many raptors. Currently, some (if well-planned in the mine permit and capable of demonstrating post mine stability) can be considered.

Recent changes in geomorphic reclamation has literally changed the landscape of the post-mining topography. Considerations for stabilizing drainage features allows the reclaimed area to mimic pre-disturbance stream channels.

Proper topsoil salvage and replacement was mandated by SMCRA. Direct haul of salvaged topsoil was an important consideration for the biological viability of the topsoil, as well as mine planning. Variable topsoil replacement depths are being used by some mines to create more plant diversity in the landscape and assist shrub growth.

Alleviating compaction by ripping of backfilled areas, travel paths for equipment, etc. is an important consideration. Contact between the overlying topsoil and underlying compacted layer enhances water movement and allows for plant roots to further break up those compacted zones.

Depending on the desired post-mining plant community, substrate replacement, both quantity and type, is critical. In order to get trees and some types of shrubs reestablished, it is necessary to use rockier substrate as replacement material on the reclaimed area. The uniform replacement of topsoil requirement of SMCRA did not foresee such need.

1.3 New technology

No one could have foreseen how new technology has changed reclamation. The application of Geographical Information Systems (GIS) query capability in evaluating landscape problems through imagery is extremely valuable. Until recently, crude topographic and aerial maps that guided our thinking have given way to those produced with the use of Light Detection and Ranging (LIDAR), satellite, digital high altitude aerial imagery and drones or unmanned aerial systems (UAS).

Global Positioning System (GPS) is used in heavy equipment to replace backfill, as well as guiding topsoil salvage and replacement operations. Post-mining topography recreation is now at the dozer-level. In conjunction with the use of extensive databases, undesirable material can now be tracked in ways never understood in 1977.

2 What haven't we learned yet

Regulations are often written with broad brush implications but implemented on a site-specific or regional basis. The one-size-fits-all mentality is often engrained in regulation simply as a result of the process. Complex problems deserve complex solutions and a regulatory framework that can adapt to finding those solutions. When SMCRA was devised, it was meant to address many of the problems in the Eastern USA. Mines west of the 100th Meridian (or essentially the Western USA) had similar categories of items to address such as topsoil salvage and stream protection but the environment was vastly different. Similarly, regions throughout the world have similar categories but their environments can be vastly different too. Fixing past problems or planning for new disturbance must take into account those regionally-specific conditions. Understanding the overall ecology of any site, its limitations and its challenges are essential to achieving success.

Site specific considerations may include:

- Are there limited resources such as seed, water, topsoil etc.?
- How can those limited resources be used or preserved most effectively?
- How can organic matter be incorporated into the system to increase soil fertility?
- Is stability a priority?
- Is another land use, such as wildlife habitat, a priority?

There is much continuing debate whether *reclamation* or *restoration* is the proper word for what SMCRA should accomplish. Ecosystem function and land uses can be restored. The restoration of late successional plant communities, however, is something that only time can provide. What then is the best means for success comparison (for bond release purposes) with a young reclaimed area? Is it a similar undisturbed vegetation community, ecological site (as defined by the NRCS), or some type of technical standard? What is a reasonable expectation of trend given the 5 and 10 year bonding time frames? That debate continues.

3 Summary

Most interviews agreed that technology has advanced sufficiently to provide the tools and resources needed to properly and cost-effectively reclaim a surface coal mine in Wyoming. These include, in part, available quality and quantity of seed, soil amendments, and the proper equipment. Potential variables impacting success include the pre-planning needed to maximize the efficiency of effort, the money allocated to the necessary resources, the regulatory environment that allows flexibility in implementation, and the managerial and engineering commitments to consider reclamation a vital part of the mining process. As many western mines, such as in Wyoming, have gone from being "new" in 1977 to "mature" in 2017, how are the lessons learned over 40 years recognized and applied today?

As the nations of the world utilize their natural resources for the betterment of their people, how can impact be minimized while maintaining long-term sustainable use? Reclamation definitely has a role. Mining operators and regulators in Wyoming and the semi-arid/arid Western USA have learned much in the 40 years since SMCRA was passed and implemented. According to Schroeder et al. (2016), important positives during the course of a reclamation career include "dedication to the advancement of the sciences, professionalism, inspirational mentoring, and partnerships". Wyoming's reclamation practitioners have been a resilient group of people that have met many challenges over the 40 years since the passage of SMCRA. We likely have more to learn. Our charge is to pass on the knowledge we have gained.

References

Gasch CK, Enloe SF, Stahl PD, Williams SE (2013) An aboveground-belowground assessment of ecosystem properties associated with exotic annual brome invasion. Biol Fertil Soils 49:919–928

Hejda M, Pyšek P, Jarošík V (2009) Impact of invasive plants on the species richness, diversity and composition of invaded communities. J Ecol 97(3):393–403

Ingram LJ, Schuman GE, Stahl PD, Spackman LK (2005) Microbial respiration and organic carbon as indicators of soil quality and nutrient cycling in reclaimed soils. Soil Sci Soc Am J 69:1737–1745

Kettenring KM, Adams CR (2011) Lessons learned from invasive plant control experiments: a systematic review and meta-analysis. J Appl Ecol 48:970–979

Rimer RL, Evans RD (2006) Invasion of downy brome (*Bromus tectorum* L.) causes rapid changes in the nitrogen cycle. Am Midl Nat 156:252–258

Schroeder S, Toy T, Vories K (2016) Challenges and opportunities: reflections on a life in mined land reclamation. Reclamation matters. Spring 2016, pp 10–18. http://www.asmr.us/Publications/Reclamation-Matters

Stiver SF, Rinkes ET, Naugle DE (2010) Sage-grouse Habitat Assessment Framework. U.S. Bureau of Land Management. Unpublished Report. U.S. Bureau of Land Management, Idaho State Office, Boise, Idaho

Thilenius JF, Glass GB (1974) Surface coal mining in wyoming: needs for research and management. J Range Manag 27(5):336–341

5

Key technologies and equipment for a fully mechanized top-coal caving operation with a large mining height at ultra-thick coal seams

Jinhua Wang[1] · Bin Yu[2] · Hongpu Kang[1] · Guofa Wang[1] · Debing Mao[1] · Yuntao Liang[3] · Pengfei Jiang[1]

Abstract Thick and ultra-thick coal seams are main coal seams for high production rate and high efficiency in Chinese coal mines, which accounts for 44 % of the total minable coal reserve. A fully mechanized top-coal caving mining method is a main underground coal extraction method for ultra-thick coal seams. The coal extraction technologies for coal seams less than 14 m thick were extensively used in China. However, for coal seams with thickness greater than 14 m, there have been no reported cases in the world for underground mechanical extraction with safe performance, high efficiency and high coal recovery ratio. To deal with this case, China Coal Technology & Engineering Group, Datong Coal Mine Group, and other 15 organizations in China launched a fundamental and big project to develop coal mining technologies and equipment for coal seams with thicknesses greater than 14 m. After the completion of the project, a coal extraction method was developed for top-coal caving with a large mining height, as well as a ground control theory for ultra-thick coal seams. In addition, the mining technology for top-coal caving with a large mining height, the ground support technology for roadway in coal seams with a large cross-section, and the prevention and control technology for gas and fire hazards were developed and applied. Furthermore, a hydraulic support with a mining height of 5.2 m, a shearer with high reliability, and auxiliary equipment were developed and manufactured. Practical implication on the technologies and equipment developed was successfully completed at the No. 8105 coal face in the Tashan coal mine, Datong, China. The major achievements of the project are summarized as follows: 1. A top-coal caving method for ultra-thick coal seams is proposed with a cutting height of 5 m and a top-coal caving height of 15 m. A structural mechanical model of overlying strata called cantilever beam-articulated rock beam is established. Based on the model, the load resistance of the hydraulic support with a large mining height for top-coal caving method is determined. With the analysis, the movement characteristics of the top coal and above strata are evaluated during top-coal caving operation at the coal face with a large mining height. Furthermore, there is successful development of comprehensive technologies for preventing and controlling spalling of the coal wall, and the top-coal caving technology with high efficiency and high recovery at the top-coal caving face with a large mining height. This means that the technologies developed have overcome the difficulties in strata control, top-coal caving with high efficiency and high coal recovery, and enabled to achieve a production rate of more than 10 Mtpa at a single top-coal caving face with a large mining height in ultra-thick coal seams; 2. A hydraulic support with 5.2 m supporting height and

✉ Hongpu Kang
kanghp@163.com

1 China Coal Technology & Engineering Group, Beijing 100013, China

2 Datong Coal Mine Group Co., Ltd., Datong 037003, China

3 CCTEG Shenyang Research Institute, Shenyang 110016, China

anti-rockburst capacity, a shearer with high reliability, a scraper conveyor with a large power at the back of face, and a large load and long distance headgate belt conveyor have been successfully developed for a top-coal caving face with large mining height. The study has developed the key technologies for improving the reliability of equipment at the coal face and has overcome the challenges in equipping the top-coal caving face with a large mining height in ultra-thick coal seams; 3. The deformation characteristics of a large cross-section roadway in ultra-thick coal seams are discovered. Based on the findings above, a series of bolt materials with a high yielding strength of 500–830 MPa and a high extension ratio, and cable bolt material with a 1 × 19 structure, large tonnage and high extension ratio are developed. In addition, in order to achieve a safe roadway and a fast face advance, installation equipment for high pre-tension bolt is developed to solve the problems with the support of roadway in coal seams for top-coal caving operation with a large mining height; 4. The characteristics of gas distribution and uneven emission at top-coal caving face with large mining height in ultra-thick coal seams are evaluated. With the application of the technologies of gas drainage in the roof, the difficulties in gas control for high intensive top-coal caving mining operations, known as "low gas content, high gas emission", are solved. In addition, large flow-rate underground mobile equipment for making nitrogen are developed to solve the problems with fire prevention and safe mining at a top-coal caving face with large mining height and production rate of more than 10 Mtpa. A case study to apply the developed technologies has been conducted at the No. 8105 face, the Tashan coal mine in Datong, China. The case study demonstrates that the three units of equipment, i.e., the support, shearer and scraper conveyor, are rationally equipped. Average equipment usage at the coal face is 92.1 %. The coal recovery ratio at the coal face is up to 88.9 %. In 2011, the coal production at the No. 8105 face reached 10.849 Mtpa, exceeding the target of 10 Mtpa for a top-coal caving operation with large mining height performed by Chinese-made mining equipment. The technologies and equipment developed provide a way for extracting ultra-thick coal seams. Currently, the technologies and equipment are used in 13 mining areas in China including Datong, Pingshuo, Shendong and Xinjiang. With the exploitation of coal resources in Western China, there is great potential for the application of the technologies and equipment developed.

Keywords Ultra-thick coal seams · Top-coal caving mining · Large mining height · Mining method · Mining equipment · Roadway support · Safety guarantee

1 Introduction

In China, there is approximately 1341 Bt of proven coal resource, including 279.58 Bt of basic coal resource for coal mine construction. Among them, coal resource in thick coal seams accounts for 44 % of the total available. The annual underground coal production from thick coal seams exceeds 45 % of total coal production. Since the Eighties in the last century, there have been many research projects focusing on mechanical coal extraction from thick coal seams. The mining technologies and systems for thick and ultra-thick coal seams have been gradually developed in China. Until recently, there have been three major mechanical mining methods for thick coal seams, i.e., a fully mechanized longwall top-coal caving mining method (thereafter LTCC mining method), a fully mechanized sub-level coal mining method, and a fully mechanized mining with a large mining height method. There are some limitations with the fully mechanized sub-level coal mining method, including the thickness variation of coal seams, spontaneous combustion, unit mining cost and efficiency. Due to the low production efficiency (Qian and Shi 2003; Wang 2005; 2006a; Mao and Yao 2010; Wang 2010a), the fully mechanized sub-level coal mining method is currently replaced by the other two mining methods. Compared with the fully mechanized sub-level coal mining method, the fully mechanized coal mining with a large mining height method can reduce the amount of roadway development, mitigate the tension between mining and development, and simplify the mining process. With the progress of mining technologies and equipment in China, the fully mechanized coal mining with a large mining height method has seen widely used (Yan and Yin 2008; Yan 2009; Yan et al. 2011), and has achieved significant mining profit and coal recovery. The method is only applicable to the coal seams of low dip, relative stable, and good roof conditions, and is also limited by the shearer's cutting height and coal wall spalling. For this reason, the mining height for one mining pass is 7.0 m (Wang et al. 2012), and cannot meet the requirement for safe and efficient mining in ultra-thick coal seams above 7 m. In terms of ultra-thick coal seams above 7 m, the top-coal caving mining method is the first choice. For coal seams with thicknesses ranging from 7 to 14 m, the top-coal caving mining method can be applied as the technologies and the equipment are well established and manufactured. In order to achieve safe and efficient coal extraction in ultra-thick coal seams with thicknesses ranging from 14 to 20 m, the authors propose a fully mechanized large mining height longwall top-coal caving mining method (thereafter LLTCC mining method), developed a full range of technologies and equipment for a LLTCC mining operation. The technologies and equipment developed have been successfully applied in the Tashan coal mine in Datong, China, and has achieved a production of 10 Mtpa at a LLTCC mining operation in the ultra-thick coal seam. In the early twentieth century, the top-coal caving mining method was applied at steep seams in France, Spain and former Yugoslavia. Due to the low mechanical levels in mining, the method was actually an alternative of the high caving mining method, which was only used as a special mining method under complex geological conditions. At the end of forties and early fifties of the twentieth century, the top-coal caving mining method was used to extract the thick coal seams in the former Soviet Union, France and former Yugoslavia. With the progress of hydraulic support technology, the fully mechanized top-coal caving technology was improved and successfully applied in some countries.

In the late eighties of the last century, the top-coal caving technology was not widely applied and its potential was not fully achieved due to the competition from other energy industries, the changes of social and political systems, and no solutions for the difficulties of its application in difficult geological conditions. Up to 1992, fully mechanized top-coal caving faces had been stopped in all countries, with the last site in Russia. By that time, China was the only country using the LTCC mining method to extract coal from thick coal seams (Wang 2009). Currently, China is a world leader in developing and conducting the LTCC mining technology, which results in several large, safe, high production rate and high efficiency mining areas including Luan, Yanzhou, Yangquan and Suozhou (Fan et al. 2003; Wang 2011). In recent years, with the development of the LTCC mining technology in China, the technology and equipment is starting to be introduced in some other countries, such as Australia, and is successfully used in underground coal mines.

Although China was late to develop and apply top-coal caving mining technology, its development and applications are progressing very fast, showing significant economic potential and attracting attentions from other countries. In China, the top-coal caving mining method means the fully mechanized longwall top-coal caving mining method. A longwall panel is laid out at the bottom of the thick coal seam and extracted using a fully mechanized longwall mining method. At the same time, the remaining coal above the longwall face are fractured by ground stress and strain (manually fractured in some special cases), and caved through the sliding window at the support canopies into an additional scraper conveyor, and then transported out of the working face. There are approximately four stages for the development of LTCC technology in China.

1.1 Stage 1: Exploration stage (1982–1990)

Since 1982, the LTCC technology has been introduced and investigated in China. In 1984, the first industrial trial of the technology at a gentle inclined seam was conducted at Northern Area 3 in Puhe mine, Shenyang Province, China (Yan and Fu 2003). Due to the problems with support design and equipment, and lack of experience in production management, the trial was not successful. However, the subsequent trials at steep inclined coal seams in the No. 2 mine of Yaojie, Meihekou mine of Liaoyuan, and Liudaowan mine in Urumqi, China, were all successful. Up until 1990, although the trials on the technology at gentle inclined coal seams continued in Northeastern China and Pingdingshan, there were no breakthroughs. In the second half of 1990, the LTCC face 8603 at the No. 1 coal mine in Yangquan produced 140,000 t/m, doubling the production from a sub-level coal face and exceeded 80 % of coal recovery from the face. The dip of the LTCC face was 3°–7°, the thickness was 6 m and the face width was 120 m. The trial at Yangquan found a way to achieve high production, high efficiency at a LTCC face, demonstrated the potential with LTCC technology, and laid a solid foundation for the development of LTCC technology. The feasibility of high production capacity and high efficiency of the technology has been proven at this stage.

1.2 Stage 2: Gradual maturation (1990–1995)

The key achievement at this stage was that the LTCC face at Xinglongzhuang mine in Yanzhou, China, reached the high production of 3 Mtpa. In addition, there was a breakthrough for LTCC technology at “three soft” coal seams, “large dip” coal seams (approximately 30°), and “high gassy” coal seams, which will enable a wide application of the technology in China. At this stage, the LTCC technology developed quickly and made some coal mining companies in China realize that the LTCC technology was a technological step-change for the coal mining industry. In addition, the research and studies were very active in the areas of ground control, support-strata relationship, cavability of top coal, and methods of top-coal caving, which provided various theories and practices.

1.3 Stage 3: Technical maturation and promotion (1995–2005)

Due to the huge technical advantages demonstrated, LTCC technology has attracted great attention from mining companies in China. Also with the removal of some technical difficulties in using LTCC technology, LTCC technology was widely and quickly applied in the coal mining industry in China. In 1995, LTCC technology was selected as one of five key technologies for further research and promotion during the period of the ninth five-year national economic plan by the Ministry of China Coal Industry. Several major technical issues associated with the

LTCC mining method were named as key topics for research. Furthermore, in 1997, China National Nature Science Foundation launched a research project of "Basic study of whole seam coal extraction in thick coal seams". Various achievements were obtained at this stage. In some coal mines that meet appropriate geological conditions i.e., simple geological conditions, large coal reserves, fewer natural hazards, 6–9 m coal seams, and medium hard coals, the annual coal production and profit at LTCC faces increased substantially. For instance, in 1997, the coal production rate at the LTCC face in Dongtan mine in Yanzhou, China, achieved 4.1 Mtpa, and 208 t/manpower. In 1998, at the same face, the coal production rate reached 5.4 Mtpa, and the manpower efficiency was at 235 t/manpower, which was regarded as a world level in that time. In addition, at this stage, the LTCC technology was developed to extract some difficult coal seams, such as "three soft", "two hard", "large dip", "high gassy", "spontaneous combustion-prone", and "ultra-thin seams", etc. The various technologies were obtained from the above development.

During the period of the tenth five-year national economic plan, a project called "the research and experiment on LTCC technology and equipment with a production capacity of 6 Mtpa" was completed in Yanzhou, China. The study enabled a higher than 6 Mtpa production rate of the LTCC face for six consequent years, which showed that LTCC technology in China had reached a new high level.

1.4 Stage 4: Improvement (2005 to present)

In 2005, during the study of safe and high efficiency mining technologies for high gassy ultra-thick coal seams, the causes of problems that stagnated the production of high gassy LTCC faces were analyzed. In addition, the feasibility of further increasing the coal cutting height was investigated for coal extraction using LTCC technology at ultra-thick coal faces. In 2006, Li et al. (2006) presented some characteristics and potential of designing top-coal caving hydraulic support with large mining heights. In terms of equipment capacities, Lu et al. (2006) conducted a study on a complete set of mining equipment with a cutting height of 3.5 m. Currently, in order to overcome the difficulties in mining high gassy and ultra-thick coal seams, some mining companies in China, such as Tunliu mine and Wangzhuang mine of Luan Group, and Tashan mine of Datong Group, conducted various LTCC mining experiments with large mining heights and studies on the ground control at LTCC faces. Because LTCC mining technology with large mining height is a new coal mining method, however, there is still a lack of related theories and studies; for instance, the applicable conditions for a large mining height LTCC mining operation, the movement of the top coal and the roof strata, stress distribution around the face, and a complete set of equipment for a large mining height LTCC face.

In terms of LTCC mining operations, the characteristics of ground pressure, the movement of the top coal and roof strata, and the relationship between the support and surrounding rock are the main topics. They are also the main criteria for the selection of face equipment, mining parameters and support parameters.

The studies on ground pressure and the movement of the top coal and roof strata are limited in countries other than China. Limited literatures exist that describe top coal movement and top coal cavability (Hu 1995), while there is less description on roof structure, face pressure, and roof control, etc. Since the application of LTCC technology, there have been many studies on the movement characteristics of overlying strata, which has resulted in a large amount of achievements.

Kang Lijun (Kang 1990; 1995; 1996; 1998; 1999; Kang and Qi 1998; Kang et al. 1998; 1999) divided the roof movement into two stages, the non-complete extraction stage and the complete extraction stage. At the non-complete extraction stage, the periodic pressure from the main roof is mainly caused by the combination impact or individual impact from the broken cantilever beam roof in the irregular caving zone, and the broken or overturned overlying beam roof fixed at the two ends. But at the complete extraction stage, the periodical pressure from the main roof is mainly caused by the periodical broken, overturn and unstable arch beam roof in the regular caving zone. The strength of periodic pressure at the non-complete extraction stage is higher than that at the complete extraction stage. A case study on the roof movement of a LTCC face at the No. 4 mine in Yangquan, China, further described the roof movement at the two stages (Kang et al. 1997). Based on the movement characteristics of overlying strata at the LTCC face, Yan (1996) introduced a mechanical theory of limited deformation and proposed a balance structure of macro continuous "compressed arch beam" of overlying strata at the upper location.

Deng Guangzhe et al. from the Xi'an Mining Institute (Deng 1994) investigated the characteristics of "arch structure" of overlying strata movement and ground control law based on the observation and measurement at LTCC face and the analysis of 3D analogue modelling. Using mechanical analysis of the shell structure, the arch structure of the overlying strata at LTCC face was preliminarily analyzed. The analysis concluded that the formation of the arch structure at overlying strata relies on some external conditions. When there are no stable skewbacks, the arch structure becomes unstable. The caving of the arch structure of strata shows an abruptness. Zhang (1999) believe that after the immediate roof caved, an unstable "arch similarly" structure may be formed within 1.0–1.2 times of mining height above the coal face, and its stability is random. On the other hand, after the caving of the immediate roof above the unstable range, a stable "arch similarly" structure may be formed due to limited space. This type of "arch similarly" structure may be formed and caved with the advance of the coal face, resulting in periodical ground pressure.

Jiang Fuxing et al. proposed a concept of spatial structure of the overlying strata. Based on the different mining boundaries at different mining areas, the main roof and overlying strata may form four types of 3D structures after caving, including type "θ", "O", "S" and "C". The spatial structure type of overlying strata in mining regions or mines is initially determined by design, and then varies with the mining operation. It is determined by roof property, buried depth, coal pillars between the working face and the mining area, and fault protection pillars. In addition, it varies at different mining stages (Ma 2005; Jiang 1995; 2006; Wang 2006b; Hou et al. 2009; Liu 2010).

In terms of the movement characteristics of overlying thick hard strata, Shi Hong (Shi 2005; Shi and Jiang 2004; 2005; Shi et al. 2005) regard the hard strata as a simple "beam fixed at two ends" structural model and analyzed the broken characteristics of thick hard strata before the initial fracture of the main roof. Using the stress distribution in the strata, Shi Hong, et al., proposed a mechanical criterion for three broken modes of thick hard rock, and predicted the initial fracture and the ground pressure interval of the main roof.

Based on the "elastic plate and articulated plate" mechanical model, Jia Xirong of Taiyuan University of Technology et al. (Jia et al. 1998) extended the ground pressure calculation method in medium thick coal seams into the calculation of ground pressure at LTCC face roof. According to the structural properties of roof strata and mining conditions, the strata with no self-support capability after fracture (no self-support strata) are defined as immediate roof. The strata with support structure, but without sufficient self-support capability after fracture (transition strata) are defined as main roof. The strata with self-support capability after fracture (complete support strata) are defined as roof overlying strata. In addition, they presented the basic assessment criteria for complete support strata, transition strata and no self-support strata.

Xie Guangxiang (Xie 2005; 2006; Xie and Luo 2005; Xie et al. 2006a, b; 2007a, b; 2009) believe that there is a "stress shell" composed of high stress beams in the surrounding rock space of mines. The maximum principal stress in the "stress shell" is greater than the principal stress in the strata inside and outside the "stress shell". The stress shell bears and transfers the load and pressure from the overlying rock, and is a major support body while the main roof bears only partial load. Based on the findings on the "stress shell" and its mechanical properties, the 3D mechanical properties of the surrounding rock at LTCC faces were found, including the effect of mining height, the mechanism of pillar width and the response of advance rate. Furthermore, the mechanical properties of ground pressure abate and the mechanisms of dynamic hazard mitigation were obtained.

Some researchers have completed various studies on the theory of ground control and the law of roof movement at LTCC face (Buddery and Oldroyd 1992; Zhang 1995; Wu 1998; Wang et al. 2001; Yasitli and Unver 2005; Ren 2005; Liu et al. 2006; Nan et al. 2008; Liu and Huang 2009; Xie and Zhao 2009a, b; Altounyan and Taljaard 2001; Alehossein and Poulsen 2010; Chen et al. 2010; Miao and Lai 2011; Yu et al. 2011; Zhang et al. 2011), including seismic monitoring (Liu 2008a, b; Kong et al. 2009; 2010a, b, c), analogous material (Kang et al. 2007a, b), numerical modelling (Wang et al. 2007; Zhu and Yan 2011), theoretical analysis (Meng 2007; Yang 2007; Fu 2008; Wu 2008; Zhang and Liu 2008; Zhou and Su 2008; Li and Xu 2009; Shi et al. 2009; Xing et al. 2009; Liu 2010; Meng et al. 2010; Wang 2010b) and site observations (Cheng 2008; Duan et al. 2008; Liu 2008a, b; Su 2008; Song et al. 2010).

The research results above are all based on the characteristics of ground pressure, and the models of "beam", "arch"and "shell". In addition, theoretical analysis and analogous (numerical) modelling were used to establish the model of overlying strata structure. Subsequently the models were validated on site. As a stable structure is formed in the strata at higher locations, its deformation and fracture will not affect or have little effect on the ground pressure at working face. Although there are some differences in studying methods and aspects, the ultimate purposes of the models derived are to explain the ground pressure during the LTCC operation and maintain a high efficient safe production. In recent years, however, the strong ground pressure at a large mining height LTCC face in ultra-thick coal seam cannot be explained by previous experiences and knowledge accumulated at a normal LTCC face. Therefore, based on the LTCC operation in ultra-thick coal seam of more than 14 m, this paper proposes a "cantilever beam-articulated rock beam" structural model. Based on the model proposed, the resistance force of support was calculated and the top coal and roof movement were monitored and analyzed.

In the study on the relationship between support and surrounding rock, to overcome the difference between the actual measurement and the traditional theory, Qian Minggao (Liu 1996; Qian et al. 1996a, b; Liu et al. 2002) regarded the key blocks of the Voussoir beam as the top boundary condition of the immediate roof, and believed that the support resistance force is mainly from "the pressure of loose blocks" and "the deformation pressure of overturned blocks" caused by the "S–R" movement of the "Voussoir beam". The deformation of immediate roof results in the overturned deformation load of the immediate roof inserted by the Voussoir beam absorbed by the fractured immediate roof (including top coal) and the non-existence of double curve relationship in the "P–$\triangle L$" relationship. Based on the mechanical properties and effect of various components in the support and surrounding rock system, the mechanical model of support and surrounding rock was established and provided a basis for further investigation on the relationship between the support and surrounding rock.

With the development of LTCC technology, the comprehensive mechanical model for LTCC mining face has been greatly improved. Liu Changyou, et al. used site observation, finite element numerical calculation, and the analogous material modelling method to investigate the comprehensive mechanical properties of the immediate roof and their effect on the relationship between support and surrounding rock. By considering the immediate roof as a variable body, the characteristics of deformation and fracture of the immediate roof and the structural mechanical and stiffness properties of the immediate roof were studied. They proposed that the load of the immediate roof is regionalized, and provided a calculation method for the stiffness of the immediate roof. In addition, based on the stiffness characteristics of the immediate roof with a rectangle shape face, the immediate roof is divided into three categories as stiffness similar, zero stiffness and medium stiffness. When the immediate roof stiffness is zero, the support is in the status of a "given load" or a "limited load". When the immediate roof is in a stiffness similar status, the support is in a "given deformation" status. When the immediate roof is in a medium stiffness status, the "P–$\triangle L$" curve shows a typical double curve relationship. When the immediate roof is in a zero stiffness status, the rational resistance force of support is mainly determined by its effect on the stability of the face ends.

The thickness variation of the immediate roof (including top coal) also affects the load bearing capacity of the immediate roof. Cao Changsheng et al. (Liu et al. 1997; 2002; Cao and Liu 1997; Cao et al. 1998; 1999; Cao et al. 1999) presented a concept on the critical thickness of the immediate roof and believed that the "P–$\triangle L$" double curve flattens when the immediate roof thickness exceeds a critical thickness. In other words, the variations of support resistance and main roof overturn angles have little effect on the roof subsidence.

Shi Yuanwei et al. (Kang et al. 1991; Shi and Chen 1993; Shi et al. 1993) presented and analyzed the critical thickness of the immediate roof and its effecting factors. They state that when critical thickness is greater than actual thickness, or when the immediate roof in front of the face enters a plastic zone, the immediate roof would lose its load bearing capacity after it enters the roof control zone. Therefore, the mechanical relationship of support and surrounding rock will show a variation. In contrast, the immediate roof will maintain a certain deformation capacity.

Fang (2002) simulated the effect on the roof stability at face ends by support type using discrete element software and believes that under soft coals, the critical issue is to control the distance from the support tip to the coal wall, which must be controlled within the distance limit. Under the conditions of medium hard coals, the constraint distance from the support tip to coal wall is relatively large and thus there will be a rational arrangement for the tip to wall distance and support resistance force. For hard coals, the study focus should be on improving the top coal caving ability. In the study on the mechanical relationship of support and surrounding rock in a soft coal extraction, Huang (2002) found that the increase of support resistance can reduce the tensile stress range in the top coal between support tip and coal wall, increase the self-support capacity from the top coal between support tip and coal wall, and suppress the movement and fracture of top coals. Nevertheless, the increase of support resistance has a limited effect on the improvement of subsidence of unsupported zone between the support tip and coal wall.

Jin et al. (2002) state that under the condition of hard roof and hard coal, the LTCC face shows the characteristics of hard roof, and demonstrates the characteristics of top coal cushion effect at LTCC face. The rational support resistance force should be based on the cantilever beam model with a modification by the coefficient of cushion effect.

Wu Jian et al. (Wu and Yan 1995; Wu and Fu 1998; Wu and Zhang 2001) suggest that the traditional ground control theory could not explain the interaction relationship between the support and surrounding rock. The key support parameters in the support and surrounding rock relationship includes the vertical support force, the horizontal support force, the distance between the supporting point of resultant force and the coal wall, and the distance between the support tip and coal wall. There are obvious effects from these parameters, sometimes obvious only when they interact.

Yang et al. (2009) state that the key to the interaction system of the support and surrounding rock is the interaction between the support and top coal. Under hard coal conditions, the overturn of the immediate roof is affected by the support and top coal system, and the turning angle of the lower strata is less than the turning angle of the main roof. Under the medium hard coals and soft coals, the immediate roof can equally transfer the overturn deformation of the main roof to top coals and the support system. With the consideration of the immediate roof as the top boundary of support-top coal systems, the mechanical model for two-leg shield support and surrounding rock can be established.

The studies on the relationship between roof support and surrounding rock provide theories for the design and selection of LTCC roof support, the improvement of support-surrounding rock status and effective roof control (including top coals). In this paper, based on the relationship between roof support and surrounding rock, the authors propose a 3D dynamic optimal design method with hydraulic support and surrounding rock coupling, and subsequently the determination of the working resistance of hydraulic support at the No. 8105 coal face in Tashan mine in Datong, China, and test on site.

In terms of the study on mining equipment for LTCC operation, the improvement of the production rate at a fully mechanized coal face mainly depends on the increase of longwall width and shear cutting depth, fast advance and the extension of applicable range. In particular, the application of electromechanical heavy equipment enables the increase of the production rate, including high powered electric traction shearer, large capacity conveyor, electric-hydraulic controlled hydraulic support, non-stop self-moving stage loader and conveyor end, and high voltage power supply for face equipment. Currently the mining equipment manufactured in China for the fully mechanized longwall face and the LTCC face can meet the requirements and demands for high production and high efficiency in the Chinese coal industry.

In the early years, China imported a large amount of shearers including chain haulage type and hydraulic haulage type. Since the 1970s, China has started to manufacture shearers. In the eighties, the hydraulic haulage shearers met the requirements for extracting coal seams with heights from 1.5 to 4.5 m. In addition, the difficulties in integrating all mining equipment had been overcome. There were about 50 types of shearers. Up to the 1990s, large power hydraulic haulage shearer developed in China has gone into mass production. Furthermore, the high performance electric haulage shearer had been developed, and is currently at an advanced level in the world, with a cutting height of 5.5 m. In China, the application of large power electric haulage shearers enables the increase of coal production at a fully mechanized face from 1 Mtpa to 4 Mtpa or even 10 Mtpa. Currently, several series of electric haulage shearers are manufactured in China. In addition, the electric shearers are equipped with intelligent monitoring, control and protection systems, and applied with information processing and sensor technologies to achieve electro-mechanization.

Top-coal caving hydraulic support is a key equipment for LTCC operations. During the last 20 years, top-coal caving hydraulic support was gradually developed from high position top-coal caving support, middle position top-coal caving support, to low position top-coal caving support. Currently, the low position top-coal caving support is widely used, which is a chock-shield hydraulic support with positive four-link bar, reverse four-link bar and single swing link bar.

In terms of the scraper conveyor, there has been significant improvement every 10 years in China. In the 1980s, there were various types of scraper conveyors with two series of chute width of 730 mm and 764 mm. In 1994, the SGZ880/800 cross-side unloading scraper conveyor with an entirely cast-welded chute was developed by the Taiyuan Branch of China Coal Research Institute and Northwestern Mining Equipment Manufactory, which had a capacity of 7000 t/d. During the same period, Zhangjiakou Mining Equipment Manufactory introduced some casting and processing equipment and manufactured scraper conveyors with entirely cast-welding chutes.

During the period of the ninth five-year national economic plan, with the cooperation of Zhangjiakou and Northwestern Coal Mining Equipment Manufactory, using CAD technology, the Taiyuan Branch of China Coal Research Institute developed the first scraper conveyors for LTCC face with retractable end chain adjusting device. They were the SGZ960/750 front scraper conveyor and the SGZ900/750 rear scraper conveyor, and also stage loader and crusher. The equipment developed met the requirements from Yankuang Group for a coal production of 10,000–13,000 t/d. During the period of the tenth five-year national economic plan, Yankuang Group, the Taiyuan Branch of China Coal Research Institute and Northwestern Benniu Group jointly developed the SGZ1000/1200 and the SGZ1200/1400 long distance high reliability front scraper conveyor and rear scraper conveyor. Currently, the SFZ1200/1400 long distance high reliable rear scraper conveyor is an ultra-heavy scraper conveyor for LTCC faces, which is characterized by its highest power level, widest chute and longest length. Furthermore, it integrated various advanced technologies, including automatic extension of the conveyor end, hydraulic motor chain tensioner, speed adjustment hydraulic coupler and compact chain. In the 1990s, its main specifications reached an advanced level accepted worldwide. The SGZ1000/1200 front scraper conveyor is currently the largest front scraper conveyor, and can be integrated with various powerful shearers and hydraulic supports. Its installed power is 2×600 kW. The development of the SGZ1000/1200 and the SGZ1200/1400 scraper conveyors enables not only the technical advance in scraper conveyor technology, but also in LTCC technology.

In order to promote the development of large mining height LTCC technology and equipment for ultra-thick coal seams, and to achieve safe, high efficiency and high recovery mining of ultra-thick coal seams, in 2008, the Ministry of Science and Technology of China launched a significant R&D project called "the study of large mining height LTCC technology and equipment" The project was supported by the national science and technology plan. With 14–20 m thick coal seams, the project has developed a large mining height LTCC mining method (LLTCC mining method), a complete set of mining equipment, roadway drivage and support technology and equipment, and mine safety warrant technology and equipment (Wang 2013a, b). In 2011, the technologies and equipment developed were tested at the No. 8105 coal face of Tashan mine in Datong, China, with a production of 1.0849 Mtpa. The success of the project marked a new level for large mining height LTCC technology and equipment for ultra-thick coal seams in China. In this paper, the major research achievements, case studies and technical economical results from this project are presented.

2 Geological and mining conditions at the Tashan coal mine

2.1 Brief description of the mine

The Tashan coal mine is a large modern coal mine in Datong extracting the coals from Permo-Carboniferous system in the Datong coal field. The mine is located at the middle edge of Eastern Datong coal field in Shanxi Province, China, Fig. 1. The strike length of the mine is 24.3 km, the dip width is 11.7 km and the total area is 170.9 km^2. The mine extracts five coal seams including seam 2, 3, 4, 5 and 8. The mine has 5070 Mt of geological coal reserve, 4760 Mt of industrial coal reserve and 3070 Mt of mineable coal reserve. In 2007, the mine had a nominal production capacity of 1.5 Mtpa, and a life of 140 years.

Fig. 1 Locality of Tashan coal mine

2.2 Geological conditions of the mine

The main coal seams extracted at Tashan coal mine are coal seam 3–5; their buried depth ranges from 347 to 447 m; average seam thickness is 16.8 m and maximum is 20 m; and average seam dip is 4°. The protodyakonov coefficient F of the coal seams ranges from 2.7 to 13.7. There are 4–14 plies partings, and their lithologies consist of grey brown kaolinite, kaolinitic mudstone, grey and black carbonaceous mudstone, sandy mudstone, and locally dark grey siltstone.

The top of coal seam 3–5 is a 0.33 m thick false roof of grey black carbonaceous mudstone. The top part of the immediate roof consists of brown grey kaolinitic mudstone, magmatic rock and the seam 2. The middle part of the immediate roof is composed of shed coal and silicified coal. The bottom part consists of grey black carbonaceous mudstone, kaolinite and magmatic rock. The thickness of the immediate roof ranges from 2.4 to 6.3 m. From top to bottom, the upper roof consists of seam 4, magmatic rock, silicified coal, siltstone, fine sandstone, kaolinitic mudstone and sandy mudstone. The thickness of the upper roof varies from 11.31 to 27.36 m. The immediate floor consists of grey brown kaolinitic mudstone and sandy mudstone. It is brittle, fragile and contains coal bits. The lower floor is dark grey siltstone with local fine sandstone. The length of the No. 8105 LLTCC face is 207 m. The face is affected by igneous rock. Thus, there are some metamorphism and silicification in the seam 3-5. The seam structure is complex. The coal with metamorphism and silicification is hard, broken and with well-developed cleats. Mixed coal is lose and fragile. A lower coal seam is stable and hard, and forms a structure like a sandwich, shown in Fig. 2.

Fig. 2 The upper seam structure at No. 8105 working face

Table 1 Geotechnical measurement of surrounding rock at Tashan coal mine

Order number	Measurement contents	Measurement instrument	Number of measurement points
1	Ground stress	SYY-56 ground stress monitor with hydraulic fracture	3 holes
2	Rock strength	WQCZ-56 measurement instrument of rock strength	8 holes
3	Rock structure	KDVJ-400 borehole camera	13 holes

Table 2 Test results of ground stress

Order number	Roadway name	Buried depth (m)	Vertical stress (MPa)	Maximum horizontal principal stress (MPa)	Minimum horizontal principal stress (MPa)	Direction of maximum horizontal principal stress
1	1070 auxiliary transport roadway	467	11.44	12.00	6.40	N26.7°E
2	5102 gate road	467	11.44	12.40	8.22	N24.8°E
3	2102 gate road	467	11.44	12.90	7.24	N19.0°E

The relative gas emission at coal seam 3–5 in Tashan coal mine is 1.95 m^3/t, which is classified as a low gas mine. However, the coal dust has the possibility of a dust explosion, with an explosion index of 37 %. The period of spontaneous combustion is 68 days. There is no high temperature heat hazard area. The geothermal gradient is 2.41 °C/100 m.

2.3 Geotechnical measurement and analysis of coal and rock masses

In order to fully understand the geotechnical properties of the surrounding rock, the measurement of geotechnical properties was systematically and comprehensively conducted, including the underground measurement of ground stress, rock strength test and rock structure observations. The measurement contents are listed in Table 1.

2.3.1 Measurement results of in situ stresses

Using the hydraulic fracturing method, the in situ stress of rock in the roof of the seam 3–5 at Tashan coal mine was measured. The measurement results are shown in Table 2. These results show that the maximum horizontal principal stress is greater than the vertical principal stress. The type of stress field is the $\sigma_H > \sigma_V > \sigma_h$ type. Therefore, the stress field tested is mainly a structural stress field, and maximum horizontal principal stress is the principal stress.

2.3.2 Test results of surrounding rock strength

The surrounding rock strength was directly measured in the borehole underground using the WQCZ-56 instrument. There were eight boreholes at four test stations, including four roof holes and four wall holes. Figures 3 and 4 show

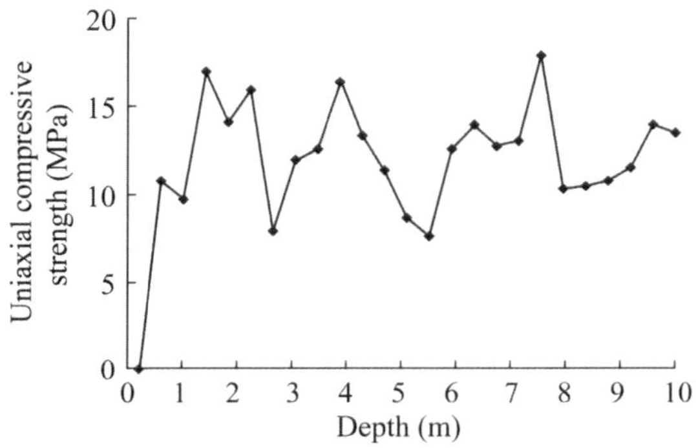

Fig. 3 Test results of surrounding rock strength at roof of No. 2102 roadway

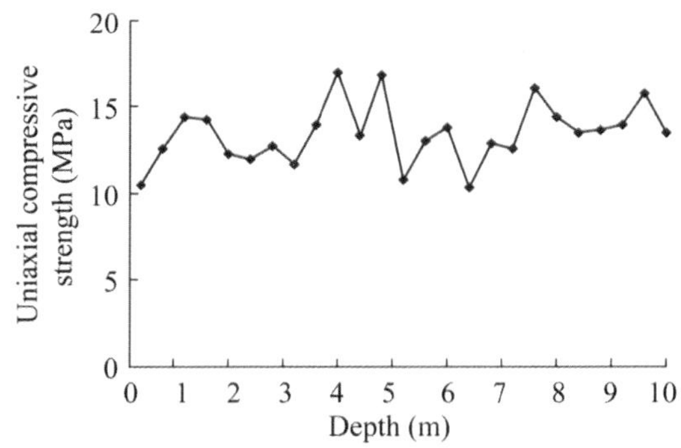

Fig. 4 Test results of surrounding rock strength at wall of No. 2102 roadway

the test results in the No. 2102 roadway that was the closest station to the No. 8105 working face.

The test station was located in the No. 2102 roadway and 500 m away from the intersection of the No. 2102 roadway and the No. 1070 return airway. At this location, the roof holes and wall holes were all in coal seams and the coal strength was measured in detail. A 0.4 m fractured

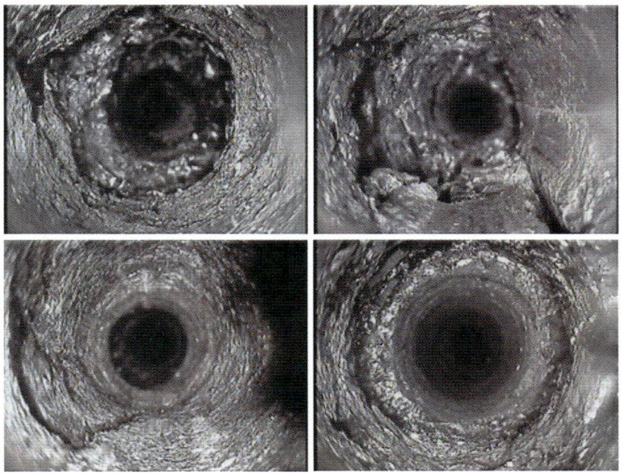

Fig. 5 Typical photos inside boreholes at roof of roadway in Tashan coal mine

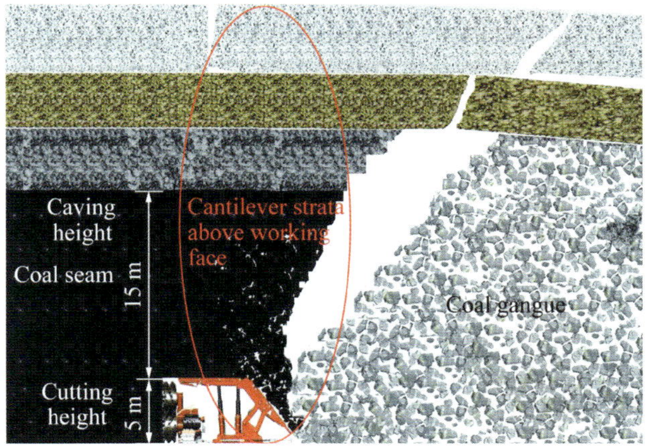

Fig. 6 LLTCC mining method in ultra-thick coal seam

zone at roof holes was ignored and thus the maximum strength of the coal was 17.85 MPa, the minimum was 7.65 MPa, and the average was 12.43 MPa. The average coal strength in wall holes was 13.43 MPa.

2.3.3 Observations of the surrounding rock structure

The development of cleavage and fracture of the surrounding rock was directly observed in the boreholes underground using a KDVJ-400 borehole camera. The observations of 13 boreholes demonstrated that the cleavage and fracture in the top coal and immediate roof at the Tashan coal mine were well developed. In addition, there were various degree of layer separation at shallow roof strata and there was a trend to extend the layer separation into the upper strata. Figure 5 shows the photos of cleavage and fracture in boreholes.

3 Mining method and ground control technology for LLTCC operation in ultra-thick coal seams

3.1 LLTCC in ultra-thick coal seams

In a conventional fully mechanized top-coal caving longwall face, the coal cutting height is normally 3.0–3.5 m. Based on the rule of cutting to caving ratio 1:3, the conventional top-coal caving height is 9–10.5 m. In other words, the maximum thickness of coal extraction is 12–14 m. In terms of ultra-thick coal seams with thicknesses greater than 14 m, if the cutting height remains as 3.0–3.5 m, there will be many problems. For instance, the cutting to caving ratio will be too low (1:4.7 if cutting height is 3.5 m in a 20 m thick coal seam); it would be difficult to cave top coals, and the coal recovery will be low. At the same time, the increase of coal production will result in the increase of gas emission and the difficulties in removing gas due to limited cross-section area at the coal face.

In order to overcome the problems mentioned above, the authors propose a LLTCC mining method, that is to increase the coal cutting height to 5 m, the top coal caving height to 15 m, and the total coal extraction height to 20 m, shown in Fig. 6. The LLTCC mining method involves a large coal cutting height, a large coal caving height and a large extraction space, that will result in some additional problems including coal wall spalling, difficulties in top-coal caving, and strong ground pressure. Those problems require comprehensive, in-depth and systematic research and testing.

3.2 3D spatial structural model of the roof in LLTCC mining

Because of the large thickness of the coal seam extracted through LLTCC mining in ultra-thick coal seams, the range of ground control should also be extended. After the completion of top coal caving operation, it is not possible for the roof rock to cave and fill in the goaf in a short time. Therefore, the immediate roof will form a "cantilever beam", and the upper roof will form a stable "articulated beam". In other words, a "cantilever beam–articulated beam" will be formed in a LLTCC coal face. In a general case, the overlying strata does not simply follow the movement from lower layer to upper layer, and instead forms various combinations of structures and moves with some laws. Figures 7 and 8 show the structural model of the roof.

Based on the study discussed above, a "cantilever beam–articulated beam" structural model for overlying strata at a LLTCC face has been developed, shown in Fig. 9. The strata that need to be controlled are the strata

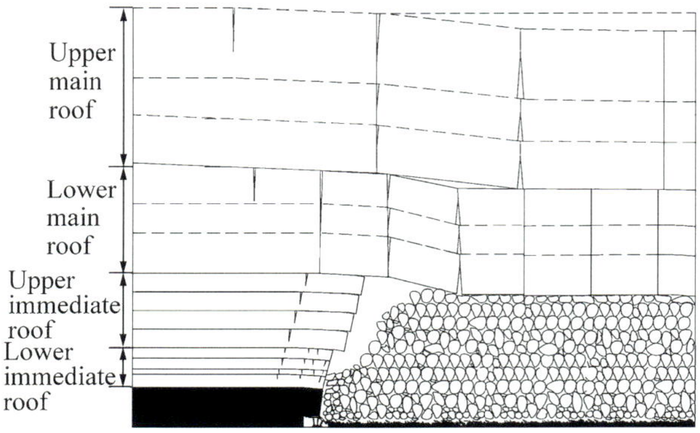

Fig. 7 Strike cross-section of a LLTCC face

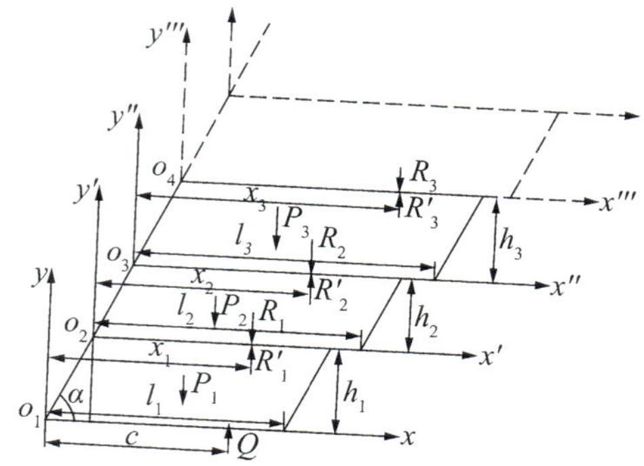

Fig. 10 Force analysis and calculation of the immediate roof

Fig. 8 Dip cross-section of a LLTCC face

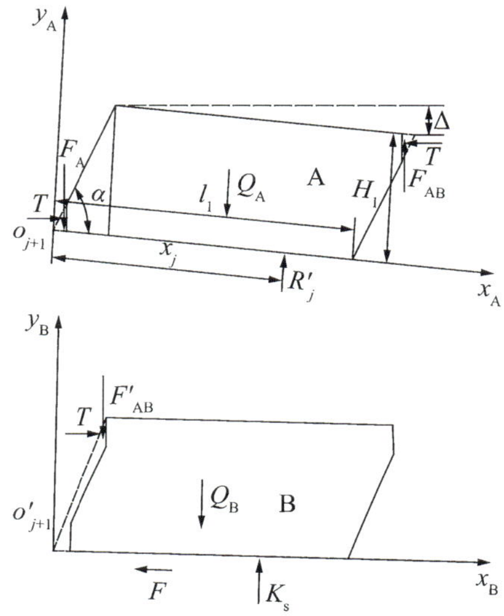

Fig. 11 Force analysis of rock in main roof

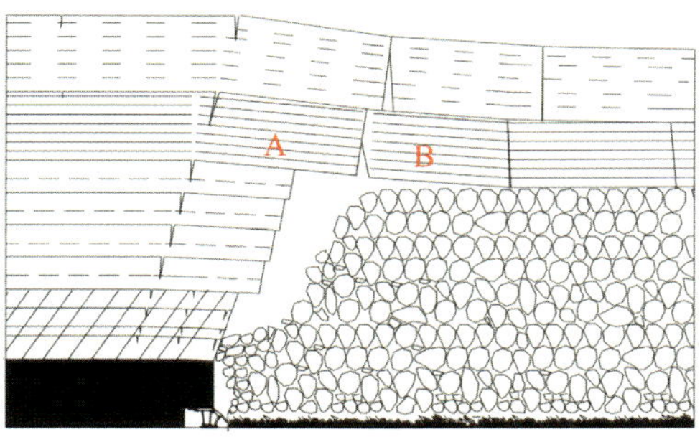

Fig. 9 Structural model of "cantilever beam–articulated beam"

that cannot be supported by caved rocks before they cave. They are in the form of a "cantilever beam" and defined as immediate roof. Above the immediate roof, the strata that are supported by caved rocks are called the upper roof with an "articulated beam" structure. The roof strata satisfying the $\Delta_j - \Delta_m \leq 0$ condition are defined as immediate roof while the roof strata satisfying the $\Delta_j - \Delta_m > 0$ condition are defined as upper roof. Δ_j is the ultimate subsidence.

$$\Delta_j = h - \frac{ql^2}{2kh\sigma_c} \tag{1}$$

where h is the strata thickness analyzed, k is a dimensionless coefficient ($k = 0.1\ h$), q is the line load, l is fracture length of the strata analyzed, $\sigma_c = (0.30 - 0.35)R_c$ (R_c is compression strength), Δ_m is the possible subsidence (Yan 2009),

$$\Delta_m = (h_c + h_f)(1 - p_1) + (1 - k_P)h_m \tag{2}$$

where p_1 is the coal loss rate; h_c is the coal-cutting height; h_f is the coal caving height; k_p is the expansion coefficient; h_m is thickness of the immediate roof.

3.3 Determination of working resistance of the LLTCC hydraulic support

The top coal body of the LLTCC face in ultra-thick coal seams can be regarded as a "semi-rigid top coal body" that can transfer the roof deformation. The deformation effect of the part of the immediate roof enables the top coal body becoming "semi-rigid top coal body". In terms of pressure on the support, this part of the strata can be called "no deformation pressure strata". Through the "semi-rigid top coal body", the pressure generated by the deformation of immediate roof and upper roof is transferred to hydraulic support. This part of the strata can be called "deformation pressure strata" (Yan 2009). The loads on the support are the combination of the weight of the top coal and the pressure generated by roof deformation. The force analysis of the immediate roof with deformation pressure is shown in Fig. 10. The force analysis of the upper roof is shown in Fig. 11.

(1) Force analysis of the immediate roof with deformation pressure.
According to $\sum M_{oi} = 0$ ($i = 1, 2, 3, \ldots, j$), get:

$$P_1\left(\frac{l_1}{2}+\frac{1}{2}h_1\text{ctan }\alpha\right)+R_1x_1-Qc=0$$

$$P_2\left(\frac{l_2}{2}+\frac{1}{2}h_2\text{ctan }\alpha\right)+R_2x_2-R_1'(x_1-h_1\text{ctan }\alpha)=0$$

$$P_3\left(\frac{l_3}{2}+\frac{1}{2}h_3\text{ctan }\alpha\right)+R_3x_3-R_2'(x_2-h_2\text{ctan }\alpha)=0$$

$$\cdots\cdots\cdots\cdots$$

$$P_j\left(\frac{l_j}{2}+\frac{1}{2}h_j\text{ctan }\alpha\right)+R_jx_j-R_{j-1}'\left(x_{j-1}-h_{j-1}\text{ctan }\alpha\right)=0$$

The j formulas above are combined, and which will get:

$$Qc=\frac{1}{2}\sum_{i=1}^{j}P_i(l_i+h_i\text{ctan }\alpha)+\sum_{i=1}^{j-1}R_ih_i\text{ctan }\alpha+R_jx_j \tag{3}$$

where Q is the deformation pressure on support, kN; c is the distance from resultant force of support to coal wall, m; P_i is the rock weight of immediate roof of ith layer with deformation pressure, kN; h_i, l_i are the rock thickness and length of the immediate roof of ith layer with deformation pressure respectively, m; α is the rock fissure angle, (°); R_j is the additional force of upper strata, kN; x_j is the distance from additional force of upper strata to caving point of jth layer immediate roof, m.
During the analysis of the large roof structure, the roof with deformation pressure is regarded as an entire body, and the interaction of internal layers can be ignored. Therefore, the Eq. (3) can be simplified as follows:

$$Qc=\frac{1}{2}\sum_{i=1}^{j}P_i(l_i+h_i\text{ctan }\alpha)+R_jx_j \tag{4}$$

(2) Analysis of rock A in main roof with deformation pressure.
During the analysis of rock A in main roof, assume that the broken distance of main roof A exceeding immediate roof P_j is $h_j$ctan α, thus, $\sum M_{j+1} = 0$ to get:

$$R_jx_j=Q_A\left(\frac{L}{2}+\frac{1}{2}H\text{ctan }\alpha\right)-T(H-\Delta)$$
$$-F_{AB}(L+H\text{ctan }\alpha) \tag{5}$$

where Q_A is the weight of rock A in main roof with deformation pressure, kN; T is the compression force between rocks, kN; H, L is the strata thickness and broken distance of main roof with deformation pressure respectively, m; Δ is the subsidence of rock A in main roof with deformation pressure, m; F_{AB} is the friction force between rock A and B, kN.
Substitute Eq. (2) into Eq. (3), to get:

$$Qc=\frac{1}{2}\sum_{i=1}^{j}P_i(l_i+h_i\text{ctan }\alpha)+Q_A\left(\frac{L}{2}+\frac{1}{2}H\text{ctan }\alpha\right)$$
$$-T(H-\Delta)-F_{AB}(L+H\text{ctan }\alpha)$$

Assume $Q_A = P_{i+I}$, $L = l_{j+1}$, $H = h_{j+1}$, the equation above is simplified as:

$$Qc=\frac{1}{2}\sum_{i=1}^{j+1}P_i(l_i+h_i\text{ctan }\alpha)-T(h_{j+1}-\Delta)$$
$$-F_{AB}\left(l_{i+1}+h_{j+1}\text{ctan }\alpha\right) \tag{6}$$

(3) Analysis of rock B in main roof.
Force balance in vertical direction:

$$F_{AB}'=Ks-Q_B \tag{7}$$

In Eq. (5), $F_{AB}' = Tf$, $F_{AB}' = F_{AB}$, therefore:

$$T=\frac{Ks-Q_B}{f} \tag{8}$$

where K is the rigidity of gangue in goaf, N/m; s is the compression of gangue in goaf, m, ($s=(k_1-k_2)\sum_{i=1}^{j}h_i$, k_1 is the expansion coefficient, k_2 is the residual expansion coefficient); f is the friction coefficient among rocks; Q_B is the weight of key block B in main roof (assume $Q_B = Q_A$), kN.

Substitute Eq. (8) into Eq. (6), to get:

$$Q = \frac{f\sum_{i=1}^{j+1} P_i(l_i + h_i\text{ctan }\alpha) - 2(Ks - Q_B)(h_{j+1} + fl_{i+1} + fh_{i+1}\text{ctan }\alpha - \Delta)}{2cf} \quad (9)$$

In summary, the calculation formula of hydraulic support for LLTCC mining is as follows:

$$P_z = K_d B(G_d + Q) = K_d B\left[L_d h_d \gamma_m + \frac{f\sum_{i=1}^{j+1} P_i(l_i + h_i\text{ctan }\alpha) - 2(Ks - Q_B)(h_{j+1} + fl_{i+1} + fh_{i+1}\text{ctan }\alpha - \Delta)}{2cf}\right] \quad (10)$$

where P_z is the working resistance force of LLTCC support, kN; K_d is the dynamic load coefficient; G_d is the weight of top coal, kN; B is the distance between the centers of two supports, m; L_d is the distance of roof control, m; h_d is the thickness of top coal, m; γ_m is the bulk density of top coal, kN/m^3.

The equation for calculating the support external load obtained from the theoretical study discussed above was tested and validated at the No. 8105 coal face in Tashan coal mine in Datong, China. The borehole analyzed was WEI 1403 at the No. 8105 coal face in Tashan coal mine. Table 3 shows the structure of the strata in borehole, and Table 4 shows the physical and mechanical parameters.

The determination on the working resistance force of hydraulic support for LLTCC mining in ultra-thick coal seams:

(1) Based on criteria of immediate roof.
Δ_{mi} is the possible subsidence of ith layer, $i = 1, 2, \ldots, 12$.

Table 3 Structure of rock strata in WEI 1403 borehole

Layer No.	Layer thickness (m)	Rock name	Layer No.	Layer thickness (m)	Rock name	Layer No.	Layer thickness (m)	Rock name
C1	4.49	Mudstone	C22	2.40	Siltstone	C43	7.25	Coarse sandstone
C2	1.33	Coal	C23	2.24	Kaolinitic mudstone	C44	1.70	Fine conglomerate
C3	2.97	Sandy mudstone	C24	1.40	Siltstone	C45	10.15	Sandstone
C4	3.17	Sandstone	C25	14.39	Kaolinitic mudstone	C46	20.52	Fine conglomerate
C5	15.67	Sandy mudstone	C26	4.10	Siltstone	C47	10.18	Fine sandstone
C6	1.12	Mudstone	C27	2.40	Fine conglomerate	C48	4.95	Siltstone
C7	1.78	Coal	C28	3.70	Coarse sandstone	C49	4.58	Fine sandstone
C8	5.37	Sandy mudstone	C29	3.00	Siltstone	C50	2.16	Fine conglomerate
C9	1.10	Coarse sandstone	C30	14.13	Sandy mudstone	C51	9.34	Fine sandstone
C10	1.3	Siltstone	C31	4.10	Siltstone	C52	1.40	Siltstone
C11	1.55	Kaolinitic mudstone	C32	1.70	Fine sandstone	C52	8.60	Fine conglomerate
C12	5.03	Siltstone	C33	7.96	Sandy mudstone	C54	5.90	Fine sandstone
C13	1.00	Coarse sandstone	C34	2.66	siltstone	C55	1.70	siltstone
C14	1.10	Siltstone	C35	1.65	Fine sandstone	C56	1.00	Fine sandstone
C15	5.37	Kaolinitic mudstone	C36	3.50	siltstone	C57	12.28	Sandstone
C16	1.25	Siltstone	C37	5.35	Sandy mudstone	C58	10.20	Siltstone
C17	2.93	Coarse sandstone	C38	1.60	Kaolinitic mudstone	C59	5.34	Sandy mudstone
C18	0.90	Siltstone	C39	0.65	Coarse sandstone	C60	10.68	Fine sandstone
C19	1.10	Coarse sandstone	C40	13.64	Siltstone	C63	122.56	
C20	4.43	Kaolinitic mudstone	C41	7.72	Fine sandstone	to		
C21	2.70	Fine sandstone	C42	2.00	Sandy mudstone	surface		

Table 4 Lithological parameters of rock strata

Rock strata	Volume force (kN/m^3)	Tensile strength (MPa)
Coal	14.50	2.50
Mudstone	25.10	2.90
Sandy mudstone	26.20	3.10
Kaolinitic mudstone	25.13	3.30
Siltstone	25.80	5.10
Fine sandstone	26.50	4.30
Sandstone	24.10	3.70
Fine conglomerate	26.50	4.23
Coarse sandstone	25.80	4.10

$$\Delta_{m1} = 10.73\,\text{m}$$

$$\Delta_{j1} = h - \frac{ql^2}{kh\sigma_c} = 3.87\,\text{m} < \Delta_{m1}$$

Thus rock strata C1 is in the immediate roof.

......

$$\Delta_{m11} = 1.89\,\text{m}$$

$$\Delta_{j11} = h - \frac{ql^2}{kh\sigma_c} = 0.56\,\text{m} < \Delta_{m11}$$

Thus, rock strata C11 is in the immediate roof.

$$\Delta_{m12} = 0.63\ \text{m}$$

$$\Delta_{j12} = h - \frac{ql^2}{kh\sigma_c} = 4.32\ \text{m} > \Delta_{m12}$$

Thus, rock strata C12 and strata above are in the upper roof.

(2) Determination of rock strata without deformation pressure.

$\sum_{i=1}^{n} (\Delta_j)_i = \Delta_d$, Δ_d is a total deformation of coal body in roof control zone, $\Delta_d = \eta h_d(1+\lambda)$ (Yan 2009); η is the porosity of top coal, $\eta = (1-\gamma_d/\gamma) \times 100\,\%$, γ_d is the dried bulk density of coal seam, γ is the bulk density of coal seam, λ is the coefficient of horizontal pressure (measured value at Tashan coal mine is 1.13).

$$\Delta_d = \eta h_d(1+\lambda) = 0.3 \times 10.61 \times (1+1.13)$$
$$= 6.78\,\text{m}\ \sum_{i=1}^{4}(\Delta_j)_i = 5.50\,\text{m} < \Delta_d;$$

$$\sum_{i=1}^{5} (\Delta_j)_i = 12.48\ \text{m} > \Delta_d;$$

Rock strata C5 is divided into two parts, $C5^{-1} = 10.17$ m, $C5^{-2} = 5.50$ m.

$$\sum_{i=1}^{5^{-1}} (\Delta_j)_i = \Delta_d$$

Thus rock strata C1–$C5^{-1}$ is rock strata without deformation pressure.

(3) Determination of working resistance force.

To maximize Eq. (10):

$$P_z = K_d B\left[L_d h_d \gamma + \frac{f\sum_{i=1}^{j+1} P_i(l_i + h_i \cot\alpha)}{2cf}\right] \qquad (11)$$

Analyze Table 3, refer to the mechanical property of each rock stratum, and take C12–C16 as a whole. Substitute the basic data of C5-2, C6, C7,...,C16 into Eq. (11), get:

$$P_z = K_d B\left[L_d h_d \gamma + \frac{f\sum_{i=1}^{j+1} P_i(l_i + h_i \cot\alpha)}{2cf}\right]$$
$$= 1.6 \times 1.75 \times (923.07 + 4378.11)$$
$$= 14843.3\,\text{kN}$$

Therefore, based on the geotechnical conditions of rock strata at the Tashan coal mine, the working resistance force of hydraulic support for the No. 8105 LLTCC face at ultra-thick coal seam in Tashan coal mine was determined as 15,000 kN. The ZF15000/28/52 hydraulic support for LLTCC face was selected.

3.4 Analysis on the applicability of hydraulic support type

The dynamic change characteristics of the ZF15000/28/52 fully mechanized high strength LLTCC hydraulic support in a shift is shown in Fig. 12. The figure shows that the variation of working resistance was not large, implying that the support had a very high reliability.

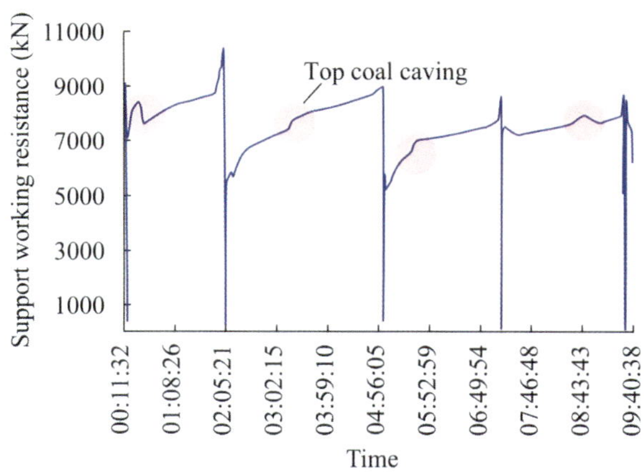

Fig. 12 Periodical change of support working resistance at the middle of the face

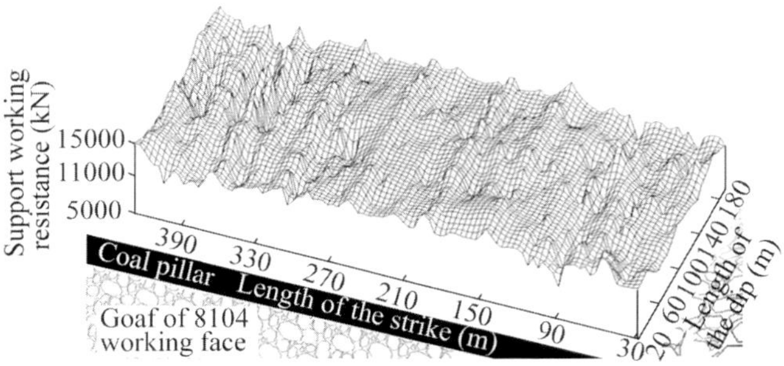

Fig. 13 Distribution of ground pressure at the working face

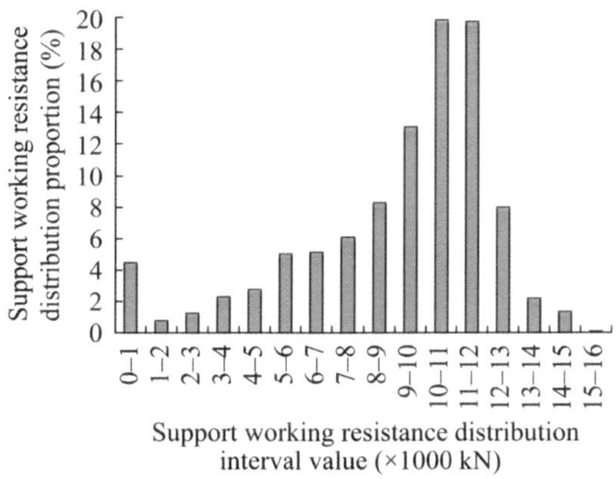

Fig. 14 Histogram of support working resistance

Figure 13 shows the 3D support resistance variation at the end of the circle in the No. 8105 face. The impact range of the initial caving and periodical caving of the upper roof was extensive and covered the whole working face. In a normal working period, the average working resistance is 9073.06 kN. Table 5 shows that there was almost no effect on supports, although the dynamic load coefficient is large during loading period, indicating that the impact resistance of support is high. When load is high, and the working resistance force exceeds nominal resistance force, the relief valve can be activated in a timely manner, showing the stability of the support is good.

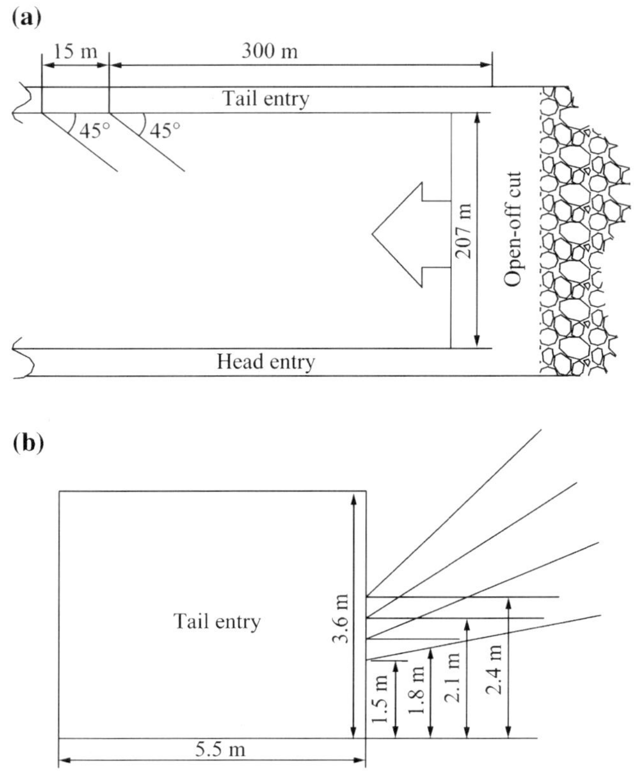

Fig. 15 Boreholes layout. **a** Top view of boreholes layout, **b** vertical cross-section diagram

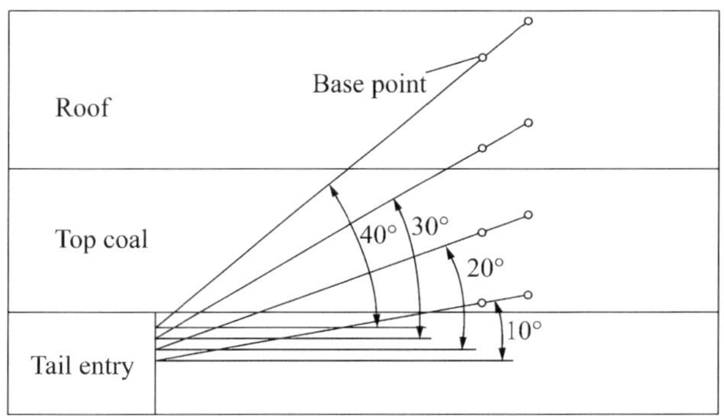

Fig. 16 Layout of deep base points

Table 5 Statistics of dynamic loading coefficient of upper roof

Support No.	20	21	63	64	95	96
Dynamic loading coefficient at initial loading	1.51	1.57	1.56	1.58	1.62	1.49
Average dynamic loading coefficient at initial loading	1.56					
Dynamic loading coefficient at Ith periodical loading	1.43	1.58	1.58	1.58	1.49	1.50
Dynamic loading coefficient at IIth periodical loading	1.54	1.58	1.57	1.56	1.55	1.46
Dynamic loading coefficient at IIIth periodical loading	1.30	1.51	1.66	1.41	1.41	1.40
Dynamic loading coefficient at IVth periodical loading	1.38	1.51	1.64	1.52	1.59	1.37
Dynamic loading coefficient at Vth periodical loading	1.44	1.52	1.63	1.56	1.57	1.46
Average dynamic loading coefficient at periodical loading	1.43	1.55	1.61	1.54	1.54	1.45
Total average dynamic loading coefficient at periodical loading	1.52					

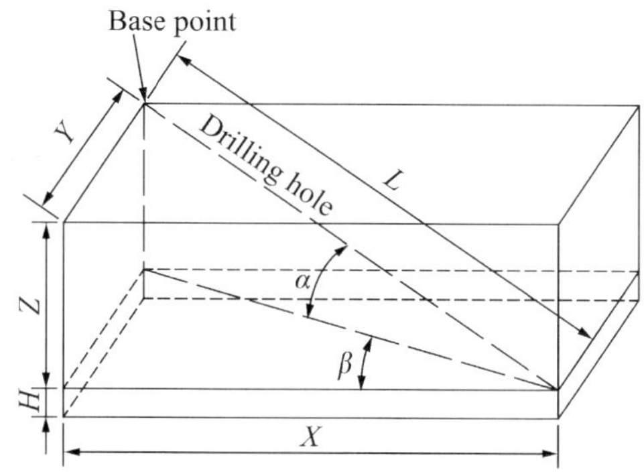

Fig. 17 Geometry of deep base point parameters

The percentage of support resistance at different intervals was described as a histogram (Fig. 14). Figure 14 shows that the distribution of working resistance of hydraulic support was rational, and appeared as a distribution of sine wave. The support resistance was mostly within 8000–13,000 kN. In summary, the selection of support was reasonable and their utilization was high. The support selected was suitable for the roof conditions in the working face.

3.5 Analysis of the characteristics of top coal and roof movement in a LLTCC coal face

In order to study the movement law of top coal and roof at a LLTCC coal face, some boreholes were drilled into the top coal and roof of the tailgate at the No. 8105 LLTCC coal face in Tashan coal mine. Basic anchors were installed at different layers inside the boreholes. GUW300 tracking instruments were installed at the opening of the boreholes. The first monitoring station was setup at the location 300 m away from the box cut of the No. 8105 face. The second monitoring station was 15 m away from the first station. At each station, four boreholes were drilled into the top coal and roof, two deep base points were installed inside each hole, eight deep base points in total were installed, and four GUW300 deep base point movement tracking instruments were installed. The intersection angle of borehole and tailgate was 45°. The angles of from No. 1 to No. 4 boreholes with horizontal level were 40°, 30°, 20° and 10° respectively. The heights of borehole positions from No. 1 to No. 4 with floor were 2.4, 2.1, 1.8 and 1.5 m respectively. Figure 15 shows the layout of the boreholes. Figure 16 shows the layout of eight deep base points in each monitoring station with four points in the top coal and four points in the roof (Table 5).

Figure 17 shows the basic parameter geometry of the boreholes layout. Tables 6 and 7 show the parameters of deep base points. Basic geometry is described as Eq. (12).

$$X = \sqrt{(L\cos\alpha)^2 - (L\sin\beta)^2} \quad (12)$$

where L is the depth of base point, m; α is the angle of elevation, (°); β is the angle between borehole and tailgate, (°); X is the horizontal length from borehole opening to deep base point and parallel to tailgate, m; Y is the horizontal length from borehole opening to deep base point and perpendicular to tailgate, m; Z is the vertical height from floor to deep base point, m; H is the vertical height from floor to borehole opening, m.

From Figs. 18, 19, 20 and 21 show the movement of top coal and roof from two monitoring stations at the No. 8105 working face.

As shown from Figs. 18, 19, 20 and 21, among the initial displacement positions of eight top coal deep base points at two monitoring stations, the distance ahead of the coal wall at the working face ranged from 34 to 48 m with an average of 38 m; among the initial displacement positions of roof deep base points, the maximum distance ahead of the coal wall at the working face ranged from 22 to 34 m, with an average of 28.6 m.

The average measured displacement at the top coal base points in front of the coal wall was 114.88 mm, accounting for 26.6 % of total displacement. The average measured displacement at the top coal base points behind coal wall was

Table 6 Parameters from first monitoring station

Hole No.	Height from floor to hole opening (m)	Base point No.	Angle of elevation (°)	Height to floor (m)	Height to hole opening (m)	Depth to base point (m)	Horizontal projection of base point (m)	Distance of tailgate projection (m)
1	2.4	11	40	32.1	29.7	46.2	35.4	25.0
		12		26.1	23.7	37.0	28.3	20.0
2	2.1	21	30	22.5	20.4	40.9	35.4	25.0
		22		18.4	16.3	32.7	28.3	20.0
3	1.8	31	20	14.7	12.9	37.7	35.4	25.0
		32		12.1	10.3	30.1	28.3	20.0
4	1.5	41	10	7.7	6.2	35.9	35.4	25.0
		42		6.5	5.0	28.7	28.3	20.0

Table 7 Parameters from second monitoring station

Hole No.	Height from floor to hole opening (m)	Base point No.	Angle of elevation (°)	Height to floor (m)	Height to hole opening (m)	Depth to base point (m)	Horizontal projection of base point (m)	Distance of tailgate projection (m)
1	2.4	11	40	32.1	29.7	46.2	35.4	25.0
		12		26.1	23.7	37.0	28.3	20.0
2	2.1	21	30	22.5	20.4	40.9	35.4	25.0
		22		18.4	16.3	32.7	28.3	20.0
3	1.8	31	20	14.7	12.9	37.7	35.4	25.0
		32		12.1	10.3	30.1	28.3	20.0
4	1.5	41	10	7.7	6.2	35.9	35.4	25.0
		42		6.5	5.0	28.7	28.3	20.0

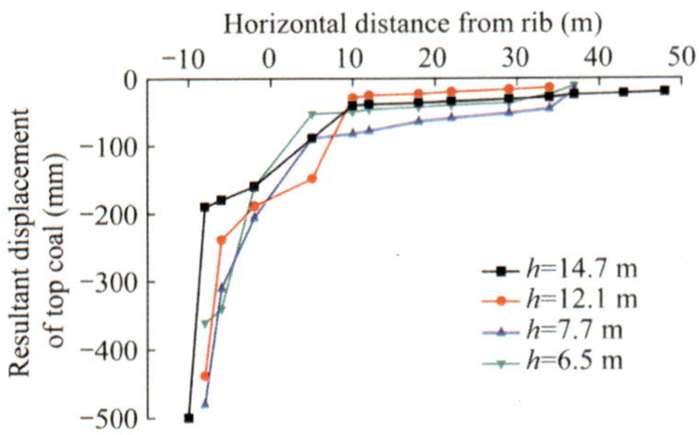

Fig. 18 Top coal displacement at first monitoring station

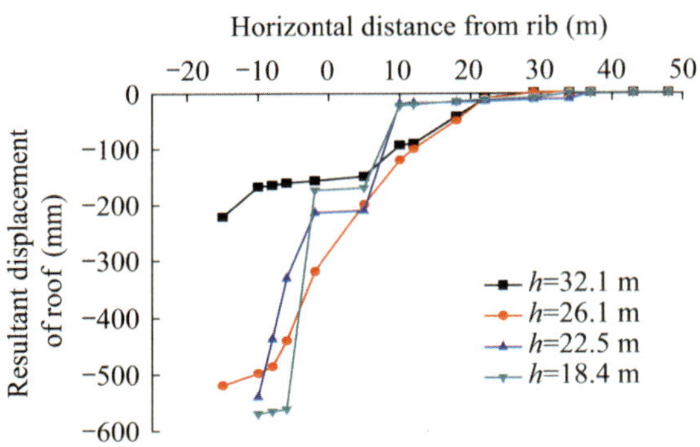

Fig. 19 Roof displacement at first monitoring station

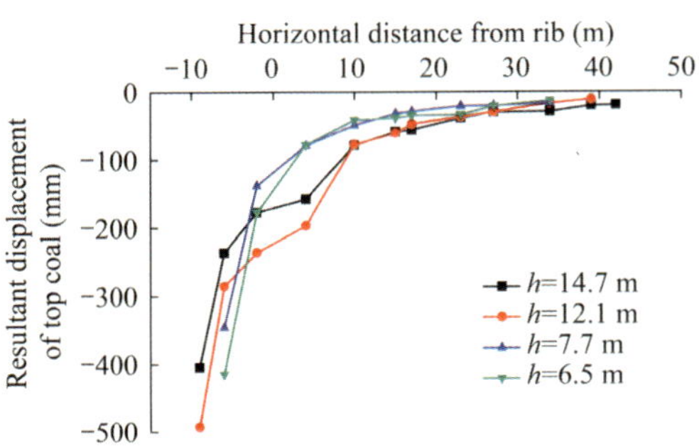

Fig. 20 Top coal displacement at second monitoring station

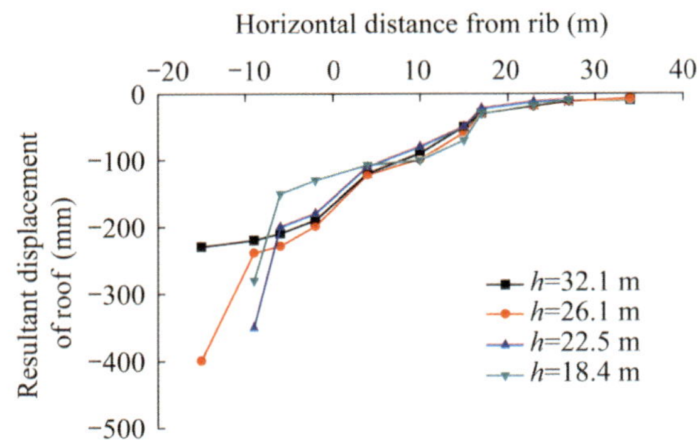

Fig. 21 Roof displacement at second monitoring station

317.63 mm, accounting for 73.4 % of total displacement. The total displacement was 432.5 mm. Average measured displacement at the roof base points in front of the coal wall was 149 mm, accounting for 38.3 % of the total displacement. Average measured displacement at roof base points behind coal wall was 240.25 mm, accounting for 61.7 % of total displacement. The total displacement was 389.25 mm. Although from initial displacement position, the top coal base points were affected by abutment pressure in front of the face, that lasted a long period and affected a long distance, the accumulated displacement of the top coal and roof was relatively small due to the constraint from limited space before entering the roof control zone. After entering the roof control zone, the movement of the top coal and roof followed the movement law of vertical subsidence after caving.

In front of the coal wall at the working face, the movement of the top coal and roof mainly showed a horizontal deformation under the abutment pressure due to the vertical constraints of the top coal and roof. At the rear of the coal wall which was under the repeated loading and unloading actions, the vertical constraints on the lower part of the top coal exhibited subsidential weakening with the advance of coal cutting resulting in a large space. Therefore, the movement of roof appeared as fracture subsidence and the movement of the top coal appeared as vertical displacement caused by the roof fracture subsidence. There was a large difference in accumulated top coal and roof displacement in the front and in the rear of the coal wall. In addition, the difference of displacement per unit advance was relatively large. The displacement rate of the top coal in front of the coal wall was 2.63 and 42.29 mm/m at rear of the coal wall. The displacement rate of the roof in front of the coal wall was 5.4 mm/m, and 21.35 mm/m at rear of coal wall. It demonstrated that the horizontal deformation rate of the top coal and roof caused by abutment pressure in front of the coal wall was much less than the vertical subsidence rate of the top coal and roof at rear of coal wall.

At eight base points in the top coal in two monitoring stations, from the coal wall to the gulf, the distance between the caving position to the coal wall ranged between 6 and 10 m, with an average of 7.8 m. Assume the average distance of the roof control was 6.5 m, the average caving position of the top coal was greater than the average distance of the roof control. In particular, the average caving position was 9 m in the middle or upper part of the top coal ($h = 12.1$ m, or $h = 14.7$ m), that indicated that there was a delay of the middle or upper part of top coal caving, the average distance of delay was 2.5 m, and the top coal was caved in a form of inverted step. From the point view of eight base points at two monitoring stations, the roof base points at $h = 18.4$ m, $h = 22.5$ m and $h = 26.1$ m caved in time with the top coal caving. The displacement of the roof started to increase at the roof base point of $h = 32.1$ m and 15 m at the rear of the coal wall. Therefore, the immediate roof was not completely caved or caved on time, and formed an immediate roof in the form of a cantilever beam above the coal seam. Based on the position of the top coal caving, the caving angle of the top coal can be approximately calculated, i.e., 64.3° for the first monitoring station, 70.1° for the second monitoring station, and 67.2° for an average caving angle.

3.6 Spalling of coal wall at a LLTCC coal face

3.6.1 Factors of coal wall spalling at a LLTCC coal face

At different stages of non-periodical ground pressure and periodical pressure, as the differences in mining height, advance speed and distance from the support tip to the coal wall, the spalling depth and spalling ratio were different at a LLTCC coal mining face (Yan 2008; Yin et al. 2008; Zhang et al. 2010). In order to better understand the spalling mechanism at a LLTCC coal mining face, the coal wall spalling were examined under different conditions including mining heights, advance rates, distances from the support tip to the coal wall, and ground pressure stages.

Using the numerical simulation method, the horizontal displacements of the coal wall with mining heights of 3.5, 4.5, 5.0 and 5.5 m were simulated and the results are shown in Fig. 22.

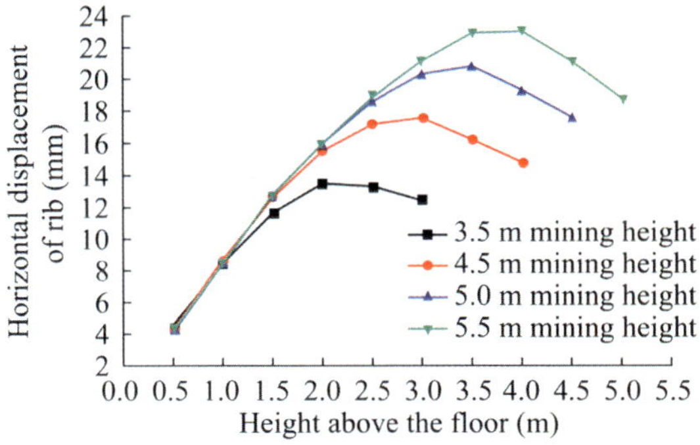

Fig. 22 Horizontal displacements of rib with different mining heights

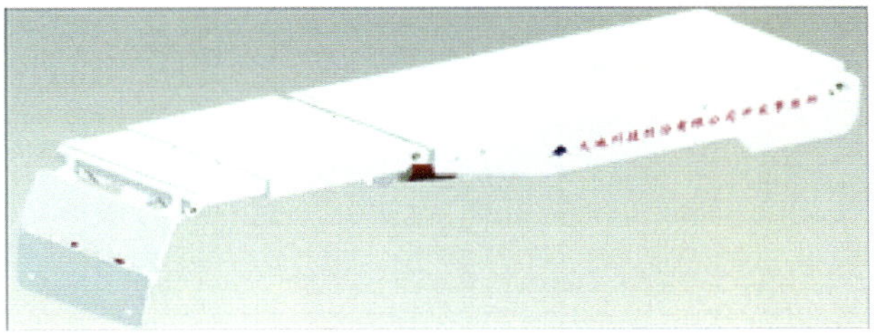

Fig. 23 Structure of the second level wall guard plate installed on the extendable beam

Figure 22 showed that the peak horizontal displacement increased with the increase of mining height. Considering the normal mining height was 4.0–4.5 m, the point of peak horizontal displacement to floor was approximately 2.5–3.0 m. Therefore, the rational height of the wall guard plate of support was 1.5 m. Based on simulated results, the second level wall guard plate was installed on a high reliable extendable beam and articulated front beam to prevent the spalling of the coal wall. Its structure is shown in Fig. 23.

The face advance speed also significantly affected wall spalling. The observations at the No. 8105 coal face demonstrated that with the increase of face advance speed, there was a reduction of damage area around the coal wall, and a reduction of wall horizontal displacement. In other words, the spalling of coal wall was suppressed. It is therefore suggested that the face advance speed should be rationally increased to reduce the exposure time and spalling of coal wall. Nevertheless, the effectiveness of controlling coal wall spalling was limited through the increase of face advance speed. With a large amount of observations on site, the effect of increasing face advance speed on coal wall spalling was found. If the advance speed is too high, the coal spalling might increase.

With the increase of the distance from support tip to coal wall, the peak abutment pressure in front of the face decreased. Subsequently, the displacement at support tip increased significantly, and the damage zone in front of the face also increased, redulting in significant spalling. Undoubtedly, the distance from the support tip to the coal wall should be reduced appropriately.

Coal cutting height, face advance speed and distance from support tip to coal wall were three factors that had substantial effects on coal wall spalling. Their weighted effects might be ordered as coal cutting height > face advance speed > distance from support tip to coal wall.

The effect on coal wall spalling from the caving and displacement of the top coal and the roof had an obvious regular pattern. The coal wall spalling clearly appeared before ground pressure commenced, increasing to maximum with ground pressure, and decreased to a minimum after ground pressure disappeared. During periodical ground pressure, the average spalling ratio of the coal wall increased to 24.58 %. With a low face advance speed, the effect on coal wall spalling would be obvious from periodical ground pressure. Therefore, the maintenance of the coal wall should be enhanced during the period of periodical pressure occurring.

The seriousness of coal wall spalling is different at different positions. Coal wall spalling was found to be most serious at the middle part of the face, less serious at the lower part of the face and better at the upper part of the face.

According to the observation at the No. 8105 LLTCC face, the position of peak abutment pressure in front of the coal face was between 10 and 25 m, the relative stress concentration coefficient k was 1.3–6.5 during the peak period, and the average was 2.13. From the 10–25 m zone in front of the coal face to the coal wall was the stress release zone, and the effected zone by abutment pressure was 80 m.

With the analysis on the characteristics of coal wall spalling at the LLTCC face under different conditions, including mining height, face advance speed, distance from support tip to coal wall, ground pressure, and different positions at the coal face, the authors have developed a comprehensive spalling control technology of controlling the face advance speed, increasing initial support power, and improving the support capacity for wall protection. The technology developed has been tested and applied at the No. 8105 coal face.

3.6.2 Observation plan of coal wall spalling at the No. 8105 face

The observation of coal wall spalling consisted of two aspects:

(1) During the non-periodical pressure period, under different face advance speeds, the coal wall spalling ratio and depth are at different positions of the face.

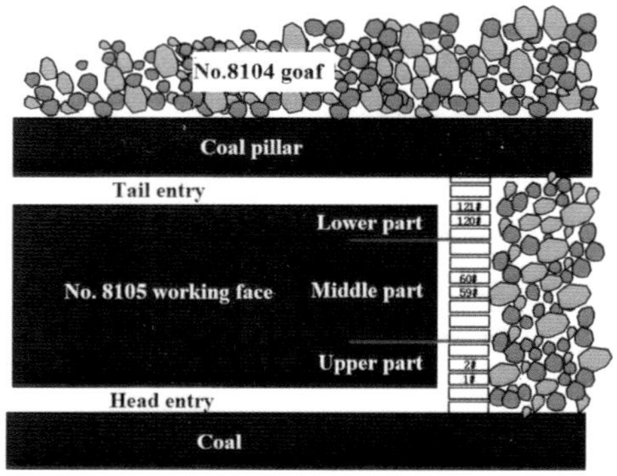

Fig. 24 Observation plan

(2) During the periodical pressure period, under different face advance speeds, the coal wall spalling ratio and depth are at different positions of the face.

The observation plan of coal wall spalling was as follows: the length of the No. 8105 face was 207 m, and the number of supports was 121. The face was divided into three observation zones including upper part, middle part and lower part (upper part: No. 1–No. 30 supports, middle part: No. 31–No. 89 supports and lower part: No. 90–No. 121 supports). Figure 24 shows the observation zones.

3.6.3 Analysis of coal wall spalling during non-periodical pressure period

(1) Coal wall spalling at the upper part of the face.
Observation results of coal wall spalling at the upper part of the face (non-periodical pressure period) are shown in Table 8 and Fig. 25.
In the non-periodical ground pressure period, the depth of coal wall spalling increased with the face advance speed at the upper part of the face, and eventually maintained at a certain depth range. When the face advance speed reached 3.05 m/d, the coal wall spalling length and depth reached their maximum values of 35 m and 0.6 m respectively, and the maximum spalling ratio was 66.67 %. The relationship between spalling length, depth, ratio and the face advance speed displayed a significant range. But generally, the values tended to increase with the face advance speed.

(2) Coal wall spalling in the middle part of the face.
Observation results of coal wall spalling at the middle part of the face (non-periodical pressure period) are shown in Table 9 and Fig. 26.
During the non-periodical pressure period, being similar to the coal wall spalling at the upper part of the face, the depth of coal wall spalling decreased with the increase of face advance speed in the middle

Table 8 Characteristics of coal wall spalling at the upper part of the face (non-periodical ground pressure period)

Face advance speed (m/d)	Average spalling depth (m)	Spalling length (m)	Spalling ratio (m)	Maximum spalling depth (m)
2.05	0	0	0	0
2.25	0	0	0	0
2.25	0.40	31.50	60.00	0.50
2.40	0.35	31.50	60.00	0.50
3.00	0.20	14.00	26.67	0.30
3.05	0.40	35.00	66.67	0.60
3.10	0.35	22.75	43.33	0.60
3.15	0	0	0	0
3.40	0.35	7.00	13.33	0.40
3.60	0	0	0	0
3.85	0.30	1.75	3.33	0.30
3.90	0.30	17.50	33.33	0.40
4.00	0.40	17.50	33.33	0.40
4.50	0.25	17.50	33.33	0.40
4.60	0	0	0	0
4.75	0.25	1.75	3.33	0.30
5.00	0.30	28.00	53.33	0.30
5.40	0.30	26.25	50.00	0.50
5.50	0.25	14.00	26.67	0.30
5.65	0	0	0	0
5.75	0.20	12.25	23.33	0.30
6.00	0.25	14.00	26.67	0.30
6.15	0	0	0	0
6.25	0	0	0	0
6.50	0	0	0	0
7.00	0	0	0	0
8.50	0.30	15.75	30.00	0.50

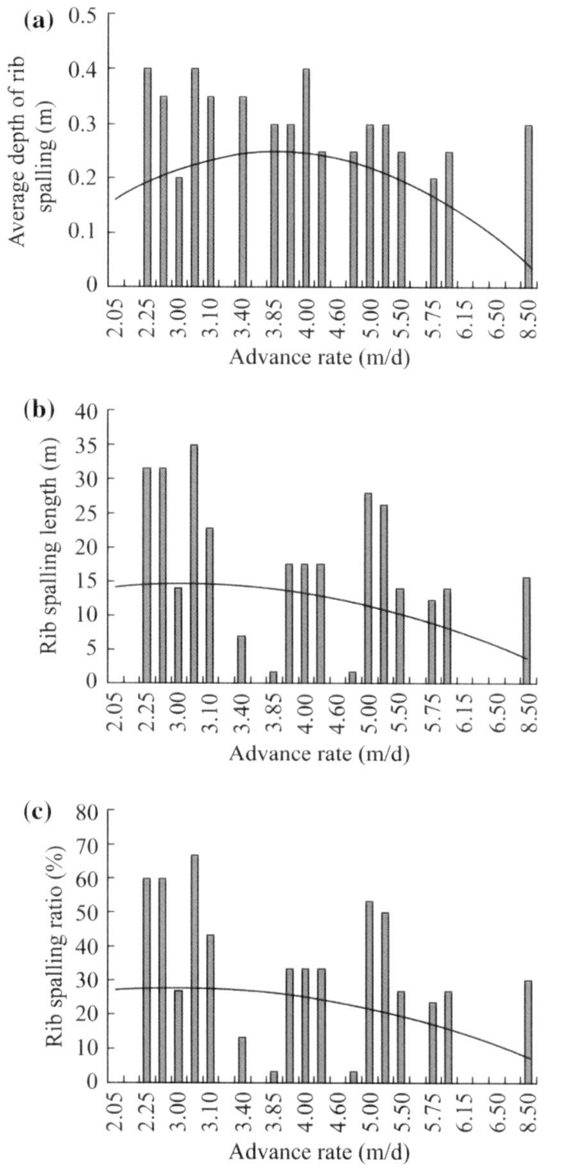

Fig. 25 Relationship *curves* between coal wall spalling and face advance speed at the upper part of the face (non-periodical ground pressure). **a** Relationship between average rib spalling depth and face advance rate, **b** relationship between rib spalling length and face advance rate, **c** relationship between rib spalling ratio and face advance rate

part of the face, the spalling ratio showed a trend of non-linear decrease with the increase of face advance speed.

(3) Coal wall spalling at the lower part of the face.

Observation results of coal wall spalling at the lower part of the face (non-periodical pressure period) are shown in Table 10 and Fig. 27.

During the non-periodical pressure period, at the lower part of the face, the relationship between coal wall spalling and face advance speed showed a certain degree of randomness. When the face advance speed reached 2.4, 3.05, 5, 5.5 and 5.75 m/d, there was no spalling occurring. In general, the depth of coal wall spalling showed less variation with face advance speed at the lower part of the face, the average spalling depth concentrated within 0.3–0.4 m. With the increase of face advance speed, the spalling length and ratio showed a trend of decrease early, to increase at a later time.

(4) Comparison of coal wall spalling at different face positions during non-periodical ground pressure period.

Figure 28 shows the comparison of the depth of coal wall spalling at different face positions during non-periodical ground pressure period. Table 11 lists the

Table 9 Characteristics of coal wall spalling in the middle part of the face (non-periodical ground pressure period)

Face advance speed (m/d)	Average spalling depth (m)	Spalling length (m)	Spalling ratio (%)	Maximum spalling depth (m)
2.05	0.30	29.75	28.81	0.40
2.25	0.30	24.50	23.73	0.30
2.40	0.35	22.75	22.03	0.50
3.00	0.35	28.00	27.12	0.60
3.05	0.40	42.00	40.68	0.60
3.10	0.40	15.75	15.25	0.70
3.15	0.45	21.00	20.34	0.70
3.40	0.35	31.50	30.51	0.50
3.60	0.35	50.75	49.15	0.50
3.85	0.30	33.25	32.20	0.30
3.90	0.35	26.25	25.42	0.50
4.00	0.30	17.50	16.95	0.30
4.50	0.30	50.75	49.15	0.40
4.60	0.25	14.00	13.56	0.30
4.75	0.30	12.25	11.86	0.40
5.00	0.40	8.75	8.47	0.40
5.40	0.30	22.75	22.00	0.40
5.50	0.20	15.75	15.25	0.20
5.65	0.30	14.00	13.56	0.30
5.75	0.35	29.75	28.81	0.50
6.00	0.35	19.25	18.64	0.30
6.15	0.30	26.25	25.42	0.40
6.25	0.20	7.00	6.78	0.30
6.50	0.30	33.25	32.20	0.30
7.00	0.20	8.75	8.47	0.30
8.50	0.25	7.00	7.00	0.40

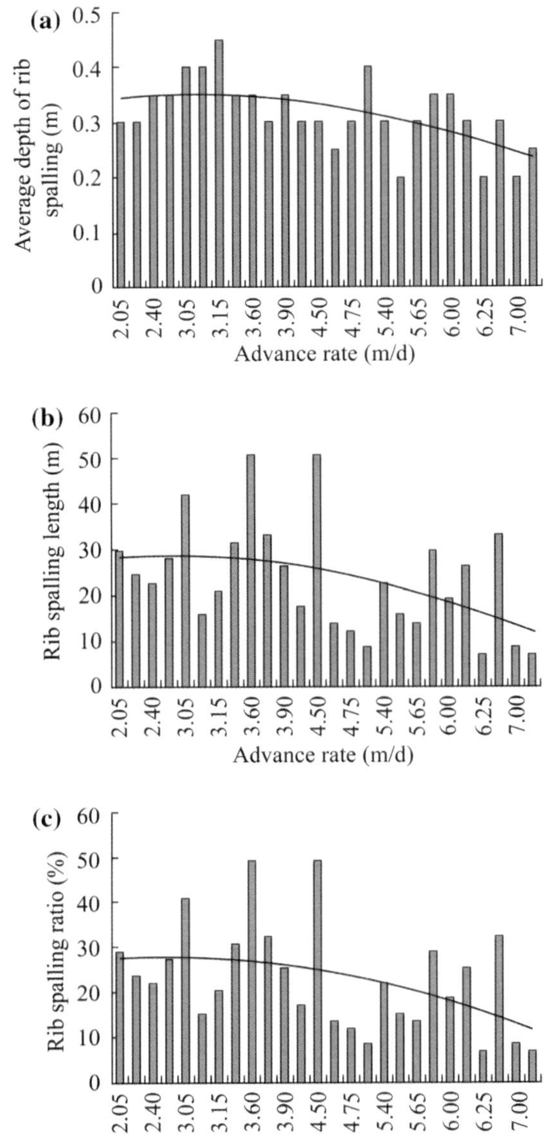

Fig. 26 Relationship *curves* between coal wall spalling and face advance speed in the middle part of the face (non-periodical ground pressure). **a** Relationship between average rib spalling depth and face advance rate, **b** relationship between rib spalling length and face advance rate, **c** relationship between rib spalling ratio and face advance rate

statistical analysis results of coal wall spalling at different face positions during the non-periodical ground pressure period. Based on the analysis and comparison, the spalling depth was 0.2 m at the upper part of the face, 0.32 m in the middle part of the face and 0.26 m at the lower part of the face, suggesting that the spalling depth was deepest in the middle part of the face, less deep at the lower part and least deep at the upper part.

Figure 29 shows the comparison of the spalling ratios at different face positions during the non-periodical ground pressure period. The spalling ratio showed a trendency of reduction with the increase of face advance speed during the non-periodical ground pressure period. As shown in Table 12, the coal wall spalling ratio was 22.56 % at the upper part of the face, 22.81 % in the middle part of the face, and 19.05 % at the lower part of the face, that indicated that the spalling ratio was largest in the middle of the face, less at the upper part of the face, and the least at the lower part of the face.

3.6.4 Analysis of coal wall spalling during the periodical ground pressure period

(1) Coal wall spalling at the upper part of the face.
Observation results of coal wall spalling at the upper part of the face (during the periodical ground pressure period) are shown in Table 13 and Fig. 30. During the period of ground pressure, the spalling

Table 10 Characteristics of coal wall spalling at the lower part of the face (non-periodical ground pressure period)

Face advance speed (m/d)	Average spalling depth (m)	Spalling length (m)	Spalling ratio (%)	Maximum spalling depth (m)
2.05	0.30	26.25	46.88	0.40
2.25	0.25	24.50	43.75	0.30
2.40	0	0	0	0
3.00	0.45	8.75	15.63	0.70
3.05	0	0	0	0
3.10	0.20	15.75	12.50	0.30
3.15	0.45	21.00	20.34	0.70
3.40	0.35	8.75	15.63	0.50
3.60	0.40	14.00	25.00	0.50
3.85	0.30	7.00	12.50	0.30
3.90	0.35	8.75	15.63	0.50
4.00	0.30	12.25	21.88	0.30
4.50	0.30	21.00	37.50	0.40
4.60	0.25	10.50	18.75	0.30
4.75	0.30	3.50	6.25	0.40
5.00	0	0	0	0
5.40	0.30	19.25	34.00	0.40
5.50	0	0	0	0
5.65	0.40	8.75	15.63	0.50
5.75	0	0	0	0
6.00	0.50	14.00	25.00	0.70
6.15	0.30	17.50	31.25	0.40
6.25	0.25	15.75	28.13	0.40
6.50	0.30	14.00	25.00	0.30
7.00	0.30	15.75	28.13	0.40
8.50	0.30	8.75	16.00	0.40

depth, spalling length and spalling ratio at the upper part of the face had a similar variation trendency with the increase of face advance speed, i.e., increase early and decrease later. When the face advance speed reached 5.50 m/d, the spalling ratio was the largest at the upper part of the face at 63.33 %.

(2) Coal face spalling in the middle part of the face.
Observation results of coal wall spalling at the middle part of the face (during the periodical ground pressure period) are shown in Table 14 and Fig. 31. During the period of periodical ground pressure, the spalling depth presented a trend of decrease early on, and increasing later with the increase of face advance speed. The spalling length and ratio presented a trend of gradual reduction with the face advance speed. In the middle part of the face, the spalling depth was 0.6 m, maximum spalling ratio was 49.15 % and maximum spalling length was 50.75 %.

(3) Coal wall spalling at the lower part of the face.
Observation results of coal wall spalling at the lower part of the face (during the periodical ground pressure period) are shown in Table 15 and Fig. 32. During the period of periodical ground pressure, the average coal wall spalling depth at the lower part of the face decreased with the increase of the face advance speed. Before the face advance speed

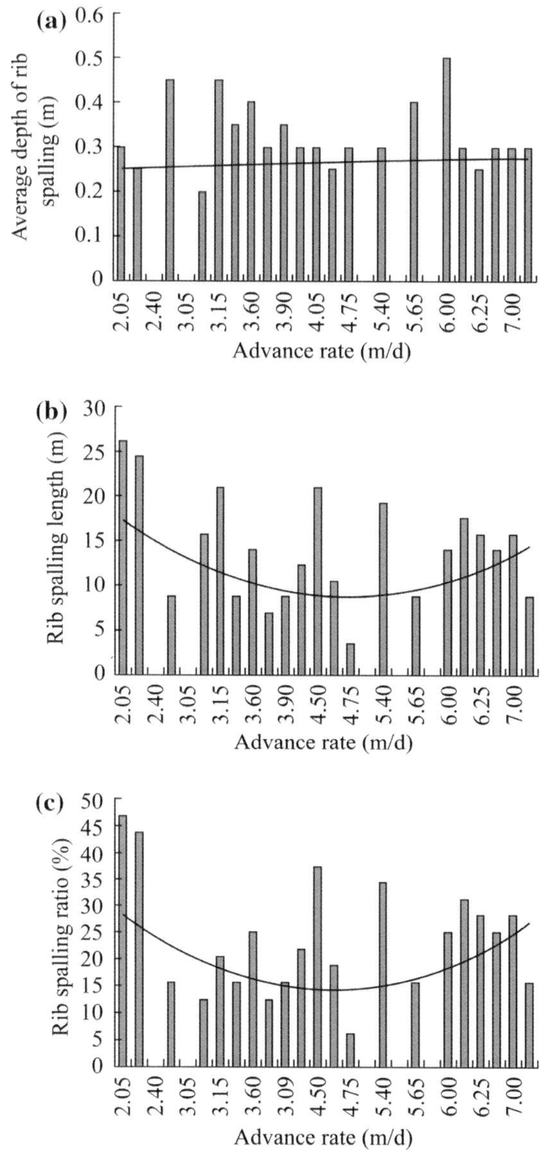

Fig. 27 Relationship *curves* between coal wall spalling and face advance speed at the lower part of the face (non-periodical ground pressure). **a** Relationship between average rib spalling depth and face advance rate, **b** relationship between rib spalling length and face advance rate, **c** relationship between rib spalling ratio and face advance rate

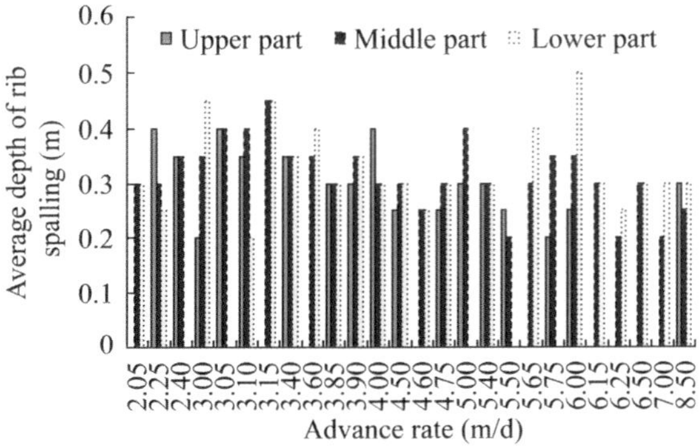

Fig. 28 Comparison of average spalling depth at different face positions (non-periodical ground pressure period)

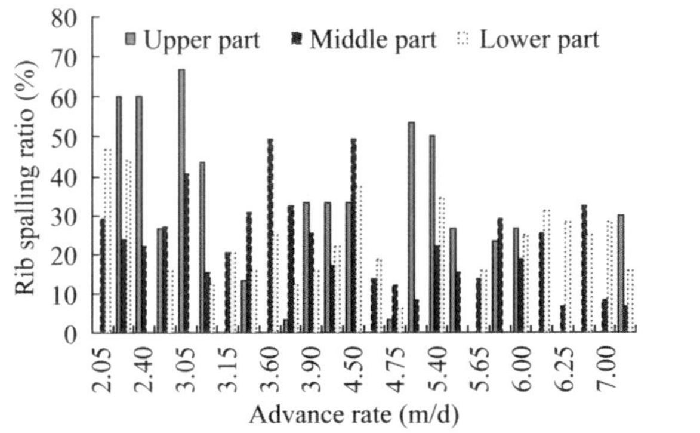

Fig. 29 Comparison of average spalling ratio at different face positions (non-periodical ground pressure period)

Table 11 Statistical analysis results of spalling depth at the upper, middle and lower parts of the face (non-periodical ground pressure period)

Face advance speed (m/d)	Average spalling depth (upper) (m)	Average spalling depth (middle) (m)	Average spalling depth (lower) (m)	Average spalling depth (m)
2.05	0	0.30	0.30	0.20
2.25	0.40	0.30	0.25	0.32
2.40	0.35	0.35	0	0.23
3.00	0.20	0.35	0.45	0.33
3.05	0.40	0.40	0	0.27
3.10	0.35	0.40	0.20	0.32
3.15	0	0.45	0.45	0.30
3.40	0.35	0.35	0.35	0.35
3.60	0	0.35	0.40	0.25
3.85	0.30	0.30	0.30	0.30
3.90	0.30	0.35	0.35	0.33
4.00	0.40	0.30	0.30	0.33
4.50	0.25	0.30	0.30	0.28
4.60	0	0.25	0.25	0.17
4.75	0.25	0.30	0.30	0.28
5.00	0.30	0.40	0	0.23
5.40	0.30	0.30	0.30	0.30
5.50	0.25	0.20	0	0.15
5.65	0	0.30	0.40	0.23
5.75	0.20	0.35	0	0.18
6.00	0.25	0.35	0.50	0.37
6.15	0	0.30	0.30	0.20
6.25	0	0.20	0.25	0.15
6.50	0	0.30	0.30	0.20
7.00	0	0.20	0.30	0.17
8.50	0.30	0.25	0.30	0.28
Average	0.20	0.32	0.26	0.26

reached 4.5 m/d, there was no distinct decrease of spalling depth, and the slope of the trend line was almost zero. When the face advance speed exceeded 4.5 m/d, there was a distinct decrease of spalling depth, thus the slope of the trend line increased.

As shown in Fig. 32b, c, during the periodical ground pressure period, the coal wall spalling ratio at the lower part of the face demonstrated a trend of increase early on, and decreasing later with the increase of face advance speed.

(4) Comparison of coal wall spalling at different face positions during the periodical ground pressure

As shown in Table 16, during the period of periodical ground pressure, the spalling depths at different face positions were almost identical, the average spalling depth was maintained at 0.29 m. The maximum spalling depth reached 0.7 m and the spalling depth exceeded 0.5 m, accounting for 30 %. During the period of periodical ground pressure, the spalling depth showed randomness, however, in general, presented a trend to decrease with the increase of face advance speed.

As shown in Table 17, during the occurrence of periodical ground pressure, the spalling ratio at the upper part of the face was the largest, reaching 38.67 %, less in the middle part at 28.81 %, and least at the lower part, 20.1 %. The average spalling ratio was 29.16 %.

3.6.5 Comparison of coal wall spalling at different ground pressure periods

Figures 33 and 34 show spalling depth and spalling ratio at different ground pressure periods respectively (average 4.9 m/d).

As shown in Fig. 33, the average difference of spalling depth with periodical ground pressure was 0.05 m (under the same face advance speed, the difference of spalling depth with periodical ground pressure or without periodical ground pressure).

Table 12 Statistical analysis results of spalling ratio at the upper, middle and lower parts of the face 8108 (non-periodical ground pressure period)

Face advance speed (m/d)	Average spalling ratio (upper) (%)	Average spalling ratio (middle) (%)	Average spalling ratio (lower) (%)	Spalling ratio (%)
2.05	0	28.81	46.88	27.05
2.25	60.00	23.73	43.75	38.89
2.40	60.00	22.03	0	26.21
3.00	26.67	27.12	15.63	24.52
3.05	66.67	40.68	0	37.20
3.10	43.33	15.25	12.50	26.21
3.15	0	20.34	20.34	20.29
3.40	13.33	30.51	15.63	22.83
3.60	0	49.15	25.00	31.28
3.85	3.33	32.20	12.50	20.29
3.90	33.33	25.42	15.63	25.36
4.00	33.33	16.95	21.88	22.83
4.50	33.33	49.15	37.50	43.12
4.60	0	13.56	18.75	11.84
4.75	3.33	11.86	6.25	8.45
5.00	53.33	8.47	0	17.75
5.40	50.00	22.00	34.00	32.97
5.50	26.67	15.25	0	14.37
5.65	0	13.56	15.63	10.99
5.75	23.33	28.81	0	20.29
6.00	26.67	18.64	25.00	22.83
6.15	0	25.42	31.25	21.14
6.25	0	6.78	28.13	10.99
6.50	0	32.20	25.00	22.83
7.00	0	8.47	28.13	11.84
8.50	30.00	7.00	16.00	15.22
Average	22.56	22.81	19.05	22.60

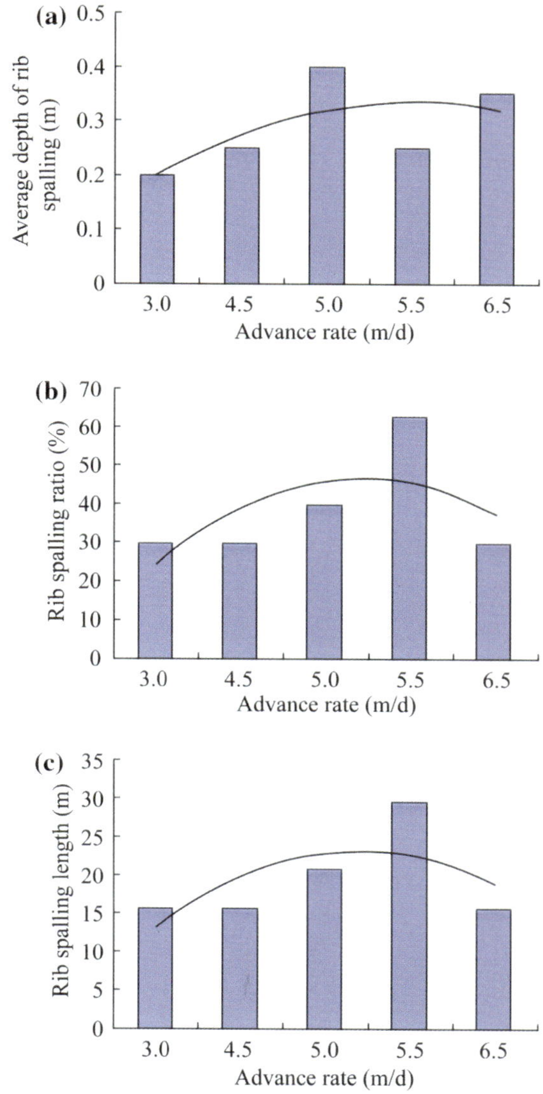

Fig. 30 Relationship curves between coal wall spalling and face advance speed at the upper part of the face (periodical ground pressure period). **a** Relationship between average rib spalling depth and face advance rate, **b** relationship between rib spalling ratio and face advance rate, **c** relationship between rib spalling length and face advance rate

As shown in Fig. 34, the periodical ground pressure has limited influence upon the coal wall spalling ratio at the No. 8105 face. The difference of spalling ratio caused by the ground pressures was 4.64 % (at the same face advance speed, the difference between the spalling ratio during periodical ground pressure period and the spalling ratio during the non-periodical ground pressure period).

Table 13 Characteristics of coal wall spalling at the upper part of No. 8105 face during periodical ground pressure period

Face advance speed (m/d)	Average spalling depth (m)	Spalling length (m)	Spalling ratio (%)	Maximum spalling depth (m)
3.00	0.20	15.75	30.00	0.30
4.50	0.25	15.75	30.00	0.50
5.00	0.40	21.00	40.00	0.50
5.50	0.25	29.75	63.33	0.30
6.50	0.35	15.75	30.00	0.50

Table 14 Characteristics of coal wall spalling in the middle part of No. 8105 face during periodical ground pressure period

Face advance speed (m/d)	Average spalling depth (m)	Spalling length (m)	Spalling ratio (%)	Maximum spalling depth (m)
3.00	0.40	29.75	28.81	0.60
4.50	0.25	50.75	49.15	0.30
5.00	0.30	19.25	18.64	0.30
5.50	0.20	19.25	18.64	0.20
6.50	0.30	29.75	28.81	0.40

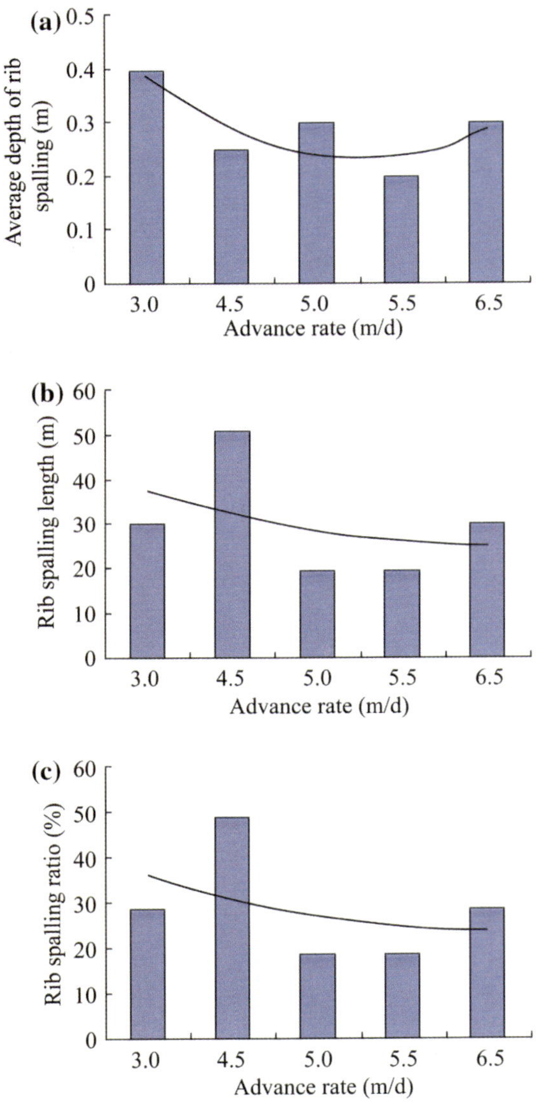

Fig. 31 Relationship curves between coal wall spalling and face advance speed in the middle part of the face (periodical ground pressure period). **a** Relationship between average rib spalling depth and face advance rate, **b** relationship between rib spalling length and face advance rate, **c** relationship between rib spalling ratio and face advance rate

3.6.6 Conclusion of the observation on coal wall at No. 8105 LLTCC face

(1) During the non-periodical ground pressure period, the relationship between the coal wall spalling ratio and face advance speed presented randomness. However, in general, the spalling ratio reduced with the increase of face advance speed. There were some differences in the spalling ratio at different face positions. The statistical analysis of the spalling ratio showed: the spalling ratio in the middle part of the face > the spalling ratio at the upper part of the face > the spalling ratio at the lower part of the face.

(2) During the periodical ground pressure period, the spalling depth and ratio showed a decreasing trend with the increase of the face advance speed. The spalling depth showed a difference at different face positions, i.e., upper part > middle part > lower part.

(3) During the period of periodical ground pressure occurrence, the spalling ratio showed a trend of increase early and decreased later on with the increase of the face advance speed. The spalling ratios at different face positions were different.

Through the comparison of coal wall spalling with the neighbouring No. 8104 face, it was found that: during the period of non-periodical ground pressure occurring, the average coal wall spalling ratio was 33.08 % and the spalling depth varied from 0.2 to 0.3 m at the No. 8104 face; the average spalling ratio was 22.06 % and the spalling depth varied from 0.18 to 0.25 m at the No. 8105 face. During the period of periodical ground pressure, the spalling ratio was 36.18 % at the No. 8104 face, and 29.16 % at the No. 8105 face. Based on the statistical analysis, the spalling depth and length were the largest in the middle parts of the No. 8104 and No. 8105 faces. In comparison with the No. 8104 face, the spalling ratio at the No. 8105 face was 33.4 % less than that at the No. 8104

Table 15 Characteristics of coal wall spalling at the lower part of No. 8105 face during periodical ground pressure period

Face advance speed (m/d)	Average spalling depth (m)	Spalling length (m)	Spalling ratio (%)	Maximum spalling depth (m)
3.00	0.45	7.00	12.50	0.70
4.50	0.30	21.00	37.50	0.40
5.00	0.40	21.00	37.50	0.50
5.50	0.25	7.00	12.50	0.30
6.50	0	0	0	0

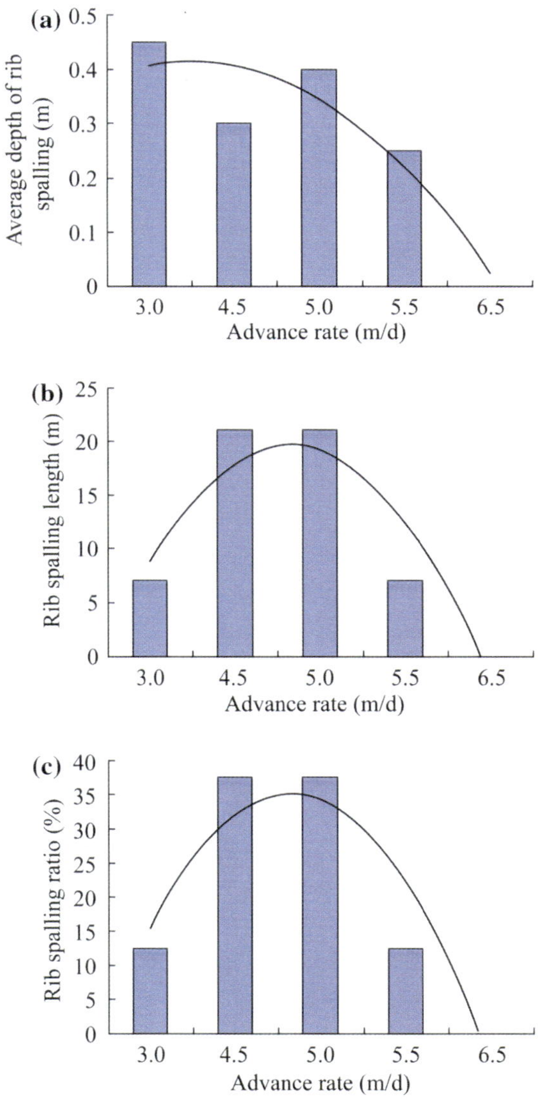

Fig. 32 Relationship *curves* between coal wall spalling and face advance speed at the lower part of the face (periodical ground pressure period). **a** Relationship between average rib spalling depth and face advance rate, **b** relationship between rib spalling length and face advance rate, **c** relationship between rib spalling ratio and face advance rate

face, and the spalling depth was also 50 mm less than that at the No. 8104 face. Therefore, it can be concluded that the coal wall spalling was significantly reduced and the face achieved a fast advance speed due to the application of proposed comprehensive control technology at the No. 8105 face in Tashan coal mine in Datong, China, including controlling face advance speed, increasing support initial abutment force and increasing the wall protection capacity of support.

3.7 High efficient top-coal caving technology at a LLTCC operation

3.7.1 Study on rational top-coal drawing interval

It is important to select a rational top-coal drawing interval to improve coal recovery and reduce the percentage of gangue. If the drawing interval is too large, the top coals might be caved into the goaf, resulting in a coal loss. On the other hand, if the drawing interval is too small, the gangue might be mixed with coals and cave into the discharge window, which may not only affect the coal quality, but also be mistaken for top coal exhaustion and result in coal loss. The factors for determining the drawing interval consists of the thickness of the top coal, the discharge window length of support, the center height of the discharge window and the shearer's cutting depth. The principal for determining the drawing interval is to fracture and loosen the top coal in the range of the discharged coal, increase top coal recovery and reduce the percentage of gangue. In addition, the rational drawing interval should match the coal mining process and form an integer-times relationship with the shearer's cutting depth.

The drawing interval must be greater than the length of discharge window to avoid mixing the gangue with coal. The small drawing interval might result in the gangue in the goaf being mixed with caved top coals. The large drawing interval might result in top coal losses. A rational drawing interval will maintain the mixture of top coal with

Table 16 Statistical analysis results of spalling depth at the upper, middle and lower parts of No. 8105 face (during periodical ground pressure period)

Face advance speed (m/d)	Average spalling depth (upper) (m)	Average spalling depth (middle) (m)	Average spalling depth (lower) (m)	Average (m)
3.00	0.20	0.40	0.45	0.35
4.50	0.25	0.25	0.30	0.27
5.00	0.40	0.30	0.40	0.37
5.50	0.25	0.20	0.25	0.23
6.50	0.35	0.30	0	0.22
Average	0.29	0.29	0.28	0.29

Table 17 Statistical analysis results of spalling ratio at the upper, middle and lower parts of No. 8105 face (during periodical ground pressure period)

Face advance speed (m/d)	Spalling ratio (upper) (%)	Spalling ratio (middle) (%)	Spalling ratio (lower) (%)	Average (%)
3.00	30.00	28.81	12.50	23.77
4.50	30.00	49.15	37.50	38.88
5.00	40.00	18.64	37.50	32.05
5.50	63.33	18.64	12.50	31.49
6.50	30.00	28.81	0	19.60
Average	38.67	28.81	20.00	29.16

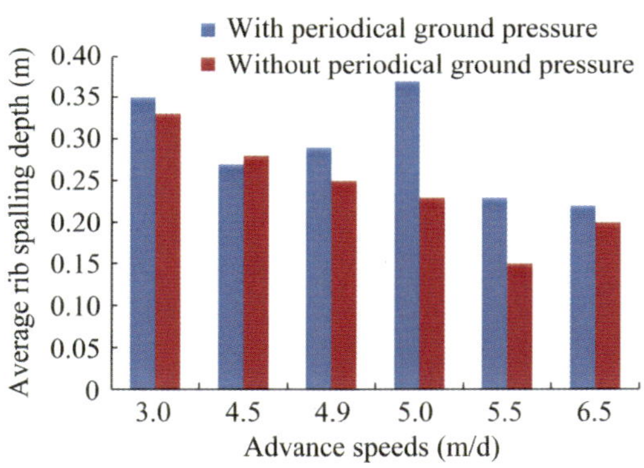

Fig. 33 Comparison of average spalling depth at different ground pressure periods under different face advance speeds

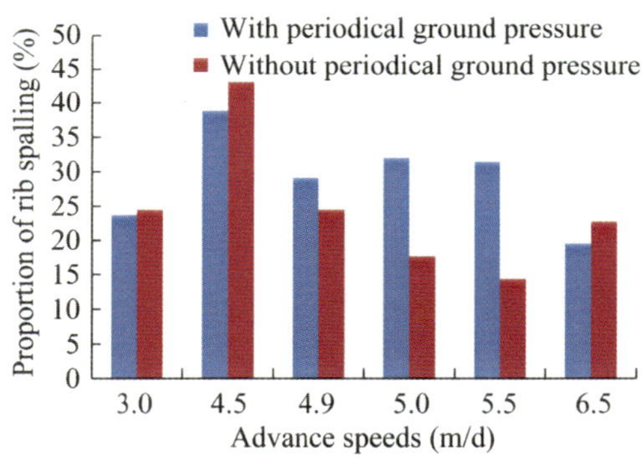

Fig. 34 Comparison of spalling ratio at different ground pressure periods under different face advance speeds

gangue and top coal losses in allowable limits. In a LLTCC face, the space for top coal fracturing becomes larger due to the increase in mining height, therefore, the effects of ground pressure on top coal fracture become greater. With the effects of ground pressure, the top coal will be fractured and broken thoroughly. The top coal fragment becomes smaller and has a better flowability.

The drawing interval has a significant influence on the coal and gangue flow and thus is a key parameter for the LLTCC process. Figure 35 shows the coal and gangue flow at different drawing intervals. The boundary of coal and gangue before and after top coal discharge is regarded as the starting line for starting coal discharge and the finishing line for finishing coal discharge. In the experimental test, three drawing intervals were tested including one cutting and one coal caving, two cuttings and one coal caving, and three cuttings and one coal caving. In the process of support shift and coal caving, the boundary of coal and gangue was moved forward and maintained in the same shape. With the increase of the drawing interval, the distance between the starting line and the finishing line also increased.

As shown in Fig. 35, with the increase in the top coal drawing interval, there were some variations in the boundary shape of coal and gangue before and after top coal caving. In other words, the initial caving point of top coal shifted forward, the distance between start line and finish line of top coal caving increased with the drawing

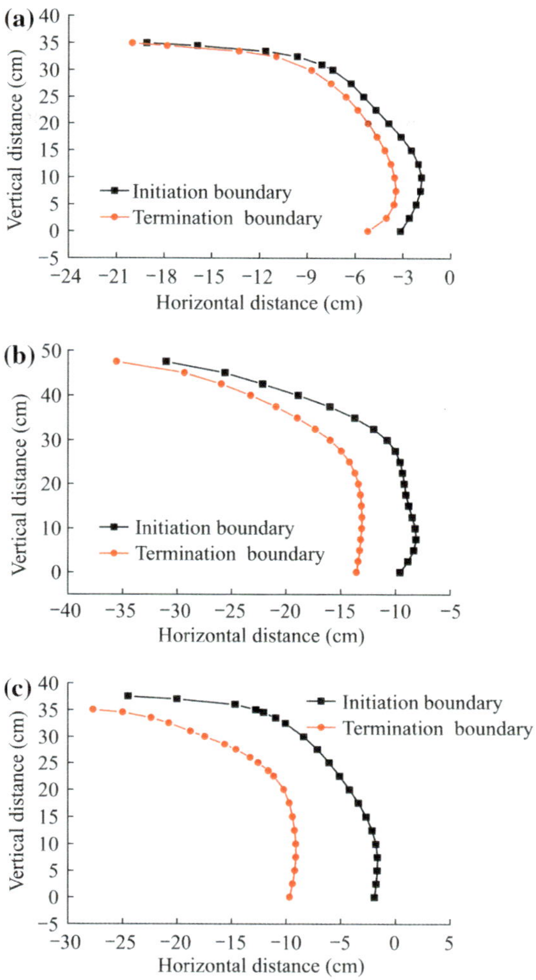

Fig. 35 Boundaries between coal and gangue with different top coal drawing intervals. **a** One cutting and one drawing, **b** two cuttings and one drawing, **c** three cuttings and one drawing

interval. This indicated that the amount of coal discharged at each discharge circle increased with the increase of the drawing interval.

There were three kinds of coal and gangue mixtures near the discharge window of support after top coal caving using different drawing intervals:

(1) When the drawing interval is small, the gangue in the direction of the goaf reaches the discharge window earlier than the top coals do, and the discharged coals mix with gangue. Before coal discharge, the boundary between coal and gangue near the discharge window is beyond the lower boundary of the tail beam. Therefore, it results in the part of middle to top coals being unable to be discharged due to the enclosure of gangue and the tail beam. In this case, the gauge might enter the discharge window from the bottom of the window, and increase the percentage of gangue in the coals obtained.
(2) With a rational top-coal drawing interval, at initial coal discharge, the gangue in the goaf and top coal will reach the discharge window of the support at the same time. Before the coal discharge, the boundary between coal and gangue is behind the lower boundary of the tail beam, thus the horizontal movement of gangue in the goaf is small. In the process of top-coal caving and discharge, the coals can smoothly reach the discharge window. With the caving of the lower and middle part of the coals, the gangue above the coal seam caves as well, then there will be some mixture of coal and gangue during the discharge of the upper layer of top coal. The mixture gangue is mainly from the roof above the seam that enters the discharge window with the discharge of top coals.
(3) When the drawing interval is too large, at the beginning of coal discharge, top coals caved reach the discharge window earlier than gangue in the goaf. Before coal discharge, the boundary between coal and gangue is obviously behind the lower boundary of the tail beam, and the gangue in the goaf cannot reach the discharge window. With a LLTCC operation, the angle of the shield beam becomes larger, and the height of the discharge window becomes higher. During the process of coal discharge, the top coal above the discharge window can be discharged quickly and the gangue above the top coal also caves quickly. In this case, the coal and gangue in the goaf near the discharge window with natural angle of depose cannot flow toward the discharge window and be discharged, resulting in coal losses and low coal recovery.

Figure 36 shows the shape of residual top coals with different drawing intervals. As shown in the figures, the one cutting and one drawing method (drawing interval is 0.8 m) can lead to a high top coal recovery.

3.7.2 Study on rational coal discharge method

Top coal discharge method involves the combination of sequence of coal discharge, the times of coal discharge at each discharge window and the quantity of coal discharge, as well as the number of discharge window opened in the same time along the face. The sequence of coal discharge can be categorized as sequential coal discharge and alternated coal discharge. The times of coal discharge at each discharge window can be categorized as single, double and multiple coal discharges. The number of coal discharge window opened in the same time can be categorized as

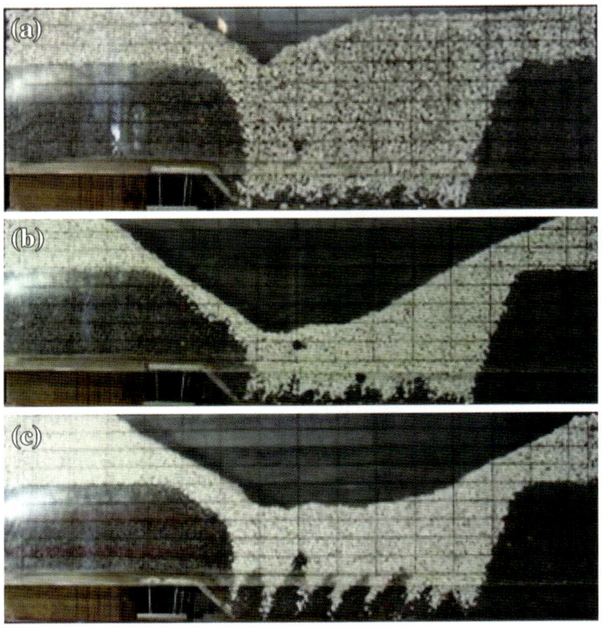

Fig. 36 Shape of residual top coals with different drawing intervals. **a** One cutting and one drawing (drawing interval 0.8 m), **b** two cuttings and one drawing (drawing interval 1.6 m), **c** three cuttings and one drawing (drawing interval 2.4 m)

single window discharge and multiple window discharge. The number of simultaneously opened discharge window can be categorized as single coal discharge and multiple coal discharge. The quantity of coal discharge can be categorized as equal quantity and non-equal quantity. Because it is difficult to manage equal or non-equal quantity coal discharge in operation, and single or multiple coal discharge can be managed in labor planning, however the determination of the coal discharge method is mainly based on the sequence of coal discharge and the times of coal discharge at each discharge window. There are four main alternative methods including single sequential coal discharge, single alternated coal discharge, multi-round sequential coal discharge and multi-round alternated coal discharge. The geological conditions for different coal seams are different, and their capabilities are also different, thus, different coal discharge methods must be used, and the recovery of top coal is also varied.

A numerical analysis was conducted to simulate the top-coal caving and discharge by using the geological parameters of coal seams and the technical parameters of hydraulic supports at Tashan coal mine. In addition, the simulated loose thickness of the top coal was 10.50–11.17 m, and the simulated cutting heights were 4, 4.5 and 5 m. The top-coal caving and discharge operation was simulated including single sequential coal discharge, single alternated coal discharge and multiple alternated coal discharge method. The numerical results showed that the recovery of coal cutted and discharged was the lowest using single-time alternated multiple windows coal discharge, and highest using discharge in multi-round & multiple windows sequence alternately. In order to increase the recovery of top coal, the method of multiple-time alternated multiple window coal discharge was selected in the LLTCC operation at the No. 8105 face, Tashan coal mine in Datong, China.

4 Key equipment for the LLTCC coal mining operation

Key equipment for the LLTCC coal mining operation involves hydraulic support, shearer, front scraper conveyor, rear scraper conveyor, headgate belt conveyor and support hauler, etc. Prior to this project, there was no equipment for LLTCC coal mining operations in China or elsewhere in the world. Therefore, the various key equipment for LLTCC mining operation have been developed in this project.

4.1 Hydraulic support with large mining heights

Hydraulic support is one of key equipment for a LLTCC coal mining operation. The requirements for hydraulic support consist of rational structure, sufficient support resistance force, stable and shock resistance, sustainability of high efficient coal discharge and high coal recovery.

4.1.1 Support and rock coupling 3D dynamic optimization design method

In the design of hydraulic support, a support and rock coupling 3D dynamic optimization design method was used. The effects of surrounding rock on support structure, support intensity, and shift capability were fully considered. As the support and rock coupling based 3D dynamic parameter design commenced from roof movement, a simulation was conducted to simulate the transfer of fracture, subsidence, crash and the caving of roof into the pressure, torque, deformation and shift speed of the support. Using a series of conditions, support working resistance, support type and control parameters were dynamically determined to satisfy the requirements of roof movement and effective support on the roof. With the consideration of coal wall and floor characteristics, the forms and parameters of the auxiliary components for coal wall protection and bottom lift were determined. The advantages of the proposed method can be stated as: the dynamic characteristics of surrounding rock were fully considered through the dynamic interaction between supports and surrounding rock, the variation of roof subsidence and fracture was transferred to the support in the forms of pressure, torque and constraints. Hydraulic support responded in terms of forms and characteristic parameters, and the response results fed back to rock in the forms of support resistance force and torque. Therefore, the roof caving and the distribution of stress and strain could be controlled.

In terms of the technique for the support design, a 3D parametric dynamic design method from top to bottom was used. In the design process, the requirements of support from surrounding rock were gradually satisfied. The flowchart of support design is shown in Fig. 37. The method proposed achieves seamless integration between the optimization of structural parameters and the 3D modelling. Furthermore, the method improves the design efficiency, provides an integrated platform for structure optimization and parametric design, and improves the applicability of support.

4.1.2 Determination of support technical parameters

Based on the geological conditions at the No. 8105 coalface in Tashan coalmine, the composition of support working resistances during their calculation was analyzed. Using the numerical modelling method, the numerical model was built based on the occurrence conditions of coal seams. Subsequently, the model was used to simulate the support intensity of 0, 0.5, 0.9, 1.2, 1.3, 1.4, 1.5, and 1.6 MPa, respectively. The analysis on the vertical stress at the coalface showed that the increase of support intensity could not change the distribution of abutment stress in the front of the coalface. The front peak abutment stress was approximately 10 m from the coal wall. With the increase of support intensity, the area of peak abutment stress in the front of the coal face decreased and became closer to the coal wall.

In order to obtain the relationship curve between support intensity and roof subsidence (P–$\triangle L$ curve) within the roof control range, the subsidence of the roof at a position 3 m

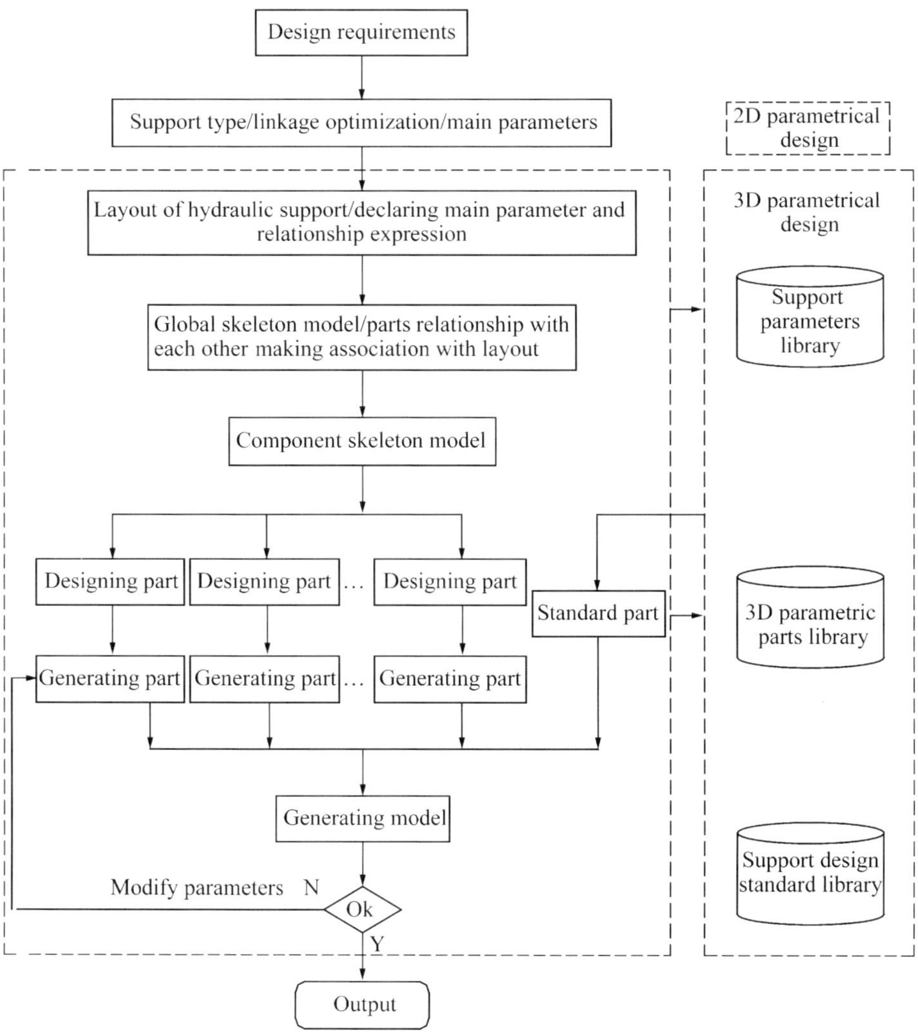

Fig. 37 Flowchart of 3D parametric dynamic design of the hydraulic support from top to bottom

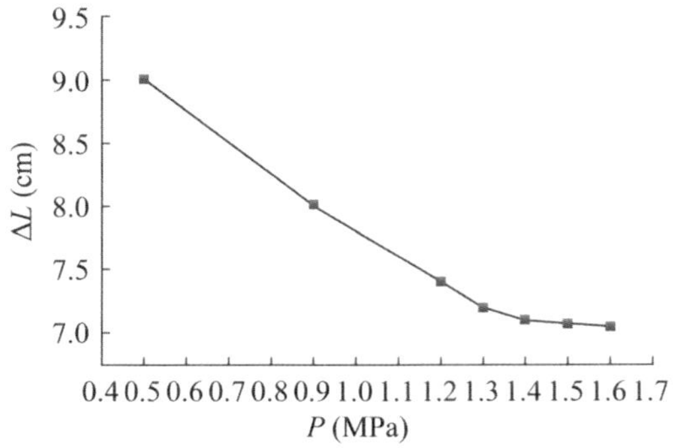

Fig. 38 Relationship curve between support intensity and roof subsidence (P–$\triangle L$ curve)

away from the coal wall were measured, and simulated with the support intensity of 0.5, 0.9, 1.2, 1.3, 1.4, 1.5 and 1.6 MPa. The curve between support intensity and roof subsidence is shown in Fig. 38.

As shown in the relationship curve between support intensity P and roof subsidence $\triangle L$, the subsidence of the roof gradually decreases with the increase of support intensity. When the increase of support intensity reaches a certain level, the increase of support intensity have little effect on the roof subsidence. Nevertheless, if the support intensity is below a given level, the reduction of support intensity will have substantial impact on roof subsidence. The given level of support intensity can be selected as rational support intensity. When the support intensity reaches 1.4–1.5 MPa, the decrease of roof subsidence will not obvious with the increase of support intensity. As a result, the rational support intensity is determined as 1.4–1.5 MPa. With comprehensive analysis on the mining and face conditions, the support intensity of the hydraulic support was determined as 1.4 MPa in Tashan coal mine.

Determination of support resistance.

Nominal working resistance F can be calculated as Eq. (13):

$$F \geq \frac{PB_cL}{\eta} \tag{13}$$

where P is the nominal support intensity at the LLTCC face, 1.4 MPa; L is the center distance, 1.75 m; B_c is the roof control distance, 5.5 m; η is the support efficiency, 0.98.

Substitute the parameters above into Eq. (13), get:

$$F \geq \frac{1.4 \times 5.5 \times 1.75}{0.98} = 13750 \text{ kN}$$

Considering the safety coefficient, the nominal support resistance was determined as 15000 kN, which is consistent with the calculation result from the "cantilever beam-articulated beam" structural mechanical model.

The support and rock coupling mechanical model (Wang 2009) was used to analyses the main technical parameters of hydraulic support at Tashan coal mine. The support parameters were optimized based on the support and rock coupling mechanical model. The structural parameters, i.e. the beam length of the support, the center distance of the support, and maximum and minimum heights, were used as optimizing variables, and the surrounding distribution of roof stress and the floor pressure of the support front were used as the optimizing objective. The optimizing expression was:

Objective functions:

$$\begin{cases} \Delta p = |p - p_j| \leq \lambda (j = 1, 2, \ldots n) \\ \min \delta \end{cases} \tag{14}$$

where $p_j = f\,(l, h_{\max}, h_{\min}, a, b)$, $\delta = k\,(l, h_{\max}, h_{\min}, a, b)$; $\triangle p$ is the absolute difference between support intensity p_j calculated iteratively based on support parameters and the result calculated based on surrounding rock parameters. When $\triangle p$ is less than or equal to a given condition of convergence λ, the support intensity is regarded as rational. The optimizing objective of floor pressure at the support front is to minimize the floor pressure. Function f and k are the mechanical status expression of the surrounding rock affected by support parameters obtained from the analysis of the interaction of support and surrounding rock, respectively. In the objective function, l is the length of top beam, $h_{\max}$ and $h_{\min}$ are maximum support height and minimum support height; a is the support center distance, being selected as 1.5, 1.75 and 2.05 m, b is the horizontal distance between the center of the top beam prop cup and the center of the base prop foot.

Before optimization, the support type should be selected. The parameters of support are used as the variables of optimization function, after which the multiple-objective optimization function is established. After the calculation results converge to a group of stable values, the rational optimization results are obtained. In the Tashan coal mine, during the process of the LLTCC mining operation in 14–20 m thick coal seams, the thickness of the top coal is large and the caving space is also large, therefore, the movement of roof increases, and the range of resultant force point of interaction between support and roof becomes large (Yu et al. 2012). According to the observation results of ground pressure at the No. 8104 LLTCC face in Tashan coal mine, the resistance alternately increased at the rear prop or front prop during the roof periodical ground pressure period. As a consequence, the positive four-link bar type of support was selected to better suit the ground pressure.

The average seam thickness at the No. 8105 face in Tashan coal mine is approximately 20 m. Based on the requirement of not exceeding the cutting to caving ratio 1:3, the cutting height reached 5 m. Based on the requirements of cutting height and the parameter optimization of support-rock coupling, the main technical parameters of support were determined as follows: the maximum support height is 5200 mm, the minimum

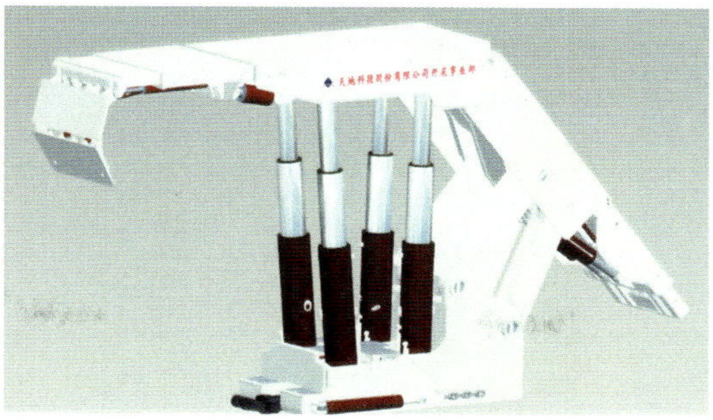

Fig. 39 LLTCC hydraulic support with a height of 5.2 m

support height is 2800 mm, the top beam length is 5615 mm, the support center distance is 1750 mm, the horizontal distance between top beam prop cup center and support prop base center is 402 mm, and the support working resistance is 15,000 kN. Figure 39 shows the LLTCC hydraulic support as developed.

4.1.3 Development of shock resistance components for LLTCC hydraulic support

In comparison to the fully mechanized longwall mining method and the fully mechanized LTCC mining method, the LLTCC mining method requires a higher standard for the shock resistance of hydraulic support. Thus, a large diameter double telescopic shock resistance prop was developed. Its main parameters are as follows: outer cylinder: cylinder diameter ϕ360 mm, cylinder outer diameter ϕ418 mm; middle cylinder: cylinder diameter ϕ 260 mm, cylinder outer diameter ϕ340 mm; small cylinder: cylinder diameter ϕ235 mm, cylinder outer diameter ϕ145 mm; working resistance ($P_1/P_2 = 43.2/82.3$ MPa): 4400 kN.

A 1000 min^{-1} large flow rate safety valve fast release device was developed to meet the requirement of a safety valve with a flow rate of 620 min^{-1} when prop was under shock force. With the development of key components above, the shock resistance capacity of the hydraulic support increased approximately 10 %.

4.1.4 High efficient coal discharge device for a LLTCC hydraulic support

At a LLTCC coal mining face in a coal seam with thickness greater than 10 m, the caving height is high after the top coals cave, and the size of coal block is relatively large. Therefore, the top coals might form an "arch" balance structure during top coal caving. To break the "arch" balance structure and enable the top coals to be easily discharged into the rear scraper conveyor is a key technology. Therefore, the technical issues discussed above must be solved for LLTCC hydraulic support, which provides the top coal control technology for a LLTCC hydraulic support.

As a result, the following method can be used for the LLTCC hydraulic support:

(1) The control of the size of the coal discharge window: The control of the discharge window is to control the coal discharge quantity at a single discharge window. Under the same flowability of caved top coals, the discharged quantity in a unit time is in proportion with the size of the discharge window. As a result, the centerline spacing of supports was modified from 1.5 m to 1.75 m to increase the size of discharge window along the inclined direction of the face (size in transverse direction). Simultaneously, the size of the discharge window in the longitudinal direction (along the strike direction of the face) was controlled.
The size of the coal discharge window is controlled by a coal discharge device (coal discharge sliding plate and tail beam). The size of the discharge window is the function of the extension length of the sliding plate and the swing angle of the tail beam. When the coal discharge window is completely opened, the maximum size of the discharge window in the longitudinal direction is equal to the coal-reception width of the rear scraper conveyor. In this case, the coal discharge rate of single support is determined by the chain speed of the rear scraper conveyor and the blockage coefficient of coal flow. The coal discharge rate of single support can be calculated using Eq. (15).

$$V_{\mathrm{f}} = 60\,K_{\mathrm{f}} F \gamma V_{\mathrm{s}} \quad (15)$$

where V_{f} is the coal discharge rate of single support, t/min; K_{f} is the blockage coefficient of discharged coal flow, $K_{\mathrm{f}} = 0.6$; F is the carrying cross-section of the rear scraper conveyor, m^2; γ is the loose coal density, t/m^3; V_{s} is the chain speed of scraper conveyor, m/s.

(2) The time of multiple coal discharge and the control of coal discharge quantity: In respect of the LLTCC mining operation at Tashan coal mine, because the ply of the top coal is very thick and the top coal is very hard, after top coal caving, the upper top coals cannot fully break due to limited caving space. Some of top coals close to the roof cannot even cave. In this case, if the caved coals are discharged completely, there will be losses of large coal blocks and uncaved top coals. As a result, there should be multiple coal discharges, e.g., two or three times. The multiple coal discharges will provide time and

space for the fracturing and caving of top coals, and enable the boundary of caved top coals and immediate roof gangue to be evenly subsided.

The purpose of multiple coal discharges is to provide time and space for the top coals and the immediate roof to fracture and cave. Therefore, the delay time of the second coal discharge cannot be too short. Generally, the second coal discharge should be more than 10 m away from the first coal discharge. It should be adjusted according to the movement of top coals and the roof.

The discharged coal quantity in a round is determined by the awaited discharge of top coals. The awaited discharge quantity of top coals can be calculated theoretically, as shown in Eq. (16).

$$Q_d = h_2 B S \gamma \eta_d \tag{16}$$

where Q_d is the awaited discharged quantity of top coals in a support, t; h_2 is the thickness of top coals, m; B is the width of support, m; S is the discharge interval, m; η_d is the recovery of top coals, %.

Because the discharged coal quantity in a unit of time at a discharge window cannot be calculated, the control of discharged coal quantity each time at a support is to control the time. After the start of a LLTCC operation, the coal discharge time at a discharge window should be calculated through several rounds of coal discharge operation. The time from the start of discharge until the gangue occurring in the discharge flow is the time of coal discharge in a round. When using two rounds of coal discharge, the time spent on one round of coal discharge should be half of the total coal discharge time, and 1/3 of the total coal discharge time for three rounds of coal discharge.

(3) During the process of coal discharge, the size of caved coal is suitable at the initial stage and the discharged coal flow is smooth. With the discharge of top coals, the large-size coals increase and form a balanced arch above the discharge window to block the coal flow. In this case, the base of the balanced arch should be broken using a coal discharge device. In addition, the large coal blocks should be crushed to enable the continuity of coal discharge. Therefore, the tail beam should be continuously swung up and down during the process of coal discharge to disturb the awaited top coals. The upswing of the tail beam can crush large coal blocks and the down swing can loosen the coal flow.

Caving and discharging the top coals and increasing coal recovery is the major objective of the present project. Based on the studies on the caveability of top coals and the laws of top coal caving, the support structure and the method of top coal caving and discharge were optimized. A new highly efficient coal caving and discharge device was developed to increase top coal recovery. Figure 40 shows the design of the coal discharge device.

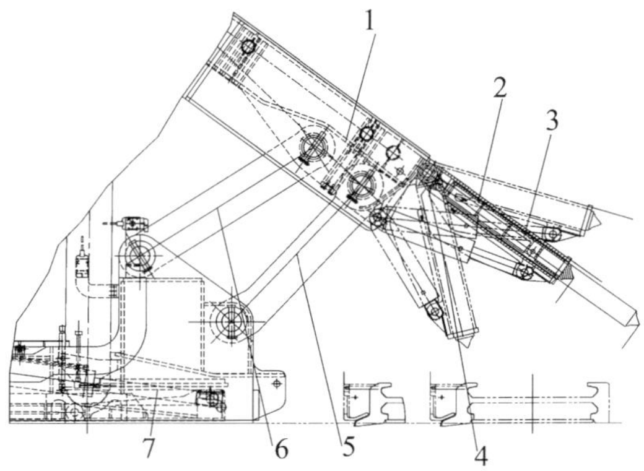

Fig. 40 Coal discharge device. *1* shield beam, *2* tail beam-sliding plate, *3* pump of sliding plate, *4* pump of tail beam, *5* rear link bar, *6* front link bar, *7* base

Through the studies on previous coal discharge devices, the following problems were found:

(1) In previous type LTCC supports, the front link bar was the double link bar, and the rear link bar was the single link bar. The rear bar was fixed in the middle of the support base, which resulted in blocking the operator's sight and difficulty in observing the effects of top coal caving and discharge.
(2) There was no passage from the front to the rear of the support. Because the rear link bar was fixed in the middle of the support base, operators could not go through two supports from the front to the rear of the support. The only way was to go to the rear of the support via face ends, which resulted in difficulties in terms of safety and maintenance.
(3) The coal discharge device of the tail beam-sliding plate was easily jammed and hard to slide. Because the sliding plate was not slid back, the sliding plate occasionally cut the chain of the rear scraper conveyor when the tail beam swung.

In order to solve the above problems, a new type of coal discharge device was designed with a double rear link bar, middle passage and observation window for coal discharge through the optimization of support structure. In addition, the high efficient and high recovery coal discharge device had a strong disturbance capacity for releasing coals.

(1) A new type of coal discharge device with double front and rear link bars, middle passage and observation window.

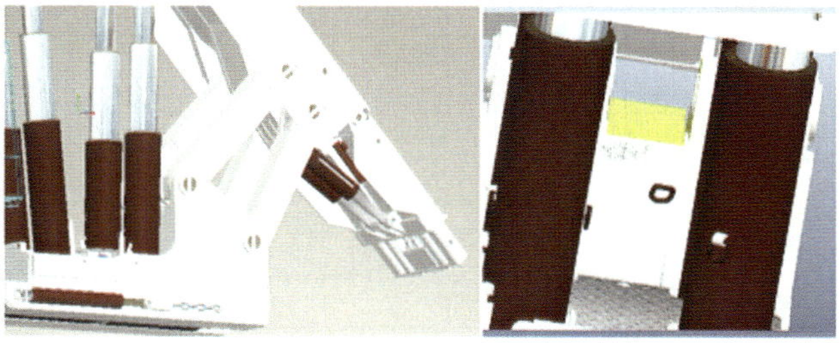

Fig. 41 Middle passage in support

Fig. 42 High efficiency and high reliability electrical haulage shearer

Through the optimization of the four-link bar on support, on the basis of meeting the requirements on support movement and limited positions, the internal components, i.e., shield beam and base, were re-designed. A new type of coal discharge device was developed with new double front and rear link bar, middle passage, and observation window, as shown in Fig. 41.

Under the ground pressure, the top coal is fractured and caved. The size of coal is different depending on the coal hardness. During the process of top coal discharge, the increase in large coals causes a balanced arch above the support discharge window and blocks the coal flow. In this case, the base of the balanced arch should be fractured using the coal discharge device. In addition, the large coal blocks should be crushed to enable the continuity of coal discharge. Therefore, a high efficient and high recovery coal discharge device was developed with a strong disturbance capacity. Two jacks with a cylinder diameter of ϕ180 mm were installed on the tail beam for operation. The force of a single jack was 1000 kN ($P = 39$ MPa) and 2000 kN with both. In order to increase the disturbance to top coals, the upswing of the tail beam can reach 19.5° to crush top coals, and the down swing can reach 41° to loosen coal flow. During the process of top coal discharge, the tail beam was swung continuously to disturb the awaited top coals and to achieve a high efficient and high recovery coal discharge.

4.2 High efficiency and high reliability electrical haulage shearer

4.2.1 Assessment of the overall reliability of electrical haulage shearer

The electrical haulage shearer is one of the key equipment in coal mining operations. High efficient continuous coal cutting is the shearer's main task, as the reliability and safety will affect coal production and operator safety. To meet the coal cutting requirements at the Tashan LLTCC coal mining operation, a new high efficiency, high reliability electrical haulage shearer had been developed. The shearer's specifications are as follows: cutting height 2.8–5.5 m, total installed power 1945 kW, coal cutting power 750 kW, maximum haulage speed 15–25 m/min, and maximum haulage power 2 × 570 kN, as shown in Fig. 42.

Table 18 Common faults of electrical haulage shearer

No.	Fault	No.	Fault
1	Not reach correct position	20	Structure crash
2	Broken	21	Jam
3	Loosen	22	False open
4	Fall off	23	False close
5	Bending	24	Unusual wear
6	Twisting	25	Oil leakage
7	No input	26	Inappropriate gap
8	No output	27	Puncture
9	Circuit open	28	Plastic deformation
10	Short circuit	29	Poor contact
11	Unable to switch	30	Blockage
12	Unable to start	31	Poor sealing
13	False action	32	Peel off
14	Flow blockage	33	Burn out
15	False indicator	34	Worn
16	Unexpected operation	35	False connection
17	External leakage	36	Burn down
18	Internal leakage	37	Fractured
19	False action	38	Rusty

The reliability of the shearer is the key factor controlling normal mining operation. As a result, the assessment method for assessing shearer's reliability has been investigated. Table 18 lists 38 common faults of the electrical haulage shearer. A fault tree analysis method was used to analyses the reliability and design. Based on the theoretical analysis on component design, installation and the environment of usage, the main causes of component failure were investigated to find the weakest link in reliability, and improve the reliability and utilization of the shearer.

The fault tree method for assessing the faults in electrical haulage shearer was divided into two stages:

(1) Establishment of fault tree for the electrical haulage shearer system.

① Collect and analyses the technical document of the electrical haulage shearer, including the drawings of the shearer, maintenance records, working environment and geological conditions, etc.
② Determine the contents and constraints of the electrical haulage shearer.
③ Determine top events based on the electrical haulage shearer system.
④ Analyze the causes and factors related with top events, the causes and factors related with each sub-system.
⑤ Map the fault tree of the electrical haulage shearer system. Deductively analyses from top events, search for causes level by level until reaching the basic event, map the fault tree according to the their logical relationship.

(2) Quantitative analysis of the electrical haulage shearer system.

① Simplify the fault tree, number the logical gates and basic events for further analysis.
② Calculate the failure rate of the basic event.
③ Calculate the occurrence rate of the top event.
④ Calculate the importance of parts of each basic event, including probability importance and criticality importance.
⑤ Establish the main control measures on key components individually according to the importance of each sub-system and elements.

4.2.2 *Shearer load database and faults database*

A remote data file submission scheme was developed to achieve a remote submission of shearer monitoring data by using the PHP language and B/S architecture. Through the automaton, the remote data was summarized and sent to database at headquarters to provide data for the simulation of shearer operation.

Using a new Browser/Server architecture, under Browser/Server three-layered architecture, the presentation,

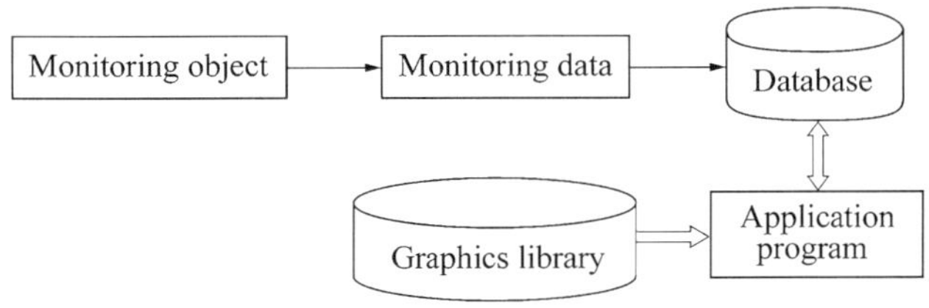

Fig. 43 Relationship of monitoring objects, data and the graphic library

business logic and data service were divided into three individual units: presentation: Web browser; business logic: Web server with extension function of application program; data service: database server.

Monitoring objects are the monitored components in the shearer, which generate monitoring data. Monitoring data is the data converted from real-time signals generated by monitoring objects, which can be recognized by the computer. The graphic library in the monitoring system is composed of the instrument graphics of monitoring objects, which can be used by an application program. The relationship of three is shown in Fig. 43.

The parameters of the monitoring system of the electrical haulage shearer were mainly used to define the environmental parameters of system operations, including the parameters displayed at the client terminal and the parameters of network transmissions, and the geological parameters, mining parameters, gateroad parameters, the parameters of the scraper conveyor, the parameters of the hydraulic support, and the parameters of the shearer and its accessories.

Based on a whole configuration electric control system, the monitoring data of the shearer system is shown in Table 19.

Four current transformers were used in the shearer's electric control system to monitor the single-phase current of each cutting motor. In this case, the loading status of the motor can be known. The DC0-10 V voltage signal output from the current transformer is sent to PLC for comparison to obtain the signals of the underload or overload. Pt100 thermal resistances are installed inside the motor windings of four cutting motors on the left and right, haulage motors on the left and right. The thermal resistances are directly connected with the RTD module of PLC. The data collected are stored in CF cards as .csv files for subsequent data analysis. The visualization system of monitoring data connects the .csv files to the application programs through the ADO, OLE, and ODBC methods. The application programs input the data into corresponding database tables to display the data in a graphic format.

Based on the analysis on the requirement for the functions of monitoring the data visualization system, the

Table 19 Data structure of the parameters of shearer operation condition

Parameters	Variable type	Variable name	Note
Left cutting motor power	Float	LCMP	Left cutting motor power
Right cutting motor power	Float	RCMP	Right cutting motor power
Left haulage motor power	Float	LHMP	Left haulage motor power
Right haulage motor power	Float	RHMP	Right haulage motor power
Oil pump motor power	Float	OBMP	Oil pump motor power
Water pump motor power	Float	WBMP	Water pump motor power
Left cutting motor temperature	Float	LCMT	Left cutting motor temperature
Right cutting motor temperature	Float	RCMT	Right cutting motor temperature
Left haulage motor temperature	Float	LHMT	Left haulage motor temperature
Right haulage motor temperature	Float	RHMT	Right haulage motor temperature
Oil pump motor temperature	Float	OBMT	Oil pump motor temperature
Water pump motor temperature	Float	WBMT	Water pump motor temperature
Haulage direction	Bolean	HD	Haulage direction
Haulage speed	Float	HS	Haulage speed
Shearer position	Int	SP	Shearer position

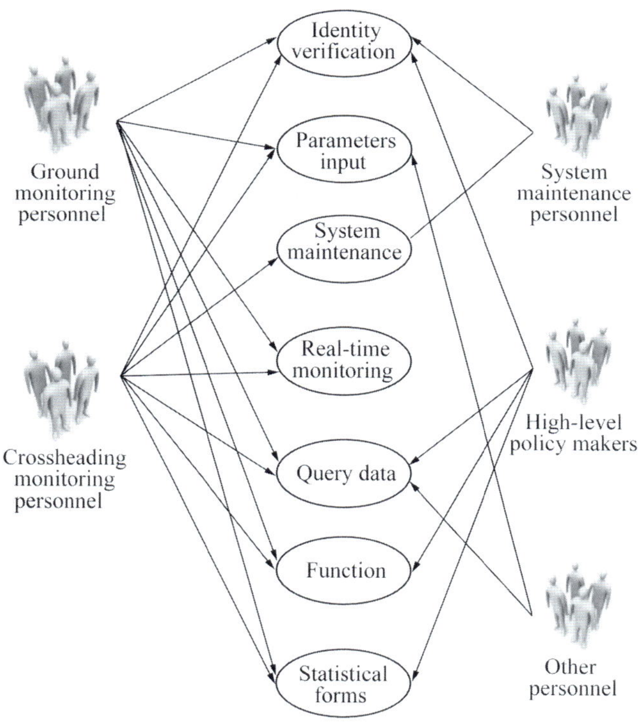

Fig. 44 System case model for shearer monitoring data

system case model was established for the system, as shown in Fig. 44.

The large amount of monitoring data generated during the shearer operation were compiled and transferred into the database. After being processed in the database, the relevant data was extracted to link with the graphic library. The visualization of data was achieved. The large amount of historical data was analyzed and mined to form a decision support system.

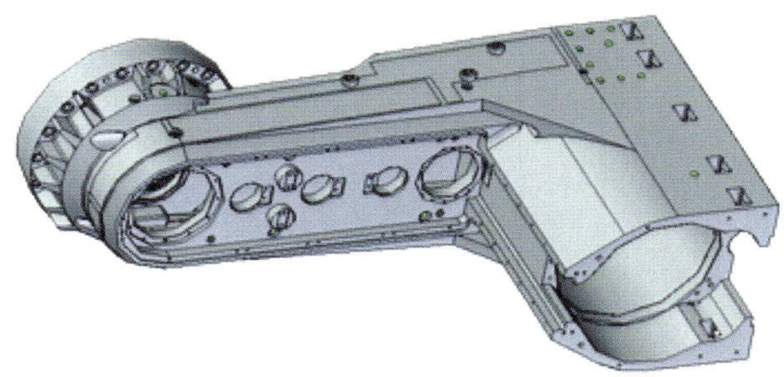

Fig. 45 3D model of shearer arm shell

4.2.3 Development of 750 kW high reliable shearer ranging arm

Shearer's ranging arm shell (Fig. 45) is a power transmission part of drum rotary cutting and a supporting part of the curving movement of the drum. To enable the shearer's ranging arm maintaining its normal operation during the period of nominal coal cutting quantity (overhaul life maintenance), the arm shell should have excellent comprehensive mechanical properties, i.e., high tensile strength, high fatigue strength and hardness, and high impact toughness. As a result, a new material CrNiMo series of hardened and tempered steel was developed. The medium carbon CrNiMo series special cast steel material was selected because the alloy steel with Cr and Ni has a high strength, in addition to high plasticity and toughness. The addition of Mo can control the secondary temper brittleness, reduce overheating, increase temper stability and obtain high strength and toughness of steel after it is hardened and tempered. Because of the addition of Cr, Ni, Mo and other minor alloy elements in the alloy, the alloy steel has a high hardenability. In addition, due to the solid solution strengthening of Ni and Mo elements, and the carbide precipitation strengthening of Cr and Mo elements, Mo and Cr elements hinder the growth of austenite grain, which will improve the strength, hardness and worn resistance of the alloy steel.

The large power ranging arm shell is made of large heterotypic cast steel, thus, it is in an irregular shape with large internal space and a complex structure. Furthermore, the shell's thicknesses vary significantly. Therefore, if the cooling rate after high temperature austenite is too fast, the shell might have cracks. At the same time, the alloy steel of the arm shell has a high hardenability, and has a high hardness under air quenching (air cooling, normalizing) with modest cooling rate. Therefore, during the process of air quenching and high temperature tempering, the rate and method of increasing temperature and the temperature of air quenching need to be appropriately controlled to meet the mechanical property requirements from the arm shell body, and prevent the occurrence of cracks.

The heat treatment of the shearer's ranging arm shell is high temperature tempering after air quenching (normalizing), which is completed before the machining of the shell. The process is: check before heat treatment—air quenching—intermediate check—tempering—final check. Figure 46 shows the normalized curve.

The shell cast outline of the split-type ranging arm is 2948 mm × 1434 mm × 961 mm, the minimum wall thickness is 45 mm, the maximum wall thickness is 119 mm, and the cast weight is 6380 kg. The position of water passage in the cast part needs to be tested in a hydrostatic test under eight standard atmospheric pressures, and the inner cavity is water leakage-proof. To make the castings with excellent mechanical properties, the density of the castings needs to be high and reaches GB 7233—87 II level in ultrasonic flaw detection. The geometry, allowance of machining, and dimensional deviation of the castings are consistent with the specifications of cast steel for engineering machinery JB/T 5939—1991. Therefore, based on ranging arm shell cast material composition based casting characteristics and structure, size, and casting

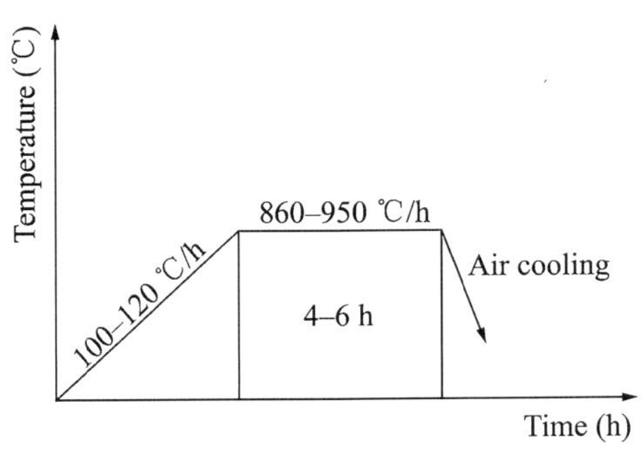

Fig. 46 Normalizing curve

weight level, the foundry program was determined, including the sand casting method and casting process (pouring position, parting surface, sand box structure and size). After the casting process was determined, the designs were conducted including the design of casting parameters, sand core design, gating system design, riser design, chill design, and vent design.

Ranging arm shell casting is a large heterotypic steel casting. A template sample manual molding was used. The molding sand is quartz sand with SiO_2 concentration greater than 96 %. The molding process used ester-hardened sodium silicate sand, core sand used 30–50 mesh of quartz sand, and binder used phenolic resin.

On the design of the outlet, sand core was used to make the exhaust channel, and the vent was reserved in sand mound. More vents were designed on the external mound upper box, in particular, on the sand mound upper box at the water jacket position of the ranging arm.

In terms of the design of feeding system, the gating and rising, and chill were appropriately laid out based on the modulus of each structural element by using the modulus method. The top insulation riser and side insulation riser were set up in the local area, and the external chill was placed. In addition, a riser subsidy was set up at necessary parts to feed castings.

With the computer numerical simulation of the process of shell mold filling and solidification, the casting process and parameters were optimized. Among them, the machining allowances were adjusted according to the JB/T5939—1991 the specifications of cast steel for construction machinery, and the shrinkage was set based on the inner cavity and external shape. In addition, using the lofting small casting test, the rationality and stability of the shell casting process was determined.

With the optimization of the casting process, the casting quality of the shell of the large power ranging arm was good in trial production. Through the process of heat treatment discussed above, the surface hardness test on the CrNiMo series casting steel of the ranging arm shell showed the hardness deviation in a reasonable arrange. The mechanical properties are as follows: Hardness HB200-240, tensile strength $\geq$700 MPa, yield strength $\geq$600 MPa, percentage elongation $\geq$18 %, area contraction $\geq$40 %, and impact toughness $\geq$70 J.

After using the technologies above, the performance of the shearer's ranging arm shells remained in good condition. The mechanical properties of the shell can meet the requirements of the high-power shearer.

In addition, the heat balance of the shearer ranging arm was investigated and a special forced lubrication and cooling device was developed. The heat of the shearer ranging arm comes mainly from the gears, bearings and stirring oil by gears. There are four main ways of heat

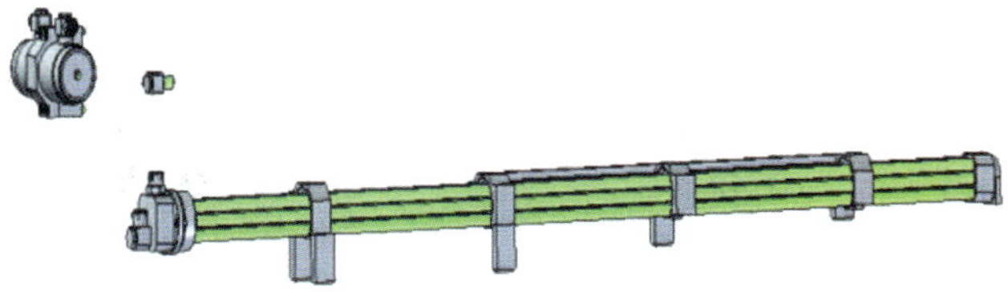

Fig. 47 Built-in cooler

dissipation for the shearer ranging arm: convective heat transfer between the ranging arm shell and air flow; heat removal by the cooling water of the ranging arms; heat absorption by the lube oil of the ranging arm; radial heat transfer between ranging arm and its connection devices.

With the significant increase of ranging arm power, and the working requirements of equipment continuous transshipment by high production and high efficiency working face, it is difficult to solve the problem of heat caused by the loss of transmission efficiency through traditional methods. Thus, a dedicated heat balance system was set for the ranging arm. The design of the water jacket for the ranging arm shell was optimized. A cooling pipe was installed in the gear chamber. The built-in cooler is shown in Fig. 47. A hydraulic gear pump system was installed for forced lubrication and cooling. In addition, the temperature sensor was installed, when temperature exceeded the limit, the alarm was triggered. The design forms a unique, advanced and reliable ranging arm heat balance system.

In order to increase the service life of gears and bearings inside the ranging arm, and increase the reliability of the shearer, the hydraulic system of the gear pump was installed for forced lubrication and cooling. The ranging arm gearbox was used as an oil tank for the automatic lubrication device. At the other side of shearer ranging arm I, screw pump 1 was mounted. In the gear oil tank, N320 middle load gear oil was injected. The gear on axial group I drives the gears on the gear pump, and the lubricating oil in the pump, under the action of centrifugal force, was pumped out through the oil tube to filter, which filters out impurities in the oil. After the cooling, the oil was transported by pipeline to the axial group III, axial group IV, axial group V, axial group VI and axial group VII. Each axial group and gear had special holes drilled, and the lubricating oil lubricated the mating surfaces through the hole as the gear turned.

In order to clearly describe the effect of heat radiation on the shearer ranging arm temperature field after optimizing the cooler, the heat radiation without the cooler was theoretically and numerically analyzed when the ranging arm angle is 0° and ambient temperature is 0°. The temperature rise curve of the ranging arm (Fig. 48) and the color cloud of stead temperature field (Fig. 49) were obtained.

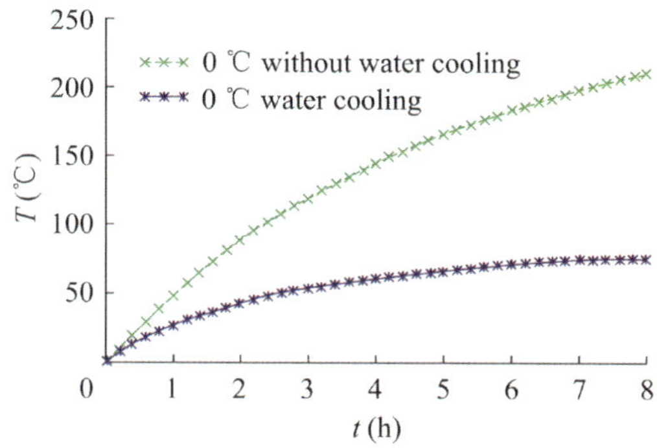

Fig. 48 Temperature rise curve with the effect of water-cooling

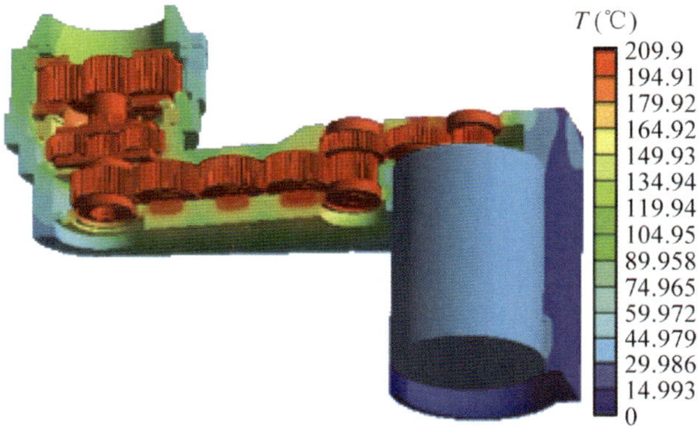

Fig. 49 Temperature field of ranging arm without water-cooling at ambient temperature 0°

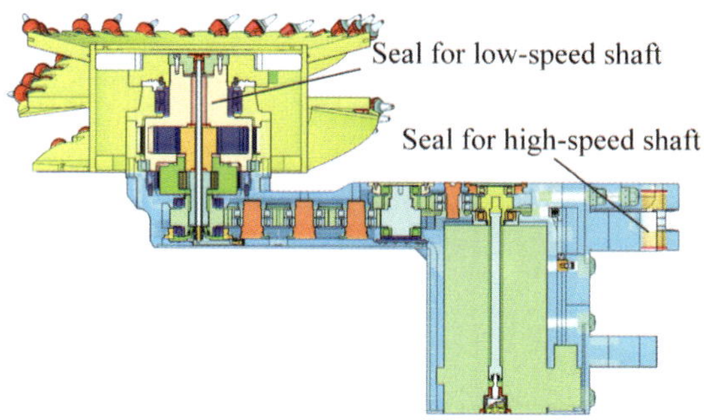

Fig. 50 Locations of high speed and low speed seals on shearer ranging arm

As shown in Fig. 48, the effect of the water cooler after optimization on the temperature rise of the ranging arm is significant. Without the water cooler, the acceleration speed of temperature increase is very large. Figure 49 shows the distribution of the ranging arm temperature field without the water cooler under working conditions. Without the water cooler, the temperature gradient generated between transmission shaft and box is very large. It indicates that the water cooler of ranging arm design has solved the overheating problem with the 750 kW ranging arm. In addition, it provides a reference for the study on the heat

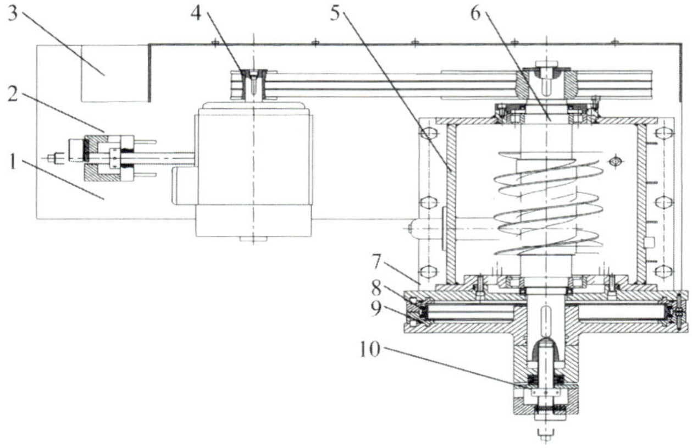

Fig. 51 Major components of floating oil-seal test device. *1* Support, *2* 2 V belt tension device, *3* electric control console, *4* moter, *5* oil tank, *6* shaft, *7* static ring fixer, *8* float oil seal, *9* moving ring fixer, *10* axial loading device

Fig. 52 Floating oil-seal test device. **a** Electric control console, **b** mechanical parts of test device

balance of large power ranging arms, and the optimization of cooler design.

In order to further improve the reliability of the shearer ranging arm, the seal performance of the shearer ranging arm was evaluated. In reality, the seal problems mainly appeared in moving seals. There are two moving seals in the ranging arm, one is the seal between drum connection sets and the gland at the output shaft of the ranging arm (low-speed shaft), the other is the seal between the ranging arm shaft I (high-speed shaft) and motor cavity body, as shown in Fig. 50.

Currently, a floating seal is used on the low-speed shaft seal of the ranging arm. The floating oil-seal is a sealing device combining dynamic and static sealing techniques. The floating oil-seal is composed of two identical metal rings (static floating seal ring and dynamic floating seal ring) and an "O" shape ring. In order to further understand the mechanism of the floating oil-sealing and optimize the assembly clearance, a test device of floating oil-seal was developed, as shown in Fig. 51. Figure 52 shows the photos of the test device.

The working principal of the floating oil-seal test device is as follows: a axial force is imposed on floating oil-seal through the axial loading device; the speed of floating oil-seal is adjusted through the electric control console, the oil temperature in floating oil-seal cavity and the test time of floating oil-seal are recorded; through the heater installed in the oil tank heat the oil in the tank, then the relationship between the axial force on floating oil-seal and assembly clearance can be measured; the floating oil-seal operation, and temperature increase and wear is monitored; and the oil leakage assembly clearance of floating oil-seal under static or dynamic status at a given temperature is detected. Main technical parameters of the test device are shown in Table 20.

Based on the main factors that affect the performance of the floating oil-seal, and the working principal of the test device, the routine test items and methods were determined as follows: hardness test on "O" type ring; test on the relationship between pressure and shift; static oil leakage test; test on the elastic resilience of the static "O" type ring; test on dynamic temperature rise; test on the operation of the floating oil-seal; test on dynamic oil leakage; test on the wearing condition of the floating oil-seal; test on service life; and the comparison of performance of floating oil-seals from different manufacturers.

The main problem for using the framework oil-seal of ranging arm's high-speed shaft is its premature failure, i.e., the failure before the service life of the framework oil-seal. The main causes of the premature failure for the framework oil-seal at the ranging arm shaft I are as follows: poor

Table 20 Main technical parameters of test device of floating oil-seal

Output (V)	Motor power (kW)	Variable frequency speed adjustment range (r/m)	Maximum output torque (N m)	Floating oil-seal size
380	3	0–200	265	ϕ293–ϕ560

Table 21 Chemical composition of cladding materials (wt %)

Sample	C	Si	Ti	V	Cr	Mn	Ni	Fe
A	1.12–2–1.25	–	0.2	–	3.6	0.7	–	Remains
B	1.0–1.2	0.15	–	0.33	5.8	0.6	–	Remains
C	1.45–1.7	0.35	0.4	–	12.3	0.88	0.21	Remains
D	1.12–2–1.25	–	0.24	0.27	3.7	1.03	1.16	Remains

quality framework oil-seal, e.g., poor quality port, and defects; poor quality of self-tightening spring; the magnitude interference between port inner diameter and shaft I outer diameter is too small, which results in low radial pressure; and the material of framework oil-seal is not good, which results in wearing out at port. Design and machining: there are some misalignments at some positions, such as ranging arm, shaft I and bearing cup; there is a low degree of finish at shaft I that amounted with framework oil-seal; and inappropriate heat treatment, etc. Assembly: damage on the port of framework oil-seal, inclination of bearing cup, loose spring, different type of grease or different quantify of grease on oil-seal port and shaft surface. To overcome the problems associated with the framework oil-seal of ranging arm shaft I, and to increase the sealing reliability of high-speed shaft of ranging arm, the following measures were taken: select high quality framework oil-seals; maintain the machining precision of key components; checking and maintenance on oil leakage of the framework oil-seal.

4.2.4 High speed ultra-powerful pin and rail haulage system

Based on the previous system, a high speed ultra-powerful chainless haulage system was developed. It has a rated traction of 2 × 570 kN and a traction speed of up to 20 m/min.

The clearance matching of chainless haulage system with a pitch of 147 mm was studied. First of all, the variable range of meshing central distance of the shearer walk wheel and scraper conveyor pin rail was determined. Its determination principles are: under the correct meshing situation, the minimum meshing central distance can warrant the no interference and no gear jam between walk wheel and pin rail (two section pin row connection), and it is usually approximately 10 mm less than the pitch radius of the walk wheel; the maximum meshing central distance must satisfy the rational meshing, the simulation shows that the increase of the meshing central distance has little effect, based on analysis and experience, the value is taken as 15 mm larger than the corresponding pitch radius of the walk wheel; a further increase of meshing central distance would cause: imprecise rolling, impact on gears from every contact and alteration, the wear of wheel gear and rack gear, etc.

The wear resistance of components in the chainless haulage system of maximum traction force 2 × 570 kN was investigated. In the study, four types of cladding materials A, B, C and D were studied; the cladding of substrate steel can dramatically improve the surface hardness and abrasion resistance of the substrate. Based on the analysis, the C type is the hardest cladding layer among the four types of cladding material, up to HRC66.9, the A cladding layer is a less hard layer, the B cladding layer is the least hard layer, with its hardness reaching HRC51. The friction and wear test showed that the four types of cladding layers have the same wearing manner, e.g., abrasive and adhesive wears. Among the four types of cladding materials, the wear resistance of A is the best, followed by B, C and D that have the same wear resistance. Cladding material is Fe-based alloy powder; its chemical composition is shown in Table 21.

The modified substrate steel after the cladding of Fe–Cr–Mn–Ti Fe-based alloy powder has an A cladding layer and a high hardness, excellent anti-wear properties, and a substrate with very good toughness, and is superior to any other alloy steel wear-resistant material. This modified substrate steel can be used as the shearer's guide sliding shoe material to meet the performance requirements of the shearer guide sliding shoe under complicated mining conditions.

4.2.5 Development of large-diameter, ultra-powerful and wear-resistance spiral drums with pick-shaped cutter and its components

The shearer's pick holder and welding process were investigated. In earlier years, the pick holder of a small power shearer was cast with medium carbon steel. The common damage of pick holders were: weld cracking; overall cracking along pin hole due to high substrate hardness, poor plastic toughness; surface wear due mainly to poor surface organization, low hardness, poor wear resistance; inner hole wear due to poor surface organization, low hardness, and poor wear resistance. Comparison of different materials for the pick holders is shown in Table 22.

Table 22 Comparison of material properties

Property	ZG40Cr	35CrMnSi	$15CrNi_3MoA$
Material property	Not compact, many defects, low strength, low plasticity	High strength, good plasticity	High strength, excellent plasticity
Processing property	Poor hardnessability, high carbon-equivalent, poor welding property	Good hardnessability, high carbon-equivalent, poor welding property	Good hardnessability, low carbon-equivalent, good wielding property
Application property	Same surface hardness. In terms of the core, hardness is too high, poor plasticity, poor impact resistance; in terms of surface, hardness is too high, poor wear resistance	Same surface hardness. In terms of the core, hardness is too high, poor plasticity, poor impact resistance; in terms of surface, hardness is too high, poor wear resistance	After carburizing and quenching, the hardness at core is appropriate, good plasticity, high surface hardness, good wear resistance

In order to better meet the working conditions of shearer's pick holders, a low carbon alloy steel 15CrNi3-MoA was used to make the pick holder which was subsequently treated by carburizing and quenching. Because the carbon content of the material is low, its carbon equivalent is also low and its welding property is good. The total alloy element content is low, therefore it is low cost, but its hardenability is high. After carburizing and quenching, the carburized layer is martensite plus carbide, its hardness exceeds 58 HRC, so the surface wear resistance is good, and substrate is a thin lath martensite. A material with a hardness of approximately 300 HB, a strong toughness, and strong impact resistance, would be a more appropriate for making pick holders.

The wear process of the shearer drum helical vane was studied. In poor production conditions, the tail end of the shearer drum helical vane was often worn out. After a comparative study of several methods, the chrome SA1750CR wear-resistant composite plate was selected. The plate is made as follows: using the heap welding method, a layer of high chrome and carbide material are surfaced on a low alloy steel plate, and form a heap welding layer with high wear property, while the back of the plate has high toughness property; the large difference in the properties of front and back side of the composite plate results in the large improvement in the properties of bending, welding and mechanical processing of composite plate and original low carbon alloy steel plate.

4.2.6 Development of the prototype shearer and performance test

Based on earlier studies in June 2009, the Shanghai Branch of Tiandi Science & Technology Co., Ltd commenced the manufacture of the prototype shearer and completed in February 2010. The key process of its main components is described as follows:

As the MG750/1915-GWD shearer developed is a heavy-duty large power shearer, large casting shells, such as ranging arms and traction gearbox, are heavy in weight, and the cavity structure is complex and the thickness is uneven. Through software simulation, the parameters of the riser, such as number, size and position, were rationally designed to effectively avoid the casting defects, e.g. cracks and loose. To overcome the problem of blowholes in casting, in terms of process, a vent was made on the core and on the sand box. In addition, a well-shaped exhaust duct was made on the sand box. Large particles of synthetic quartz sand were used as the back sand. In these ways, the casting blowhole was effectively eliminated. In addition, the measures of a one-box multi-core combined tool and a core with multiple positioning were applied to ensure dimensional accuracy. In the machining process, the benchmark superposition method was used to achieve location accuracy. Before semi-finishing machining, a baseline was mapped properly, and the base surface was machined properly. Based on the correct baseline, the surface of each hole was machined in semi-finishing or finishing. To meet the requirements of geometric tolerances of some holes, a one clamping multichannel processing method was used to ensure the requirements of finishing size. A "casting-air quenching and high temperature tempering-rough-finishing-stress removal tempering-finishing" processing route was used to ensure the strength and dimensional stability of the castings. With the development of materials and the processing method, on the basis of solderability, the hardness of ranging arm shell reaches 200–240 HB, and the tensile strength reaches σ_b - $\geq$ 700 MPa, which greatly enhances the mechanical properties.

Before forging the gear, the oxidized layer and surface defects were removed using a hand sand wheel to effectively avoid the cracking phenomenon of forging residue. In the forging process, the procedures must be strictly followed; the forging temperature must be strictly controlled with the temperature difference not exceeding 300 °C. After forging, the forging part must be normalized under more than 950 °C and tempered less than 600 °C to eliminate forging stress and improve cutting performance. After a rough-machining of the gear blank, each part underwent an ultrasonic defect check. A grinding undercut large radius hob was used for gear grinding to increase gear root fillet, and improve the impact resistance and bearing capacity of gears. In the carburizing and quenching heat treatment process, the carbon potential concentrations were strictly controlled at 0.8 %–1.0 %. A "quenching + deep freezing + low temperature tempering" process was used to ensure that the gear surface hardness reached HRC58-62, surface carbide reached Grade 1–3, the depth of effective hardened layer reached 1.5–1.7 mm range (including grinding), core hardness reached HRC38 approximately, and retained austenite content was less than 9 %. All gears were ground in high precision grinding machine NILES, and some of the key components, such as planetary gear and sun gear, were under tip relief operation. At the same time, the accuracy of gear machining was one level higher than the specified level on the drawings to ensure a good meshing of the gears. The ground gears were treated by strengthened shot blast to eliminate the burns on the tooth surface and increase residual compressive stress, thereby enhancing the gear's contact fatigue strength and bending fatigue strength.

The working conditions of the shearer walk gear are harsh, thus the shearer walk gear is one of the weakest parts in the shearer. In order to improve its work life, the electro-slag remelted steel material 18Cr2Ni4WE was used to make the shearer walk gear. The electro-slag remelting greatly reduces the harmful content of non-metallic materials (these ingredients are the causes of quenching crack and fatigue crack), increases the purity of materials, and maximizes material performance. As the gear modulus was large and there was no standard hob, the gear tooth was machined twice, the first machining cut using NC wire cutting, leaving 2–3 mm allowance. The second machining was precisely machined to its place using a processing center. In particular, in the last processes, the amount of feed and cutting speed should be strictly controlled to avoid obvious cutting marks (machining marks are the stress concentration source of gear's tooth; currently fracturing of the shearer walk gear is often caused by these marks). To reduce the gear tooth's early failure due to stress concentration, it should be treated by shot-blast after heat treatment.

High quality medium carbon alloy steel was used to cast the shearer's guiding sliding shoe. At rough machining, the shoe was hardened and tempered to improve its substrate mechanical properties. The high quality wear-resistance material was selected to heap weld the guiding surface, then, the surface was machined using a special machine to reduce the surface roughness and friction force.

The trial production of other components also strictly followed the product standard and were checked to avoid

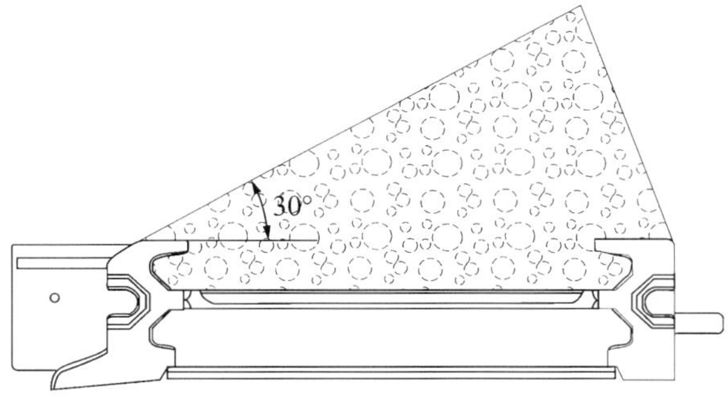

Fig. 53 Cross-section of coal transportation at rear scraper conveyor

installing non-confirming products in the prototype shearer, and can warrant the reliability of the shearer manufactured.

4.3 Rear scraper conveyor for LLTCC coal face

In order to meet the requirements of coal production of 1000 Mtpa at the No. 8105 face in Tashan coal mine, with the consideration of the effects of other uncertain factors, the calculation shows that the total transportation capacity of the front and rear scraper conveyors is 3500 t/h. Considering a cutting and caving ratio of 1:3, the individual transportation capacity of the front and rear scraper conveyors should be greater than 2625 t/h. Taking the uncertainty of caving quantity and some allowances in transportation capacity into consideration, the capacity of the rear scraper conveyor was determined as 3000 t/h.

Because of the specific environment in the underground mine and the matching characteristics of the scraper conveyor, shearer and hydraulic support, there are many constraints on many structural parameters of the scraper conveyor including the width of the pan and other important structural parameters. Balancing between economic and technical contexts, the pan width of the rear scraper conveyor was determined as 1200 mm. Its coal transport section is shown in Fig. 53.

Calculation of chain speed of scraper V:

$$V = n/2iZp/60\lambda \tag{17}$$

where n is the motor speed, 149 min^{-1}; i is the reduce deceleration ratio, 28.43; Z is the tooth number of driving sprocket, 7; P is the chain pitch of 48 × 154, 0.152 m; λ is the slip ratio of coupler, 0.95. Substitute the parameters into Eq. (17), get:

$$V = 1491/28.43 \times 2 \times 7 \times 0.152/60 \times 0.95 \approx 1.77\ \text{m/s}$$

Calculate the hourly capacity of coal transportation Q:

$$Q = 3600\,VA\rho \tag{18}$$

where V is 1.77 m/s; A is the cross-section of loads as to rear conveyor, because the top of the cross-section is open, based on the coals rest angle of 30°, the calculated coal transportation area $A = 0.63\ m^2$; ρ is the loose coal density, 0.85–1.0 t/m^3, select 0.9 t/m^3. Substitute the data into Eq. (18), get: $Q = 3600 \times 1.77 \times 0.63 \times 0.9 \approx 3613$ t/h > 3000 t/h,meet the requirements.

Based on a conveying capacity of 3000 t/h and a chain speed of 1.77 m/s, and the installed power of 2000 kW, the theoretical power consumption utilization was analyzed under the conditions of face lengths of 300, 250 and 207 m respectively (the actual length at the No. 8105 face in Tashan coal mine). The calculation results showed that: when the face length is 300 m, and face dip is 0°, the power utilization reaches up to 94.3 %, power margin is small, which is only used in a flat or downhill face. When the face length is 250 m (or the actual face length at No. 8105 face is 207 m), and the face dip is 0°, the power utilization is 78.6 % and 65 %, respectively, showing a large margin. Therefore, a total power of 2 × 1000 kW can meet the requirement of transport capacity of 3000 t/h.

In order to warrant the non-fracture of the scraper chain, the safety coefficient of the chain was calculated. When the total power is 2000 kW, scraper chain speed is 1.77 m/s and the broken load of scraper chain is 3290 kN, the safety coefficient of scraper chain should reach 1.72 to satisfy the requirements of safe coal transportation.

Based on the analysis discussed, SGZ1200/2 × 1000 large load rear scraper conveyor was selected for the LLTCC face to extract the ultra-thick coal seam at Tashan coal mine. The installed power is 2 × 1000 kW, and transportation capacity exceeds 3000 t/h. The transportation distance meets the length requirement at the No. 8105 face.

In a LLTCC mining operation, the rear scraper conveyor should have reasonable technical parameters and structure; also, the key components should have high reliability. These key elements include a heavy-duty high-strength chain drive system, speed-adjustable soft start device, integrated monitoring transmission system of transmission device, heavy-loaded 1000 kW planetary retarder, high reliability long life high-strength coal transportation trough, high strength compact end discharge frame and automatic retractable tail with electro-hydraulic control, etc.

With the development of 48 × 152 compact high-strength round chain and link chain for the mine, as well as the high-strength forged scraper, the optimization of the meshing characteristics of the chain transmission system and the adaptability and chain tension, along with the development of along life 48 × 152 drive sprocket and component, a heavy-duty high-strength chain transmission system was successfully developed.

By studying the matching ability of controlled adjustable speed soft start device-valve-controlled hydraulic coupler and motor load characteristics, a rational coupler cavity has been designed and the fluid dynamics was optimized. In addition, an electro-hydraulic control technology for the coupler has been developed. A large power valve-controlled hydraulic coupler has been developed with intellectual property rights.

A comprehensive monitoring integrated transmission system has been developed. It consists of central monitoring equipment (upper equipment) and site monitoring equipment (lower equipment), which achieves a state monitoring and control on the transmission device of the rear scraper conveyor, and enhances the operation monitoring capacity of the large power rear scraper conveyor.

Through a combination method of theoretical analysis, three dimensional modelling, mechanical analysis and simulation, virtual prototyping, dynamic testing, bench testing, and industrial trial, a 1000 kW retarder for the mine has been developed.

Through the study on the bending angle design of the middle pan, and the connection method of U-shaped port of middle pan, in addition to the large amount of research and testing on the choice of materials, welding processes and equipment, and machining technology, a high-reliability, long-life coal pan has been developed.

Through the strength design on the end discharge head frame, the connection design of the head frame and joint parts, the connection design of the head frame and transition pan was improved. A high strength, compact end discharge head frame has been developed.

With the analysis on the operating conditions of the scraper conveyor chain, and the calculation on the adjustment required by conveyor tail and the calculation of traction force of the extension mechanism, an automatic retractable conveyor tail electro-hydraulic control system and retractable extension tail frame has been developed. Furthermore, an electro-hydraulic automatic control retractable conveyor tail has been developed.

Through the research and development of the key technologies and components above, the overall reliability of the rear scraper conveyor has been improved to warrant safe, efficient and stable operation.

4.4 Large-capacity and long distance gateroad belt conveyor

In order to achieve the transportation of both the coals extracted by the shearer and the coals extracted by top coal caving from the LLTCC operations, the technologies on a large-capacity long-distance gateroad belt conveyor has been investigated.

A CST soft drive system was used, in light loads or heave loads condition, the dynamic process of the loaded belt conveyor's soft start/soft stop can effectively be controlled to achieve a balance of power among various drives.

Based on the variation of tensile force during the start, stop and operation of belt conveyor, a tension technology and equipment of high accuracy and large tensioning stroke has been developed. It can quickly respond and automatically adjust tension and tensioning stroke with the tension variation of the belt conveyor. During normal operations, the tensile force is 0.9–1.1 times of the rated tensile force, during start and stop, the tensile force is 1.3–1.4 times of the rated tensile force, and during the shift of conveyor tail, the equipment can quickly respond and guarantee non-skid. The working principle is: based on the conveyor's start-up and loading, with the feedback of tension sensor signal, the PLC controller and fuel tank can automatically adjust the output of tensile force to avoid slipping between belts and drive pulley, lower the safety factor selected for the belt, and ensure the normal operation of the belt conveyor. Installed with an automatic moved conveyor tail, the automatic tensioning can be achieved when the conveyor tail is automatically moved, thus to a certain extent, the downtime rate of the belt conveyor and the downtime rate of the three face machine can be reduced.

A new high-efficiency belt storage device was developed; the belt can be folded into six layers and stored with a length of 200 m. A belt storage system of multi-layer winding has been developed, which enables an automatic belt central alignment to overcome the deviation of conveyor belts in storage and the tensioning truck's off track.

A low speed large torque hydraulic motor and electro-hydraulic proportional control valves were used to develop a large-torque belt-speed controllable high efficiency hydraulic belt winding device, which significantly reduces the time on belt winding, is easy in operation, and reliable in working.

A high-speed low-friction larger-diameter roller was developed. Using labyrinth seals with special structure and oxidation resistance, water-resistance grease sealing method, the service life of the high-speed large-diameter roller is significantly increased.

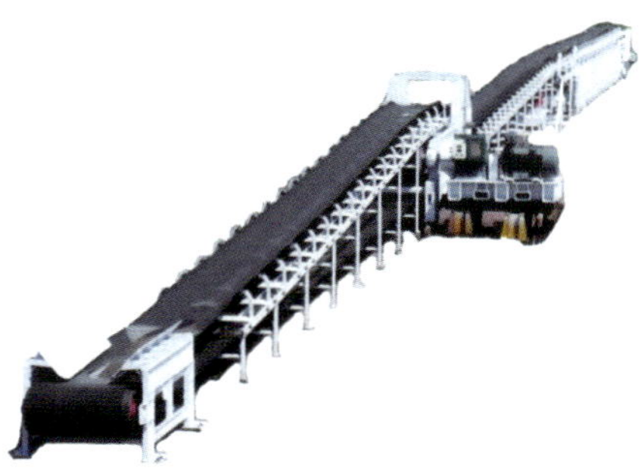

Fig. 54 The large-capacity, long-distance belt conveyor

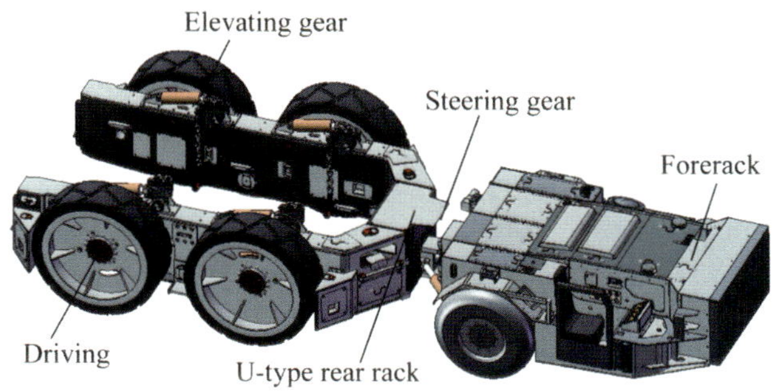

Fig. 55 WC55Y shield hauler

Using dynamic analysis software, the system was analyzed to determine a reasonable starting and braking curve. An advanced PLC programmable controller was developed. In addition, an advanced program-controlled automatic monitoring system was developed, which can conduct the operations of data collection, processing, storage, transmission, fault detection, reach and feed, etc. It can meet the requirements of the conveyor's controllable soft start and braking, power balance, automatic tensioning and self-movement of conveyor tail, as well as various functions, such as protection, communication and signal contact. In addition, it can monitor the various protection and safety devices, conveyor belt tension, the temperature of the driving drum and roller bearing, the measurement of throughput, etc.

Through the studies presented, a large-capacity, long-distance belt conveyor has been developed, as shown in Fig. 54. Its main technical parameters are as follows: transportation distance: 2800 m; transportation capacity: 3500 t/h; lifting height: 35 m; belt width: 1400 mm; installed power: 1680 (3 × 500) kW; belt speed: ≥4.5 m/s; stored belt length: ≥120 m.

4.5 Large tonnage shield hauler

The tonnage of hydraulic support for the working face in thick and ultra-thick coal seam is very large. In order to improve the efficiency of coal mining, it requires a fast relocation of coal face equipment. Therefore, a heavy-duty frame shield hauler—WC55Y shield hauler was developed, as shown in Fig. 55. The vehicle is powered by diesel, the carrying capacity is 55 t, and is an efficient equipment for a quick relocation of coalface. If equipped with a hopper, it also can be used as a multifunctional underground transport vehicle.

A V-type six-cylinder diesel engine explosion-proof equipment was used in the WC55Y shield hauler. Based on the requirements of explosion-proofing, the explosion-proof device must be set up in engine intake and exhaust systems, which will increase the resistances of engine's intake and exhaust, reducing engine efficiency, worsen the harmful emissions, and increase the pollution in underground. The difficulty in this project is how to optimize the inlet and outlet structures to minimize its effect on engine performance and reduce the emissions of harmful gases. In terms of the engine's explosion-proofing, the following four key technologies have been developed:

(1) In the explosion-proof diesel engine, exhaust turbocharged water–air cooling technology was used. Forced water circulation technology was used to solve the high temperature problem with the supercharger surface. Compared to natural air intake, the air intake increases 30 %, and lays a foundation for increasing efficiency and reducing emission.
(2) Using theoretical analysis and test methods, the fuel injection control was optimized. The oil injection method of low-speed in the early stages, high-speed in the middle stages and reduced speed in later stages was used to effectively control the emission of NO*x*, PT, etc.
(3) Intake resistance is an important factor affecting engine power. The intake spark arrestor is a key component of the air intake system in explosion-proofing the diesel engine, which directly affects engine performance. Through the experiment, after the installation of the air inlet device on the diesel engine, the losses of power and torque is up to 14 %. To improve the flow characteristics of the diesel engine intake, the original structure had been optimized. With a comparative analysis of eddy current resistance loss at inlet/outlet section in different intake spark arrestors, an inclined windward surface end grid plate was designed to substantially reduce inlet resistance.
(4) A constant pressure low resistance exhaust gas treatment device was designed. Using a water bath type cylindrical structure, the exhaust resistance was reduced by 5 %.

For the WC55Y shield hauler, a moving speed brake was developed, which integrates a planetary speed retarder and a large torque wetted multi-disc safe brake. Traditional brake concentrates on the high-speed end of wheel-side retarder, and is small in dimension. However, under a larger braking torque and repeated braking, the system generates large amount of heat and consistently fails. The working condition of the shield hauler is harsh. In the aspect of driveless axle hydraulic-driven design, there have been no successful braking studies anywhere in the world. Fully enclosed, high torque and large brake power, small in size, and high efficiency heat dissipation are the difficulties in the study. Meanwhile, the road underground is potholed, thus during the movement of heavy-loaded vehicles, the wheel-side retarder will withstand a greater shock load, and therefore must have enough strength. Otherwise, it will cause failures in the retarder's bearing, such as broken parts. In response to these technical bottlenecks, a "modular integrated technology" was used to design a new planetary gear of wetted multiple-disc brake. Also, an integrated motor-cartridge wheel drive and brake retarder was invented. It is a wheel-side planetary retarder, and integrates all brake functions. The brake was designed as a fully closed wetted-type multi-disc brake with spring braking and hydraulic release. Through the spline, friction tablets and steel tablets are coupled with fixed parts and turned parts respectively. Through the motor input, retarder part input torque, delivered via a two level planetary deceleration device, the torque will be transmitted to the wheels. The brake cavity and retarder cavity are completely insulated by a pair of high reliable floating seals to avoid the interference of two different functions. The braking torque of the integrated brake-retarder is greater than 60,000 N m, thermal equilibrium temperature is less than 90 C° under long term braking, and the average life of the friction disc is greater than 5000 h. Furthermore, it is safe, reliable, has good pollution resistance, and increases operational safety.

The main performance indicators of the WC55Y shield hauler developed are: rated load 55 t, minimum ground clearance 300 mm, full load speed 0–12 km/h and climbing ability 12°, engine rated power after explosion-proof 220 kW, turning radius 2800 mm (inside) and 7000 mm (outside). The WC55Y shield hauler was used in the relocation of the LLTCC coalface at 14–20 m thick ultra-thick coal seam in the Tashan coal mine to successfully solve the relocation problem of large-tonnage hydraulic support at a LLTCC coalface.

5 Bolt support technology of large cross-section roadway in ultra-thick coal seam

5.1 Deformation and failure characteristics of the surrounding rocks of coal roadway with a large cross-section in ultra-thick coal seam

The roadways for a LLTCC coalface include headgate, tailgate, top return airway and open-off cut. These roadways are responsible for equipment installation, transportation, ventilation, gas exhaust, personnel passage, etc. A safe, efficient coalface depends on roadway stability and smooth flow. Roadway support is a key to the LLTCC mining operation.

The roadway of a LLTCC coal face has following characteristics:

(1) The headgate, tailgate and open-off cut are developed along the coal seam. The roof is top coals with an average thickness of 12.9 m. Compared to rock, the coal seam is relatively soft, fractured, and thick. The top coal is easily subsided and even caved, which significantly increases the difficulty of roadway support.
(2) Because the power of mining equipment is large, and face production is high, a very large cross-section of the roadway is required. The tailgate and the headgate have a same cross-section with a width of 5.5 m wide, a height of 3.9 m and a size of 21.45 m^2. The open-off cut span is up to 10 m. The increase of roadway width can result in significant bending subsidence in the roof, and large height roadway can cause coal wall buckling and even spalling, which significantly affects the stability of the surrounding rock of the roadway, and brings great difficulties to roadway support.
(3) The tailgate suffers the side-abutment pressure caused by the previous longwal gob and the and the front abutment pressure caused by the current longwall panel. In addition, due to the distinct characteristics of ground pressure at a LLTCC coalface, the mining effect on the tailgate is very strong, resulting in serious damage of the surrounding rocks.

5.2 Bolt support mechanism of large cross-section coal roadway

Roadway support with bolt and cable has become one of the main methods at coal mines in China. In this project, bolts and cables were used to support the large cross-section coal roadway. However, the surrounding rock of the roadway in the coal seam is soft and fractured. Traditional bolt supporting failed to suppress the large deformation of the roadway due to the high mining-induced stresses. Bolt supporting mechanism for coal roadways should be investigated to provide a theoretical basis for support design.

Many studies have been conducted to investigate the support mechanism of bolt and cable support on coal roadways (Jia and Huang 2005; Zhang et al. 2006; Ma and Zhang 2008; Zhang and Yang 2010) in China. In recent years, it has been recognized that the prestress of rock bolts plays an important role in bolt support (Kang et al. 2007a, b, 2008; Wang et al. 2008; Zhang et al. 2010), such as the binding effect it plays on the layer separation, slippage, and join crack opening of the surrounding rock surface layer, and maintaining the integrity of the surrounding rock. The studies improve the bolting effect, in particular, providing an effective theoretical guidance for complicated roadway support.

Based on the deformation characteristics of large cross-section coal roadway in ultra-thick coal seam, a full coal roadway supporting mechanism is proposed.

(1) The deformation of surrounding rock in a full coal roadway consists of two parts: one is the dilatancy deformation including the separation of coal and rock, sliding, fissures opening, and new cracks at their structural plane; the second is the elastic deformation of surrounding rock and the plastic deformation before its peak strength, and a full deformation of anchorage zone. A reasonable full coal seam roadway support should greatly improve the stiffness and strength of supporting system at initial support, effectively controlling the dilatancy deformation of the surrounding rock, and maintaining the integrity of top coals. At the same time, the supporting system should have sufficient elongation and allow a second deformation to release high stress.
(2) The prestresses of bolts and anchors play a decisive role in roadway support. It is crucial for support design to determine reasonable prestresses according to coal roadway conditions, and enabling prestresses to effectively spread.
(3) The anchor cable in a full coal seam roadway plays its role in two main aspects: one is to connect the bearing structure formed by bolting in top coal with deep surrounding rock in order to improve the stability of load bearing structures; the other is that the large prestresses of anchor cables and the compressive zone of bolts form a skeleton mesh structure to actively support surrounding rock.
(4) In a full coal seam roadway, high prestress powerful bolt and anchor cables should be used to become the single necessary support during the service life of the roadway, avoiding second roadway support and roadway maintenance.

5.3 Material of high strength bolt and cable

The yield strength of a traditional thread steel bolt is 335–500 MPa, and the tensile strength is 470–650 MPa, which is applicable to the roadway with normal conditions. For a roadway under complex geological and geotechnical conditions, a higher strength and sufficient elongation steel should be used for bolts and anchor cables in order to achieve high prestress, high strength and high elongation.

In this project, with the cooperation of other steel mills, strong steel for bolts and cables has been developed. The key technology and difficulties in developing this type of

Table 23 High strength bolt steel

Strength level (MPa)	Yield (MPa)	Tensile strength (MPa)	Shock absorb energy (J)	Elongation (%)
500	570	745	89.0	26.0
600	650	805	42.6	22.0

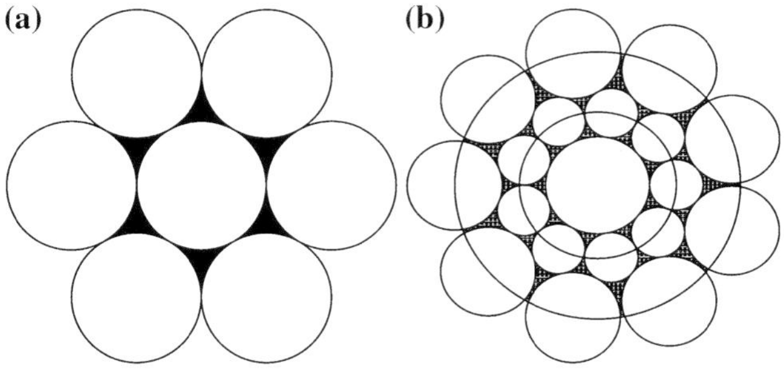

Fig. 56 Anchor cable structure. **a** 1 × 7 structure, **b** 1 × 19 structure

steel is to greatly improve the bolt strength, and simultaneously, maintain sufficient elongation and impact toughness of the bolt to prevent bolt fracture and breaking under impact loads, bending and shear loads. Starting from steelmaking, a special bolt steel formula was formulated, e.g., adding elements of V, Nb and Ti in the steel to alter the mechanical properties of steel. In addition, an appropriate rolling process was used to manufacture the bolt in accordance with its technical specifications. Table 23 lists two kinds of anchor bolt steel developed with high strength, high elongation, and high impact toughness. The yield strength of the two types of boltreaches 570 and 650 MPa respectively, and the tensile strength reaches 745 MPa and 805 MPa respectively. The elongation is 26 % and 22 % respectively, and their shock absorbing energy is more than 40 J.

5.4 Cable material with large tonnage and high elongation

The anchor cable steel strand widely used in Chinese coal mines is composed of seven wires (Fig. 56a), with a diameter of 15.2 or 17.8 mm. Its application in underground coal mines showed the following disadvantages: the cable diameter is small and doesn't match borehole diameter, thus, significantly affects anchorage force, and is prone to the anchor end sliding. Due to its small breaking load, cable fractures often occur. Because of its low elongation, it cannot adapt to the large deformation of the surrounding rock. Due to its low prestress, it is poor in controlling the separation of rock layers.

To solve the problems discussed above, a large diameter, high tonnage strong anchor cable series was developed with a diameter of 18–22 mm, and a capacity of 400–550 kN. The technological approaches used are as follows: (1) changing cable body structure. Using new type of 19 wires (Fig. 56b) to replace the original seven wires, the cable body structure is more reasonable, and the anchor cable elongation has been significantly improved; (2) increasing the diameter of anchor cable body. The cable diameter was increased from 15.2 to 18, 20 and 22 mm to improve the cable body's broken force, and match the cable diameter and drilling diameter. The anchor cable with a diameter of 22 mm has a breaking load of greater than 600 kN, and an elongation of 7 %. The comparison of different mechanical properties of anchor cable steel strand wires are shown in Table 24.

5.5 Equipment for applying high prestress on bolt

As mentioned previously, the prestress of rock bolts is one of the key parameters for bolting, and plays a decisive role on the effect of anchor support. The previous studies show that the anchor prestress should reach a level of 30 %–50 % of anchor yield strength. According to the conversion relationship, the bolts' tightening torque should be 300–500 N m. To reach such a high prestress, the existing single bolting machines in China cannot meet such a requirement. It is necessary to develop a suitable bolt

Table 24 Mechanical properties of different anchor cable strand wires

Structure of steel strand wires	Diameter (mm)	Cross-section (mm^2)	Tensile strength (MPa)	Breaking load (kN)	Elongation (%)
1 × 7	15.2	140	1860	260	3.5
	17.8	191	1860	355	4.0
	19.0	218	1860	406	4.5
1 × 19	18	215.5	1897	409	6.8
	20	262.4	1862	489	–
	22	321.6	1887	607	7.0

Table 25 Physical and mechanical parameters of coal and rock seams in a numerical model

Lithology	Density (kg/m^3)	Volume modulus (GPa)	Shear modulus (Pa)	Cohesion (MPa)	Friction angle (°)	Tensile strength (MPa)
Sandstone	2650	25.3	18.2	22	44	1.6
Coal	1500	6.5	3.9	7.9	36	0.6
Sandy mudstone	2600	8.85	5.8	9.2	41	0.8

tightening tool to significantly increase the anchor cable's prestress.

In terms of the equipment for bolting, there are two main methods for increasing bolt's prestress (Kang and Wang 2007): one is to increase the bolt-nut tightening torque; the second is to use a bolt tensioning device, which is similar to imposing large prestresses on anchor bolts. The methods for improving bolt-nut tightening torque include: (1) high torque jumbolter. With a high torque jumbolter, it can not only increase the bolt tightening torque, but also achieve an integration of bolt installation to improve installation speed. For instance, the torque of the jumbolter used with continuous miner can reach more than 300 N m. The rated torque of a single jumbolter in China is not greater than 150 N m, and unable to meet the requirement of high tightening torque of the bolt. (2) Torque multiplier. In order to improve the installation torque of the jumbolter, a torque multiplier can be used on the jumbolter to multiply the nut tightening torque when installing bolt. (3) Torque wrench. Using a large torque preload wrench to tighten the bolts and nuts is an effective way to improve the prestress of anchor bolts.

Based on the jumbolter used in large cross-section coal roadway, and the operation conditions, torque multiplier was selected to improve the prestress of anchor bolts. The bolt torque multiplier developed has an input torque of 100–150 N m and an output torque of 500–600 N m. With a single jumbolter match, the maximum torque magnification is approximately six times, which meets the high prestressing requirement of the bolts.

5.6 Stress characteristics of the surrounding rock in a full seam roadway and numerical simulation of the combined support with bolt and cable

In order to better understand the distribution characteristics of variation of surrounding rock in a large cross-section full coal roadway, as well as the support mechanism and main effect factors of anchor bolts and cables, a numerical analysis using $FLAC^{3D}$ was performed based on the data from the No. 5105 return airway of the No. 8105 coalface at Tashan coal mine The stress characteristics of the surrounding rock in a large cross-section coal roadway with different top coal thicknesses, the influence of different roadway layout locations, different roadway width-height ratios, different ground stresses, and different pre-tight forces of anchor bolts and cables were evaluated.

5.6.1 Simulation approach

The No. 5105 return air roadway was developed along the seam floor, its cross-section is in a rectangular shape with a width of 5.5 m, a height of 3.9 m. Ground stress data measured are shown in Table 2, vertical stress is 11.44 MPa, the maximum horizontal principal stress is 12.90 MPa, and the minimum horizontal principal stress is 7.24 MPa.

In the numerical simulation, the $FLAC^{3D}$ built-in cable structure element was used to simulate the rock bolts and cables. The bolts have anelastic modulus of 200 GPa, a yield strength of 600 MPa, a tensile strength of 800 MPa. The rock bolts are 2.4 m long and 22 mm in diameter. The cables have an elastic modulus of 195 GPa a length of 8.3 m, a diameter of mm, and a capacity of 600 kN. The Mohr–Coulomb constitutive model was used to describe the coal-rock body. Based on the results from the laboratory tests, the physical and mechanical parameters of coal and rock seams in the calculation are shown in Table 25.

Combining the geological conditions at the No. 5105 roadway in Tashan coal mine, based on the combined support of ancho bolts and anchor cables in coal roadway, multiple plans were compared in the simulations in this study:

(1) When the horizontal principal stress is 12.90 MPa, the vertical stress is 11.44 MPa, and the roadway was developed along the seam floor, the effect of different top-coal thicknesses on the stress distribution of the surrounding rock. Top-coal thicknesses were 0, 5, 10, 15, 20, and 25 m respectively.
(2) When the horizontal principal stress is 12.90 MPa, the vertical stress is 11.44 MPa, and the coal thickness is 15.72 m, the effect of different positions of the roadway layout on the stress distribution of

Table 26 Different ground stress combinations in the numerical simulation

Different ground stress combinations	Horizontal stress (MPa)	Vertical stress (MPa)	Lateral pressure coefficient
Combination 1	10	10	1
Combination 2	20	10	2
Combination 3	10	20	0.5

surrounding rock. In a vertical direction, the roadway was laid out along the roof and along the floor. When the coal-pillar widths in the horizontal direction were 10, 20, 30, and 40 m, the effect after the extraction of two adjacent coalfaces on the stress distribution of surrounding rock of roadway.

(3) When the horizontal principal stress is 12.90 MPa, and the vertical stress is 11.44 MPa, the effect of different width-height ratios on the stress distribution of the surrounding rock of the roadway. The height-width ratios in a full coal seam gateway are 1, 0.75 and 0.5.

(4) When the coal thickness is 16.8 m, the roadway was developed along the seam floor, the effect of different ground stresses on the stress distribution of the surrounding rock of the roadway. Different combinations of stress are shown in Table 26.

(5) The effect of different preloads of bolts and cables on the spread of stress in coal (regardless of in situ rock stress). Under the condition of the cable's prestress as 100, 150, and 200 kN, the bolt preloads are 60, 80 and 100 kN, respectively.

5.6.2 *Effect of top-coal thickness on the stress distribution of surrounding rock of full coal roadway*

In Fig. 57a, because the roadway's roof and floor are rock, and the two walls are coal, the stress concentration zones after roadway development are located at the four corners of the roadway in the roof and floor rock strata, the maximum compressed stress reaches 22 MPa. As the rock strengths of roof and floor are relative high, it enables pressure to be uniformly transferred to the coal wall bodies of the roadway to reduce the compressed stress in the coal bodies of the roadway's two sides, and reduce roadway deformation. As shown from Fig. 57b–f, with the increase of top coal thickness, the distribution zone of compressed stress in top-coal increases, and the stress in top-coal is less than that in the roadway floor. There is a distinct interface between top-coal and roof. Overall, during roadway excavation, when top-coal thickness is within 10 m, there is a strong correlation between the stress distribution of surrounding rock of roadway and the top-coal thickness. As the top-coal thickness increased, the stress concentration area in the top-coal expanded and the concentration coefficient reduced. With the continual increase of top-coal thickness, there is no appreciable variation in stress distribution of the surrounding rock.

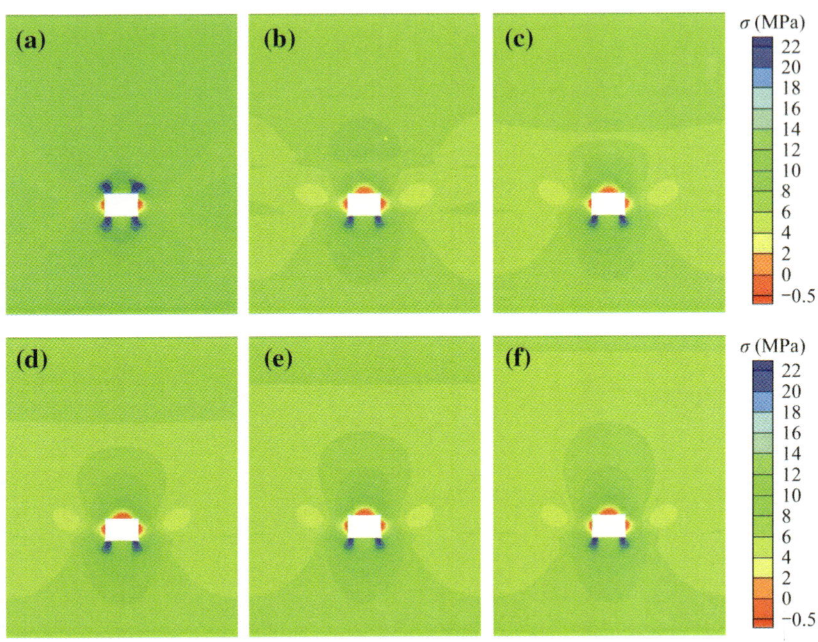

Fig. 57 Stress distribution of roadway surrounding rock at different top-coal thicknesses. **a** Rock roof **b** top coal 5 m **c** top coal 10 m **d** top coal 15 m **e** top coal 20 m **f** top coal 25 m

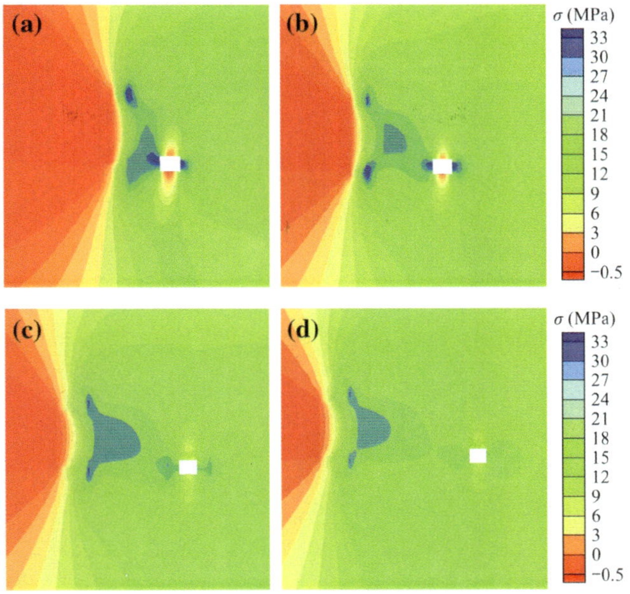

Fig. 58 Stress distribution of roadway surrounding rock with different coal pillar widths. **a** Coal pillar width 10 m **b** coal pillar width 20 m **c** coal pillar width 30 m **d** coal pillar width 40 m

5.6.3 Effect of roadway layout positions on the stress distribution of surrounding rock of full coal roadway

The coalface with full coal roadways is usually a coalface with a large mining height. The increase of mining height will have a great impact on the stress distribution in the coal pillar, thus affecting the stress distribution of surrounding rock of the roadway. In Fig. 58a, b, the stress concentration zone generated at two side walls during roadway development overlapped with the stress concentration zone in coal pillar after coal extraction. The maximum stress in the coal pillar reaches 33 MPa. In Fig. 58c, the stress concentration zone in the surrounding rock of roadway during its development is isolated from the stress concentration zone in coal pillars caused by the extraction of two adjacent coalface. However, it still has some impact on the stress distribution of the surrounding rock of the roadway, the maximum compressive stress on the side coal pillar is approximately 30 MPa. In Fig. 58d, due to the further increase of coal pillar width, the effect of the stress of the adjacent mining face on roadway decreased significantly. The stress concentration zone formed in the surrounding rocks of the roadway during its development is basically separated with the stress concentration zone formed by coalface extraction. The maximum compressive stress of surrounding rocks of the roadway is 27 MPa. Overall, due to the large thickness of coal seams with the coal roadway, after roadway development and adjacent coalface extraction, the stress concentration zone formed in the coal pillar is different to that at thick or thin coal seams. The stress is no longer distributed as standard

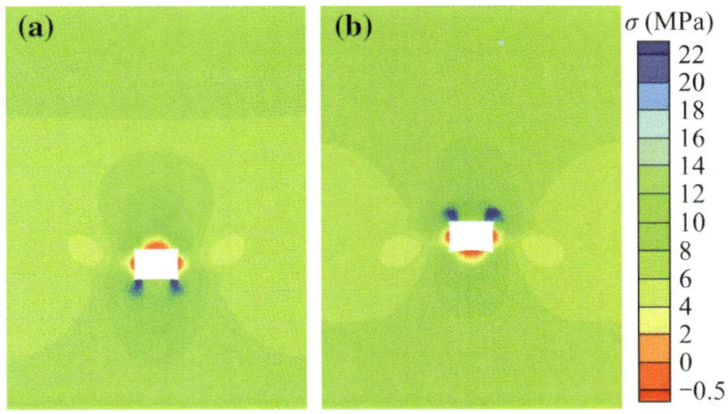

Fig. 59 Stress distribution of surrounding rock with different roadway positions. **a** Along floor layout **b** along roof layout

"arch" or "hump" distribution, but renders approximately triangular distribution.

A full coal seam roadway has two kinds of vertical arrangements, one is arranged along the floor, the other is arranged along the roof, their stress distributions have a large difference (Fig. 59). When the roadway is arranged along the floor, the stress concentration area is mainly distributed in the two corners of the roadway floor, the maximum stress reaches 22 MPa, and the maximum stress in the top coal of roadway is about 18 MPa. When the roadway is arranged along the roof, the stress state of surrounding rock of the roadway is opposite to that of roadway arranged along the floor, the stress concentration area is mainly distributed on the two corners of the roof, maximum stress is 22 MPa, and the maximum stress of the coal body at the floor is about 18 MPa. Because the stress field distributions of two kinds of arrangements are different, different measures should be taken during the roadway support. When the roadway is arranged along the floor, in addition to the top-coal support, the side walls support should be enhanced or the anchor bolt should be installed at the floor corners to prevent the occurrence of floor heave. When roadway is arranged along the roof, the support on the two corners of the roof should be strengthened, and some measures should be taken to prevent or reduce the occurrence of floor heave.

5.6.4 Effect of roadway height-width ratio on stress distribution of a full coal roadway

The stress distributions of the surrounding rock of the roadway under different roadway height-width ratios are shown in Fig. 60. With the same roadway height, the wider the roadway tunnel is, the higher the level of stress concentration in the top-coal; the greater the stress concentration area in top-coal, the higher the stress concentration factor. When the roadway's height-width ratio is equal to one, the maximum compressive stress in top-coal is

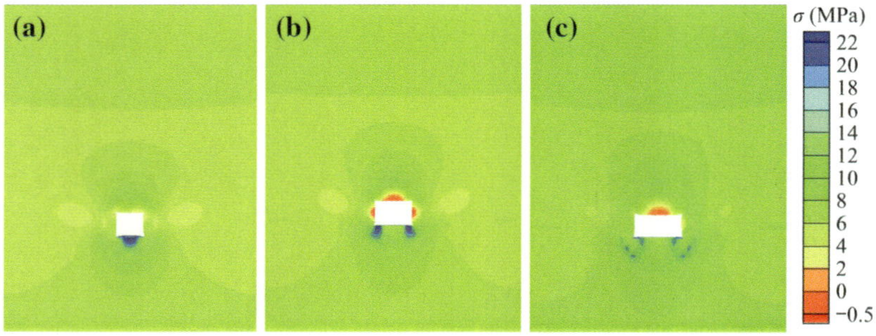

Fig. 60 Stress distribution with different roadway's height-width ratios. **a** Height-width ratio 1 **b** height-width ratio 0.75 **c** height-width ratio 0.5

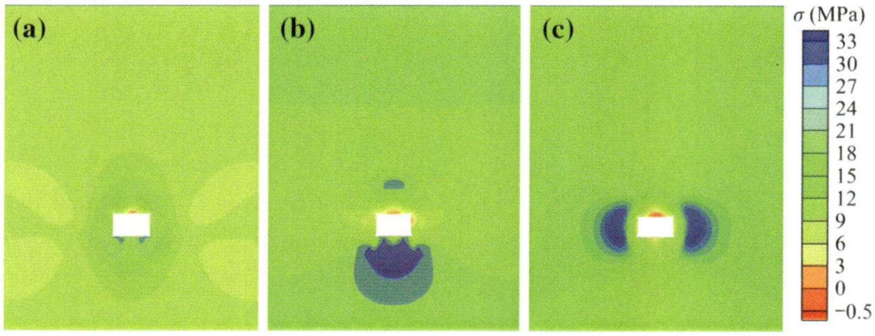

Fig. 61 Stress distribution of surrounding rock with different ground stress. **a** Combination 1 **b** combination 2 **c** combination 3

12 MPa. When roadway's height-width ratio is 0.75, the maximum compressive stress in top-coal is 16 MPa. When roadway's height-width ratio is 0.75, the maximum compressive stress in top-coal is 16 MPa, but the compressive stress concentration area increases significantly. When the roadway's height-width ratios are different, the stress distributions formed in the floor during roadway development are significantly different. When the height-width ratio is equal to one, the stress concentration zone in the floor shows an "inverted arch" distribution. When the height-width ratio is 0.75, the stress distribution zone shows an "inverted hump" shape distribution. When the height-width ratio is 0.5, an irregular stress distribution in the floor occurs, but the range increases significantly. Based on the maximum stress values in the floor, with the increase in roadway's width, the maximum stress values follow a downward trend.

5.6.5 Effect of ground stress on stress distribution of surrounding rock of a full coal roadway

The stress field distributions of surrounding rock of a full coal roadway under different ground stress conditions are shown in Fig. 61. The stress distribution of the surrounding rock of the roadway after roadway development is correlated to rock strength. When the horizontal and vertical stresses are the same, the stresses in the top-coal and side coal walls are basically the same, their values are relatively low, about 15 MPa. High stress areas are mainly in the floor with high rock strength, the maximum stress is about 21 MPa. When the horizontal stress is twice of vertical stress, the stress concentration areas are mainly in the top coal and the floor rock. In particular, due to the high rock strength in the floor, its stress and stress concentration area are significantly larger than that in top coal, the maximum stress value is up to 33 MPa. When the vertical stress is twice of horizontal stress, the maximum compressive stress of surrounding rock of roadway is approximately 30 MPa, and the stress concentration area shows an arc-shaped symmetrical distribution at two side walls of the roadway.

5.6.6 Effect of different bolt prestress on the stress distribution of surrounding rock of a full coal roadway

The increase of compressive stress in the anchorage zone and the range extension of compressive stress are the keys to improving the supporting effect. As shown in Fig. 62a, in the anchor bolt and anchor cable joint support, when anchor bolt pre-tight torque is 200 N m, a prestress overlap zone can be formed in the anchor bolt anchorage zone, but cannot cover the area between roof and anchor bolt at two side coal walls. The maximum compressive stress near anchor bolt in the coal body is about for 0.55 MPa, the compressive stress between anchor bolts is approximately 0.25 MPa. When the anchor bolt's pre-tight torque is

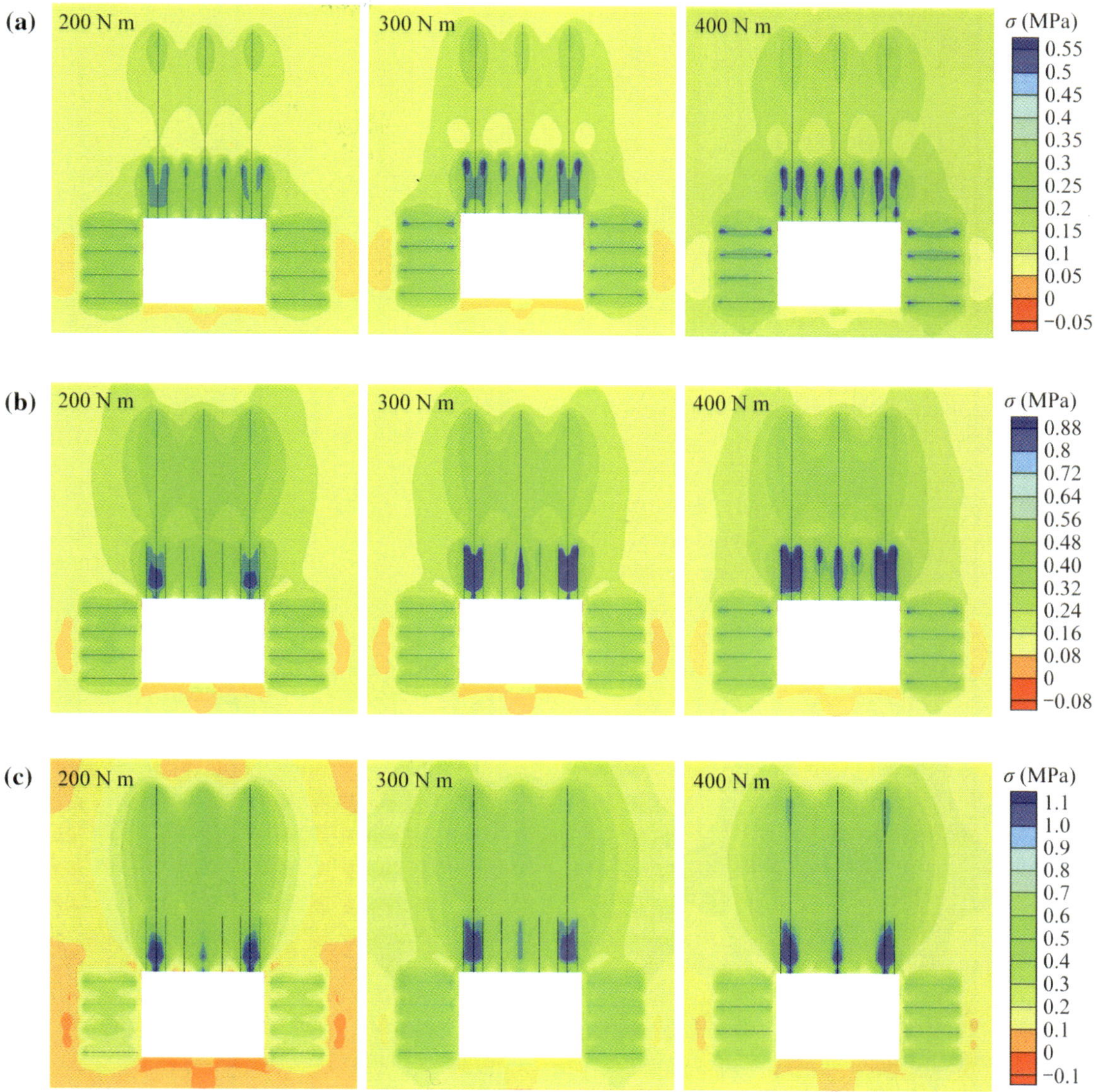

Fig. 62 Stress distribution of surrounding rock with different combinations of bolt and cable. **a** Cable prestress 100 kN **b** cable prestress 150 kN **c** cable prestress 200 kN

300 N m, with the increase of anchor bolt's pre-tight force, the anchor bolt's prestress spread range increases at roadway roof and two side coal walls, and in roadway's top coal, the spread zones can basically join and form one large zone. However, the anchor bolt's prestress spread zone is still small, and cannot cover the whole coal wall. The compressive stress between top coal and the roadway walls anchor bolt increases to around 0.4 MPa. When bolt's tightening torque increases up to 400 N m, the effect of the bolt's prestress spread within bolting range between the roof and two walls of the roadway is good. It covers the entire range of anchor zone, and the compressive stress value between bolts in the coal increases to about 0.5 MPa. Because of the lower prestress of the anchorage cable, the cable's prestress cannot spread to its anchoring range, thus, the overlapping effect of the bolt's prestress and cable's prestress is not obvious.

Compared to Fig. 62a, b shows that with the increase of the cable's prestress, there is a significant increase of prestress spread in the area along the cable length in the roadway's top coal and perpendicular to the direction of cable length. Nevertheless, in terms of the compressive stress in coal body formed by the bolt and cable's prestress, under the impact of the increase of cable's prestress and the superposition effect of the bolt's prestress, the maximum compressive stress close to the cable in the coal body reaches 0.88 MPa. When the bolt's tightening torque is 200, and 300 N m, the distribution of compressive stress

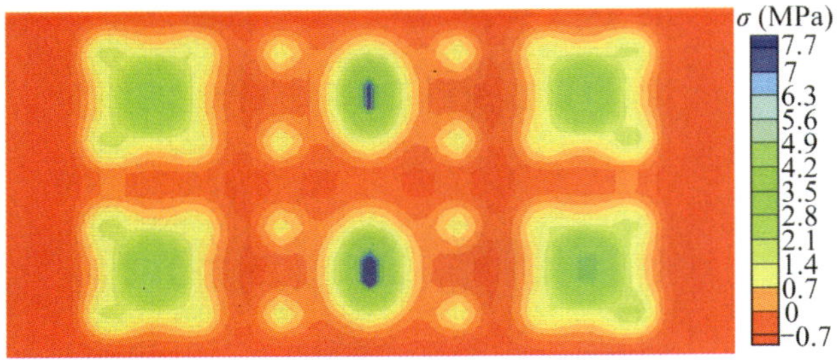

Fig. 63 Stress distribution on the surface of top coal with the bolt and cable combined support

formed in the coal body by cables and bolts is not uniform, the compressible stress between two bolts farther away from cables is 0.48 MPa. When the bolt's tightening torque is 400 N m, the effect of prestress spread formed in top coal by cables and bolts is improved to some extent. The compressible stress between two bolts farther away from cables is approximately 0.64 MPa.

As shown in Fig. 62c, when the cable's pre-tight force increases to 200 kN, the cable plays a major role in pre-stress spread in the coal body, the range of prestress spread basically covers the entire cable anchoring zone, and the maximum compressive stress formed in coal body is 1.1 MPa. The bolt's pre-tight force mainly affects the compressive stress distribution in the coal body, as the increase of bolt tightening torque, there is a distinct increase of compressive stress spread formed in the coal body and the stress values.

Figure 63 shows the compressive stress distribution on the surface of the roadway under the support of the combination of rock bolts and cables with a pre-tightening force of 400 and 150 kN, respectively. The compressive stress zones on the roadway surface formed by bolts and cables' combined support are interconnected and superimposed. The compressive stress of top coal at the bolt tray reached 2.8 MPa, at cable tray reached 7.7 MPa, the bolt's prestress and cable's prestress basically spread to the entire surface of top coal to form an effective stress arch, and improve the stress state of the surrounding rock of the roadway.

5.7 Support design of a large cross-section full coal roadway

Based on the numerical simulation results and lessons learned, a high strength bolt and cable combined support system was determined for the No. 5105 roadway at the No. 8105 coalface in Tashan coal mine. A BHRB600 bolt was selected with a diameter of 22 mm and length of 2.4 m left-lateral no longitudinal reinforcement steel rebar, extended with resin, and the bolt's tightening torque is not less than 400 N m. A 4 mm thick W steel belt and diamond-shaped metal mesh were used to protect the roof and side walls of the roadway. The roof bolt spacing was 800 mm, and the row spacing was 800 mm. Roof cable's diameter was ϕ22 mm with 1 × 19 strands high strength low relaxation prestressed steel wires, and its length is 8300 mm. Using one K2335 and two Z2360 low viscosity resin cartridge, the cable was anchored, the anchorage length is 1500 mm. With every two rows of bolts, three cable anchors were installed, the row spacing is 1600 mm, and cable spacing is 2000 mm, and all cables were vertically installed into the roof. The cable anchor's pre-tightening force was 100–150 kN. Figure 64 shows the layout of roadway support.

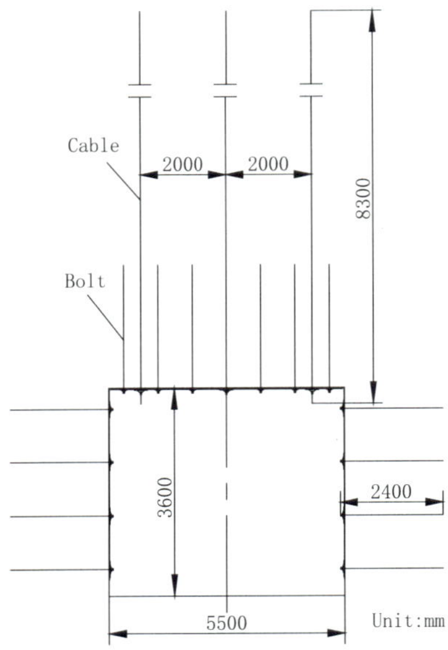

Fig. 64 Layout of roadway support

6 Safety technologies for the LLTCC coal mining operation

6.1 Gas control technologies for the LLTCC coalface

6.1.1 Characteristics of gas distribution at the LLTCC coalface

At the coal seam 3–5 in Tashan coal mine, the original gas pressure varies 0.14–0.17 MPa, the gas content in the coal seam varies 1.60–1.97 m^3/t and the average is 1.78 m^3/t. The coal seam permeability coefficient varies 171.71–428.80 m^2/(MPa^2 d), and the gas flow per hundred meters in the borehole varies 0.015–0.0212 (m^3/min hm); the gas discharge attenuation coefficient in the borehole varies between 0.602 and 0.7427 d^{-1}. The residual coal gas amount at the coal seam 3–5 is 1.17 m^3/t. The coal seams are spontaneous combustion-prone coal seams, and the shortest period of the occurrence of spontaneous combustion is 60 days; the coal seams have a tendency for coal dust explosion hazards and the explosion index is 37 %.

With the analysis of field tests, the gas emission and distribution in a LLTCC coalface have four characteristics:

(1) Low gas storage, high gas emission. Under the influence of large thickness of coal seam and high mining intensity, the LLTCC mining operation in ultra-thick coal seam shows a "low gas storage, high absolute gas emission" feature. At the coal seam 3-5, the original gas pressure was only 0.14–0.17 MPa, gas content in the coal seam was 1.6–1.97 m^3/t, and an average was only 1.78 m^3/t, however, the absolute gas emissions in the working face exceeded 40 m^3/min, and the highest absolute gas emission was up to 65 m^3/min.

(2) Unbalanced gas emission. The gas emission sources at the LLTCC coalface in Tashan mine can be divided into three parts, one is the gas at the top position of the goaf including the gas in the upper adjacent layer; the second is the gas in the deep part of the goaf including the gas of the lower adjacent layer; the third is the gas from coal extraction and caving operation carried by the airflow leakage in the working face. The gas emission in the three sources discussed above will be affected by face pressure, upper roof caving and top coal caving, and present the characteristics of imbalance, and the uneven emission coefficient might reach 1.8.

(3) Large gas emission in goaf. Local gas emission in goaf showed that the gas emission often exceeded the gas limit at the zone between the 110th support and upper corner of the coalface; at the window of top coal caving and the gap between supports, the measured gas concentration reached 2 %–3 %, sometimes even higher. At U-type ventilation coalface, the gas concentration seriously exceeded the gas limit in the upper corner of the working face. Without taking any measures in the upper corner, the gas emission concentration often exceeded 5 %. In the 10 % tail region of the coalface, the gas emission accounted for 90 % of the gas emission.

(4) The gas concentration distribution in the coalface showed a nonlinear relationship with the distance from the monitoring station. Figure 65 shows the gas concentration distribution along the vertical direction and face advance direction at the LLTCC coalface. The gas concentration distribution on the vertical face in the goaf showed an increase trend from bottom to top, increasing 2.3 % every 5 m; From the tailgate to internal displacement 32 m, the gas distribution showed an increasing trendency within the same elevation. From the support at the working face to the deep goaf, and within a 200 m range, the gas concentration distribution showed a ladder-type distribution. Within a 0–60 m range from the support to the goaf, the gas concentration increased from 1.5 % to 2.8 %; within a 60–160 m range in the goaf, the gas concentration increased from 3.6 % to 4.6 %; within a 160–200 m range in the goaf, the gas concentration increased from 5.7 % to 6.1 %, and basically coincided with the spontaneous combustion "three-zone".

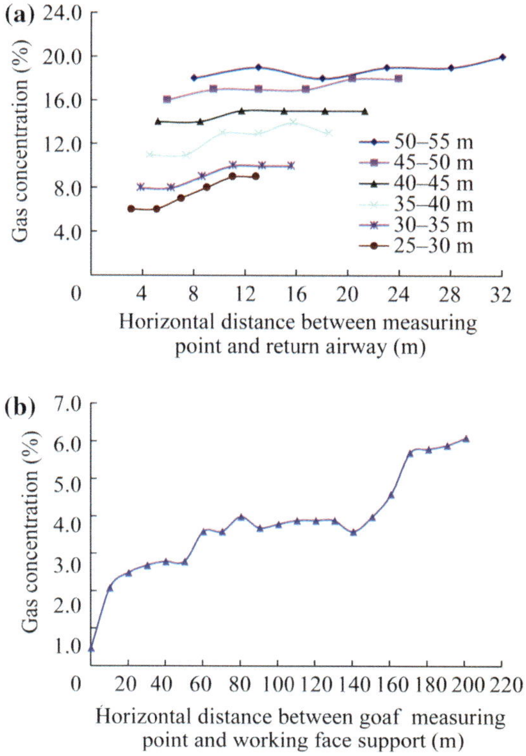

Fig. 65 Gas concentration distribution along the directions of vertical height in the goaf and face advance. **a** Along the direction of roof height **b** along the direction of face advance

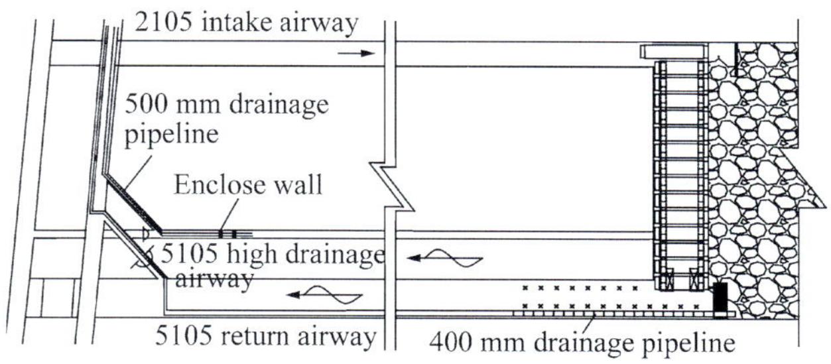

Fig. 66 Gas drainage at the No. 8105 working face

Fig. 67 Nitrogen generator in making

Fig. 68 Nitrogen generator in assembling

Fig. 69 Nitrogen generator after assembly

Fig. 70 Nitrogen generator at site test

6.1.2 Gas control technologies at the LLTCC coalface

Based on the characteristics of gas occurrence, a "mainly in strengthening gas drainage in goaf, complemented with the optimization of ventilation system in the working face" approach was used to develop a gas drainage roadway in roof-based comprehensive control technology for the LLTCC coalface. A three-step integrated program was developed: the first step was to use seals to block the gas at the upper and lower corners, and wind curtains to induce and dilute the gas for controlling the gas limit at the upper corner of the coalface at the early mining stage; the second step was to establish a large flow gas drainage system, including high-position pre-buried vertical tube, upper corner drainage tube and coalface ventilation system optimization; the third step was to developed a high drainage roadway in the roof, and use the closed roof high drainage roadway to drain the gas in the goaf. The gas drainage and mining at the No. 8105 coalface in Tashan coal mine is shown in Fig. 66.

6.2 Fire prevention technologies for LLTCC coalface

6.2.1 Large flow mobile nitrogen generator mounted on rubber tire vehicle underground

The nitrogen injection method was used to prevent fire at LLTCC coalface in ultra-thick coal seams. Therefore, a large flow mobile nitrogen generator mounted on rubber tire vehicle in the underground was developed. Figures 67, 68 and 69 show the units of large flow pressure swing adsorption nitrogen generator in the manufacturing process. Figure 70 shows the nitrogen plant in a field test.

The large flow mobile nitrogen generator mounted on rubber tire vehicle has following technical characteristics:

(1) Comparing and taking from the coarse and fine particle combined adsorption tower filling method in the US, the same method in Europe, the "blizzard" type adsorption tower absorption method, and the vibration desktop type filling method were used to develop a carbon molecular sieve filling method with the combination of vibration desktop type filling and coarse and fine grain filling, which greatly improved the service life of the molecular sieve;

(2) The recirculation calculation formula of minimum industrial pressure swing adsorption tower ω_2 (minimum) $\approx 1.15\omega_0(t_a/t_p)(p_3/P_0)$ was used to replace the calculation formula of the minimum regeneration cleaning amount required by the continuous adsorption process ω_2 (minimum) $\approx \omega_0(t_a/t_p)(p_{j0}/P_{j3})(-p_{j3} - P_{j3})/p_0$ (ω_2 is the Minimum cleaning quality flow rate of regeneration, kg/s; ω_0 is the initial gas mass flow rate, kg/s; t_a is the limiting adsorption time, s; t_p is the holding regeneration time, s; p_3 is the export airflow pressure, Pa; p_0 is the initial pressure entrance airflow, Pa), thus, the minimum amount of regeneration recirculation was determined. Combined with the control recirculation valve of the PLC controller, by reverse blowing

with the inlet direction, the fast and high efficient generation of nitrogen with the carbon molecular sieve device was achieved to increase the nitrogen amount generated by the nitrogen generator, as well as the nitrogen purity. In addition, it was possible to reduce the size of the nitrogen generator, and make the nitrogen generator more in line with the underground environment;

(3) The activated carbon filtering method was used with the combination of C, T and A three-stage fine filtrations. The activated carbon was used to remove large particles of water and oil in the compressed air, to prevent the absorption and break by the carbon molecular sieve. Using C, T, and A three-stage fine filtrations, the particles of water, oil and dust greater than 0.1 μm in compressed air will be removed to avoid large particle impurities not deposing to adsorption bed, reduce the adsorption bed resistance except maintaining the service life of the adsorbent. Furthermore, the nitrogen generator can work in a high nitrogen purity situation for a long time, ensuring that the molecular sieve service life can be extended and the efficiency of carbon molecular sieve can be improved;

(4) A dual PLC control automatic control system was developed. With the control of a dual control system, the unstable system problem caused by the interference of variable-pressure absorption in the same of twin towers was avoided during the operation of the four adsorption towers.

6.2.2 Study on the process and properties of large flow nitrogen generator

With a full consideration of the speed effect of the pressure swing adsorption tower, the size effect of adsorption bed and the adsorption equilibrium of carbon molecular sieve, the pressure swing adsorption tower was designed and the tower was filled with the vibration bench filling method and the coarse–fine combination filling method. Selecting the suitable pressure swing adsorption process, through a control system, the pressure swing adsorption nitrogen-making process was controlled, to ensure the time of pressure swing adsorption, and the time of adsorbent regeneration. Using product gas reverse blowing regeneration technologies, the nitrogen generation capacity of pressure swing adsorption and nitrogen purity were improved, and the nitrogen purity reached the requirement of ≥98 %. With the methods of activated carbon filters, and C, T and A three-stage fine filtration, the particles of water, oil and dust greater than 0.1 μm in compressed air was removed to ensure the service life of the adsorbent while reducing the adsorption bed resistance, and maintain the generation of high purity nitrogen.

Based on research and the comparison on the method of other existing models of pressure swing absorption molecular sieve nitrogen generator, through a small scale laboratory test, a comparison on the different processes, different adsorption media was conducted to determine the best process. The process is as follows: air compression—air buffer (and preliminary water removal)—activated carbon filter (removal of most of the water, and oil in compressed air)—fine filter (removal of the particles of water, and oil and dust greater than 0.1 μm in compressed air in)—pressurized adsorption (nitrogen generation)—pressure release—even pressure regeneration (reverse blowing process)—product nitrogen. In the compressed air system, two 250 kW screw-type air compressors were used to compress air, and the two sets of PLC controllers were used to control the pressure swing adsorption process. They ensure the smooth completion of adsorption and desorption process of the pressure swing adsorption nitrogen generating apparatus. After a 24 h continuous operation test, with the condition of nitrogen purity was greater than 98 %, the nitrogen flow rate was greater than 2000 m^3/h.

6.2.3 Development of fast curing stackable sealing material

The sealing on the upper corner of the coalface is an effective method for reducing the risk of spontaneous combustion in the goaf. However, in Tashan mine, the cross-section of intake and air return roadways are all up to 20 m^2. The use of traditional means, i.e. fly ash and yellow mud, would result in heavy workloads and high cost, which affects the normal transportation in the mine. Therefore, a fast curing stackable sealing material was developed.

Fast curing stackable sealing material is a polymer composite with multi-component, multi-usage, and solid gel combination, and is composed of a polymer compound, foaming agents, catalysts, flame retardants and other components. The feeding sequence for making the sealing material is: A material → B material → water-cement-sand mixture → curing agent. At the same time, the sealing material was quickly mixed to a uniform status, then pressurized and sprayed. The volume of the sealing material would instantly expand to (10–30) times of its original volume to achieve fast sealing. This material does not contain formaldehyde, is nontoxic, odorless, and non-corrosive. During the reaction expansion process, the sealing material can fill the space and cracks in coal and rock, and has a strong bonding property. It can bond coal (rock) body, concrete, and other materials to form one solid body. It has a certain level of pressure resistance and shear

Table 27 Equipment for the coalface

No.	Equipment name	Main parameters of equipment	Reference type
1	Hydraulic support	Height:2800–5200 mm; center distance: 1750 mm; support intensity: >1.35 MPa	ZF15000/28/52
2	Shearer	Cutting power: 2 × 750 kW; cutting height: 2.7–5.0 m; cutting depth: 800 mm; AC electric traction; voltage: 3300 V	MG750/1915-GWD
3	Front scraper conveyor	Length: 220 m; power: 2 × 855 kW; voltage: 3300 V; transportation capacity: 2500 t/h	SGZ1000/2 × 855
4	Rear scraper conveyor	Length:220 m; power:2 × 1000 kW; voltage: 3300 V; transportation capacity: 3000 t/h	SGZ1200/2 × 1000
4	Stage loader	Transportation capacity: 3500 t/h; power: 450 kW; voltage: 3300 V	PF6/1542
5	Crasher	Crashing capacity: 4250 t/h; power: 400 kW;voltage: 3300 V	SK1118
6	Emulsion pump	Working pressure: 31.5 MPa, Rated flow rate: 400 L/min; motor power: 250 kW, voltage: 3300 V, fluid box: 2, volume: 2500 L	BRW400/31.5
7	Spray pump	Working pressure: 12.5 MPa, Rated flow rate: 500 L/min,motor power: 132 kW, voltage: 3300 V, fluid box: 1,volume: 3000 L	BPW500/12.5

Fig. 71 Combined trial operation on the surface of complete equipment for the working face

resistance strength. It has good flexibility, is shockproof and pressure resistant, has strong elastic resilience after compression, with no cracking after curing, and no shedding. It has insulation properties, including heat insulation and flame retardant properties, suitable for filling at different locations in an underground coal mine.

The response time of rapid curing stackable material is 5–60 s adjustable at (23 ± 2) °C. The foam index is 10–30 times adjustable. The flame retardant: oxygen index is not less than 32, vertical and horizontal combustion level is V0. The content of heavy metals (Pb) is less than 0.90 mg/kg (assessment value 90 mg/kg).

7 Test in underground coal mine and result analysis

7.1 Parameters of working face and equipment

At the No. 8105 LLTCC coalface in Tashan coal mine, the face length is 207 m, the maximum cutting height is 5.0 m, and the cutting depth is 0.8 m. The installed capacity for the working face is 10 Mtpa. Based on the technical parameters and performance of fully mechanized coal mining equipment in China, the main technical parameters of the fully mechanized coalface equipment are determined as shown in Table 27.

In April 2010, the combined trial operation on surface of the complete equipment developed was conducted at Tashan coalmine (Fig. 71). From October to December in the same year, an underground industrial trial was completed.

7.2 Operation of trial equipment and production

During the period of the industrial trial at the No. 8105 face from October to December 2010, the statistics of production downtime caused by major equipment accidents are shown in Table 28. From October to December 2010, the

Table 28 Statistics of production downtime caused by major equipment failure during the industrial trial period from October to December 2010

Time	Shearer	Hydraulic support	Scraper conveyor	Belt conveyor	Power supply system
October 2010	22 h 35 min	0	0	8 h 30 min	2 h 35 min
November 2010	4 h	0	15 h 50 min	5 h 10 min	4 h
December 2010	4 h 30 min	0	1 h 25 min	15 min	0
Total	31 h 5 min	0	17 h 15 min	13 h 55 min	6 h 35 min

Table 29 Production statistics at No. 8105 coalface during the industrial test period from October to December 2010

Time	Number of working days	Cutting height average/maximum (m)	Face advance (m)	Production (t)	Average/maximum monthly production (t)	Average daily production (t)	Average manpower efficiency (t/worker)	Coal recovery (%)	Equipment usage (%)
October 2010	30	4.5/5.0	147.75	901,732	907527/916785	30057.8	362.2	88.9	92.1
November 2010	30		152.25	904,065		30135.5	363		
December 2010	30		172.5	916,785		30559.5	368.2		
Total	90	4.5/5.0	472.5	2,722,582	907527/916785	30250.9	364.5	88.9	92.1

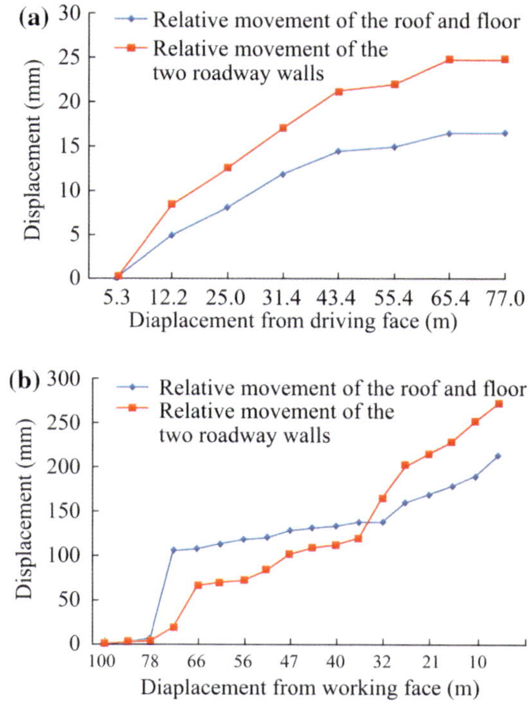

Fig. 72 Displacement curve of the roadway surface. **a** During roadway development **b** during coalface operation

production downtime caused by shearer accidents was 31 h and 5 min, the shearer usage rate was 98.45 %, and its performance was stable. The production downtime caused by hydraulic support failure was zero, its working resistance varied within a 9000–13,000 kN range, and overall the support was in good condition. The production downtime caused by scraper conveyor failure was 17 h and 15 min with an usage rate of 98.95 %. The production downtime caused by belt conveyor accidents was 13 h and 55 min, with an usage rate of 99.15 %. The production downtime caused by power system accident was 6 h and 35 min, failure-free rate was 99.61 %.

In October 2010, the production downtime caused by cumulative electrical accidents during the production period was about 57 h and 40 min. Based on four shifts with six hours each shift in the working face, the usage rate was 89.3 %. In November 2010, the production downtime caused by cumulative electrical accidents during the production period was about 53 h and 50 min, the usage rate was 91.3 %. In December 2010, the production downtime caused by cumulative electrical accidents during the production period was about 7 h and 10 min, the usage rate was 95.7 %. In the industrial test period from October 2010 to December 2011, average usage rate of equipment in use was 92.1 %.

During the industrial test period from October to December 2010 at No. 8105 coalface, the cumulative face advance was 472.5 m, average daily advance was 5 m,

Fig. 73 Roadway support at the No. 5105 roadway

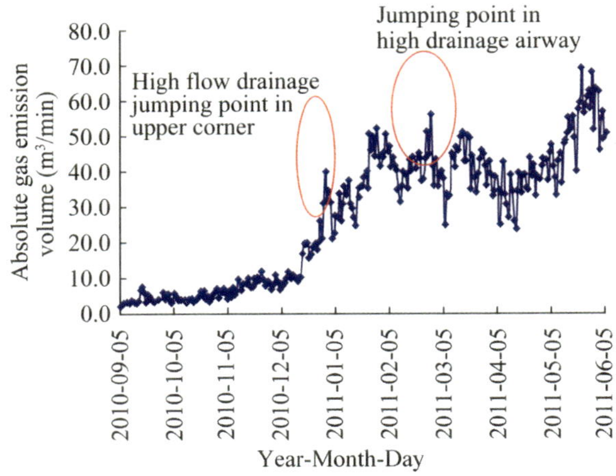

Fig. 74 Monitoring curve of absolute gas emission volume

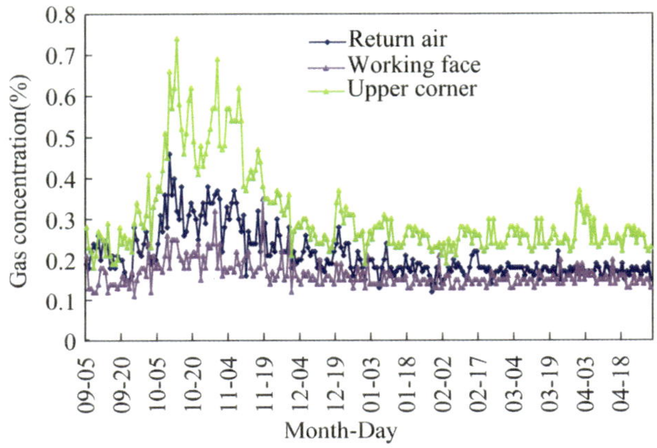

Fig. 75 Variation curves of gas concentration at the upper corner of coalface, coalface and in return air after coal extraction

maximum daily advance was 8 m; the cumulative coal production was 2.723 Mt, the average daily production was 30,300 t, maximum daily production was 38,000 t; average manpower efficiency was 364.5 t/worker. In three consecutive months during the industrial test, average monthly production reached more than 900,000 t, coal recovery rate at the working face reached 89.9 %. In 2011, the average monthly coal production was 904,000 t, the cumulative coal production for the year was 10.849 mt, achieving the goal of producing 10 Mtpa. The production statistics for the industrial test at the No. 8105 coalface from October to December 2010 is listed in Table 29.

7.3 Analysis of roadway support

During the excavation and support in the No. 5105 roadway, the deformation of the surrounding rock of the roadway was monitored, see Fig. 72a. The wall-to-wall convergence and the roof-to-floor convergence during the roadway excavation period was 25 mm and 16.6 mm, respectively. The support effect during the excavation period is shown in Fig. 73. In general, the deformation was small during roadway excavation period, and the effect of roadway support was good.

After the coal extraction at the No. 8105 coalface, under the influence of coal extraction, there were some variations in stress distribution of the surrounding rock in the roadway. The stress concentration occurred in coal and rock body in front of the coalface and in the surrounding rock of the roadway, resulting in the deformation of surrounding rocks in the roadway. The roadway surface displacement was monitored in the No. 5105 roadway within a 100 m range in front of the No. 8105 coalface, the monitoring results are shown in Fig. 72b. As can be seen from the figure, there was no change for the roadway surface displacement exceeding 78 m from the coalface, the roadway deformation affected by the coalface abutment pressure appeared from a location 78 m in front of the coalface. Within 78–66 m range in front of the coalface, large deformations occurred at the same time in the roof and two walls. Subsequently, the deformation increased as the coalface advanced slowly. Within 36 m range from coalface, with the advance of the coalface, the deformation began to dramatically increase. In the location of the monitoring station 5 m away from the coalface, the two side walls of the roadway moved closer by 280 mm, the roof and floor closer by 210 mm. However, in general there was no obvious damage to the surrounding rock of the roadway during the coal extraction period, the amount of deformation was small, and the transportation and ventilation requirements was fully able to be met.

7.4 Analysis on gas control

On the 9th of November 2010, the No. 8105 coalface and roof drainage roadway were connected. After the connection, a 2BEC80 gas drainage pump was used to drain the gas, the gas concentrations were maintained between 1.5 %

and 2.2 % in the high drainage roadway, between 0.2 % and 0.3 % at the upper corner of working face, between 0.2 % and 0.3 % at the tail of the rear scraper conveyor, and between 0.15 % and 0.25 % in the return air. Thus, the gas management at the coalface achieved very good results.

The gas emissions at the coalface were monitored before the coalface extraction, from coalface extraction to the initial caving of the upper roof, to first gas drainage at the upper corner of coalface, during the gas drainage at the upper corner of coalface, and from the gas drainage at the upper corner of the coalface to the gas drainage in return airway and high drainage roadway. The monitoring curve of absolute gas emission volumes are shown in Fig. 74.

Figure 75 shows the gas concentration variations at the upper corner of the coalface, the coalface and in the return air since the extraction of the No. 8105 coalface. As can be seen from the diagram, coalface gas emission rates gradually increased, the volume of gas drainage increased, while the gas concentration at the upper corner of the coalface, in the coalface and the return air gradually leveled off, and achieved a zero gas overlimit.

Site monitoring results showed that: the technology of roof high roadway drainage combined with other drainage can significantly reduce the gas concentration at the return air corner of the goaf, control the gas emission from the goaf to the coalface, play a role of "split-flow" of gas flow, and reduce the gas concentration in the goaf. The effect for preventing the gas accumulation at the upper corner of the coalface was significant. The drainage rate in the goaf reached 40 % of total gas emission in the goaf, achieving a zero overlimit, and ensuring a safe and smooth face advance. The gas control achieved a satisfactory result.

7.5 Analysis on fire prevention using nitrogen injection

Based on air quantity tests at different measuring points in the coalface, the amount of air leakage in the goaf was determined. With the installation of a bundle tube in the goaf and the connection of the bundle tube in the goaf with the main bundle tube monitoring system in the mine, the gas in the goaf was analyzed automatically. In addition, the airflow in the goaf was simulated in 3D to analyses the oxygen concentration in the goaf. Thus, a "three-zone" distribution was found in the goaf. Based on the finding, the nitrogen injection method and the amount of nitrogen injection were given at different face advance speeds.

Based on the actual mining conditions at the No. 8105 LLTCC coalface in Tashan mine, the nitrogen injection parameters were optimized, and the relevant nitrogen injection process and methods were developed. After calculation, an additional 3890 m^3/h nitrogen injection was required in addition to original 2500 m^3/h, to enable oxygen concentration at the spontaneous combustion zone to drop to an average of 7 %. Pipe nitrogen injection technology was used for fire prevention. The first nitrogen injection pipe along the goaf at the intake side of the coalface was buried. When the pipe was buried at a certain depth into the goaf, nitrogen injection commenced. Subsequently, a second nitrogen pipe was buried into the goaf (the shift interval of the nitrogen injection nozzle was 50 m). When the second nitrogen injection pipe was buried and started nitrogen injection into the spontaneous combustion zone in the goaf, the first pipeline of nitrogen injection stopped, and a new pipe was buried again for nitrogen injection. The process was repeated until the coalface extraction was completed.

The nitrogen injection method was determined based on fire prediction, in the case of nitrogen injection quantity of 2500 m^3/h, and coalface advance speed of 1.4 m/d, a continuous nitrogen injection method must be used. When the advance speed is less than 1.4 m/d or if the coalface has stopped, the amount of nitrogen injection must be increased. When the coalface has stopped working for more than 68 days, the nitrogen injection quantity should be not less than 6390 m^3/h; when the face advance speed is greater than 1.4 m/d, it would be appropriate to reduce the amount of nitrogen injection. Using the methods presented above, the occurrence of fire in LLTCC coalface can be prevented to achieve safe mining operation.

8 Application and benefits

The following problems are present in coal mining in China: one is that the capacity in maintaining safety in mining is relatively low; the second is that technology and equipment, technical management, and the technical innovation level are not at a high level; the third is that the waste of resources is relatively serious; the fourth is that the added-value and technological-content of products are relatively low; the fifth is that the centralization of business is relatively low.

The technologies and equipment for LLTCC coal mining operation in ultra-thick coal seam have achieved a safe and high efficiency extraction of the coal seam with a thickness of 14–20 m. The technologies and equipment developed have effectively increased the coalface production rate and top-coal recovery. The large mining height shock-resistance top-coal caving hydraulic support and the electrohydraulic control technology developed for top-coal caving greatly improved top-coal caving and discharge speed, reduced labor intensity, and increased coalface productivity. The anchor bolt support technology and material developed for the support of large cross-section full coal roadway significantly improved the effects of roadway support, and ensured the roadway is stable and unblocked. The gas control and fire prevention technology and equipment ensured safe coal production.

After adopting the technology and equipment developed, the investment on equipment in Tashan mine was decreased. Compared to the No. 8104 coalface, the investment on equipment at the No. 8105 coalface saved about 100 million yuan.

With the geological and mining conditions at the No. 8105 coalface, compared to the sub-level caving method in ultra-thick coal seam, the technologies and equipment developed enabled coal production at the No. 8105 coalface in Tashan coal mine to be doubled; the economic benefit is significant.

In addition to the Tashan mine in Datong, currently the LLTCC technologies and equipment for ultra-thick coal seam have been used in 32 coal mines in China, including 13 mining areas such as Datong, Pingshuo, Shengdong, Xinjiang, covering 8 provinces or autonomous regions in China, such as Shanxi, Inner Mongolia, Xinjiang, Shaanxi, Shandong, Gansu, Anhui, Liaoning.

Currently, 14 large coal bases have been planned in China including Shendong, Northern Shaanxi, Huanglong, Northern Shanxi, Jinzhong, Jindong, Luxi, Lianghuai, Jinzhong, Henan, Yungui, Eastern Inner Mongolia (northeast), Ningdong, and Xinjiang, where thick and ultra-thick coal seams are predominantly mined. The main coal seams in the coal mines with 10 Mtpa capacities each currently under construction are mainly thick or ultra-thick coal seams. The coal mines built with a capacity of 10 Mtpa each mainly extract the coal seams with thickness of 6–20 m, the seams above 6 m account for 50 %, such as Halagou, Shigetai, Daliuta, Buliantai, Yujialiang, Shangwan in Shengdong coalfield, Yangchangwan in Ningdong coalfield, Huangling No. 2 mine in Huangling coalfield, Anjialing mine in Pingshuo coalfield, Dafosi mine in Binchang coalfield and Sihe coalmine in Jincheng coalfield. With the construction of large-scale coal production bases in China in the future, LLTCC coal mining technologies and equipment provide technological support for the safe, high efficiency and high recovery coal production in these mines. The technologies and equipment developed will drive the upgrading of existing mechanized caving technology and equipment. The research results will be widely used and the products will have great business potential.

At the same time, the development of technologies and equipment strongly promote the development of coal mining methods and related theories, and lead the progress of coal mining technologies. It enables the capability of manufacturing the complete set of the LLTCC coal mining equipment in China, driving the development of machine processing technology, electromechanical integration technology, and automation and information technology. It curbs the occurrence of roof, gas, and fire accidents, safeguard coal mine practitioners' personal safety, improves underground mine working environment, ensures coal mine safety, efficient, integrated production, improves the efficiency of coal mining and energy supply, and provides reliable energy security for China's sustained, healthy and stable economic development.

9 Conclusions

With the study and development of the LLTCC coal mining technologies and equipment for over 10 years, the following main conclusions are obtained:

(1) In terms of coal seams with a thickness of 14–20 m, the LLTCC coal mining method has been proposed, which increases the coal cut height to 5 m, coal caving height to 15 m, and the total mining height to 20 m. The effect of different coal caving intervals and top-coal caving patterns on top-coal recovery has been investigated, which concludes that the recovery rate can be improved in long wall caving by applying the caving patterns of one-cut-one-caving and discharge in multi-round and multiple windows sequence alternately.

(2) A "cantilever beam-articulated beam" model has been established for the overlying strata at the LLTCC coal mining face. Based on the model, the working resistance of the LLTCC hydraulic support has been determined. Furthermore, the law of top coal and roof movement at the LLTCC coalface has been obtained. In front of face coal wall, the main characteristics of top coal and roof movement are mainly horizontal compressive deformation under abutment pressure; in the rear of the face coal wall, the characteristics of roof movement are mainly fractures and subsidence, the characteristics of top coal movement are mainly vertical caving movements caused by roof fractures and subsidence. The speed of horizontal deformation of the top coal and roof in front of the face coal wall under abutment pressure is much smaller than that of the vertical subsidence of the top coal and roof after fracture at the rear of the face coal wall.

(3) Comprehensive spalling prevention and control technologies have been proposed, including controlling the coalface advance speed, increasing the initial support force, and improving the ability of wall protection. These technologies reduce coal wall spalling and maintain the normal advance of the coalface.

(4) A large-diameter double telescopic shock-resistance prop and a 1000 min^{-1} large flow safety valve quick unloading device have been developed, which improve the shock-resistance of hydraulic support; an articulated front beam and high reliable telescopic beam with a secondary wall protection structure have been developed, which overcomes the wall spalling problem at LLTCC coalfaces; and a strong disturbance-type top coal discharge device with high efficiency and high top coal recovery has been developed, which improves coal recovery. On the basis of the developments presented above, a 5.2 m cutting height LLTCC hydraulic support has been developed.

(5) A high efficiency and high reliability electric traction shearer has been developed, its specifications are: coal cutting height 2.8–5.5 m, total installed power 1945 kW, coal cutting power 750 kW, maximum haulage speed 15–25 m/min, and maximum traction force 2 × 570 kN. An analysis method of the shearer reliability has been established. Furthermore, a new control system and walking system for the shearer has been developed, and the new material of ranging arm shell body and the performance of drum tooth have been optimized to improve the reliability of the shearer.

(6) A large-capacity SGZ1200/2 × 1000 rear scraper conveyor has been developed with installed power of 2 × 1000 kW, and transportation capacity of more than 3000 t/h. The transportation distance meets the requirements of the No. 8105 coalface in Tashan mine. In addition, some key accessories, including the heavy-load and high-strength chain transmission system of the scraper conveyor, controllable speed-adjustable soft start device, the integrated monitoring transmission system of transmission device, heavy-load 1000 kW planetary retarder, high-reliability and long-life coal pan, high strength and compact type end unloading headframe, and electrohydraulic control automatic telescopic conveyor tail, have been investigated to improve the performance of the LLTCC rear scraper conveyor.

(7) Various devices have been invented in the project, including a height adjustable ring chain lifting device, a new high efficient air inlet flame arrester and exhaust processing box, and an integrated motor plug type wheel side drive and brake retarder. A WC55Y support hauler with a load of 55 tons has been developed, achieving a safe, high efficiency, non-demolition, "point-to-point" transportation of the LLTCC hydraulic support, and improves the efficiency of LLTCC coalface relocation.

(8) The stress distribution of surrounding rock of large cross-section coal roadway is significantly influenced by the in situ stress and the rock strength. Under the same stress conditions, the higher coal and rock mass strength, the higher the stress magnitude of surrounding rock after the roadway excavation. The support effect of rock bolts on the roadway is mainly to generate compressive stress zones through the prestress force of bolts, and form a prestress loading structure with cables and other supporting structures, thus to strengthen the surrounding rock of the roadway, and constrain the separation, sliding, and opening of fracture in the surrounding rock, and subsequently maintain the integrity and stability of the roadway surrounding rock.

(9) A special bolt steel formula has been developed, which changes the steel mechanical properties. With suitable rolling method, a high strength (500–600 MPa), high elongation bolt material has been developed. Using new 19 structure wires instead of 7 structure wires, a large elongation, powerful cable bolt has been developed. Also, a bolt torque multiplier has been developed, which enables the output torque of bolt driller to be amplified 6 times, and solves the problems of rock bolt prestressing.

(10) In terms of the characteristics of "low gas reserve, high gas emission" in the LLTCC coalface, comprehensive gas control technology, using mainly the roof high drainage roadway, has been developed. The gas drainage amount from the goaf accounts for 40 % of the total gas emission in the goaf, achieving gas zero overlimit; a large flow underground mobile nitrogen generator has been developed to solve the fire prevention problems at the 10 Mtpa LLTCC coalface and guarantees safe coal mining.

(11) The key technologies and equipment for LLTCC coal mining operation has been successfully evaluated at the No. 8105 coalface of the Tashan coal mine. The underground test demonstrated that the match of "shield, shearer and scraper conveyor" is rational, average usage of face equipment reaches 92.1 %, face coal recovery reaches 88.9 %, the cumulative coal production was 10.849 Mtpa in 2011, achieving the target of 10 Mtpa for a LLTCC coalface in China, and safe and high efficiency coal mining in an ultra-thick coal seam.

References

Alehossein H, Poulsen BA (2010) Stress analysis of long wall top coal caving. Intern J Rock Mech Min Sci 47(1):30–41

Altounyan P, Taljaard D (2001) Developments in controlling the roof in South African coal mines—a smarter approach. International conference on coal research. South African Institute of Mining and Metallurgy, pp 33–40

Buddery PS, Oldroyd DC (1992) Development of a roof and floor classification applicable to collieries. Proceedings of the Eurock '92 conference. Thomas Telford, London, pp 197–202

Cao SG, Liu CY (1997) Study on the stiffness of the immediate roof of longwall face. Ground Press Strata Control 4:16–19

Cao SG, Qian MG, Liu CY, Miao XX (1998) New research about support and surrounding rock relationship in working face. J China Coal Soc 23(6):575–579

Cao SG, Qian MG, Liu CY (1999) Mechanical characteristics of the immediate roof in working face and working resistance of supports. Mining Science and Technology'99. vol 8, pp 309–312

Chen Y, Yu L, Lan TW (2010) Numerical simulation research on surrounding rock activities in shallow coal seam face. 2010 International symposium on safety science and technology. Science Press USA Inc, pp 1618–1622

Cheng SP (2008) Research on surface movement rules off full-mechanized caving mining with large mining height in mountain area. Coal Min Technol 13(4):37–39

Deng GZ (1994) Characteristics of movement and rupture of the roofs during longwall top coal caving process. Ground Press Strata Control 2:23–26

Duan WS, Gao XC, Dou J (2008) Law of movement and fracture of overlying strata of fully-mechanized top-coal caving face in the hard and thick coal seam. J Xi'an Univ Sci Technol 28(2):211–214

Fan YC, Kang LJ, Kang YH (2003) Fully mechanized longwall top coal caving mining technology. China Coal Industry Publishing House, Beijing

Fang XQ (2002) Study on stability of support-surrounding rocks and it's control in fully-mechanized top-coal caving face. Doctor's Thesis: China University of Mining and Technology

Fu YG (2008) Study about mining height fully-mechanized coal winning caving face gas governing technology. Coal 17(12):9–12

Hou W, Jiang FX, Wang CW, Feng ZQ, Wang DZ (2009) Pressure control in sub-level long face surrounded by three sides minedAreas of C-shaped stratas patial structure. J China Coal Soc 34(3):310–314

Hu W (1995) Roof control and selection of drawing roof-coal stope. J Xiangtan Mine Inst 10(4):7–13

Huang K (2002) Mechanical system of chock-roof and stability of soft coal longwall face during top caving process. Doctor's Thesis: China University of Mining & Technology (Beijing)

Jia AL, Huang X (2005) Research on bolt support system for full seam gateway in high stressed thick seam. Coal Sci Technol 33(7):45–48

Jia XR, Zhai YD, Yang SS (1998) Structure of roof rocks in sub-level caving face and calculation of roof weighting. J China Coal Soc 23(4):366–370

Jiang FX (1995) Structure patterns and roof-chock relationship in longwall top coal caving. Coal Mine Mod 1:33–35

Jiang FX (2006) View point of spatial structures of overlying strataand its application in coal mine. J Min Saf Eng 23(1):30–33

Jin ZM, Wei JP, Song XM, Xue Y, Niu Y (2002) Test Study on The top cracking process of large coal sample. Proceedings of the '99 international symposium on mining science and technology, Beijing, pp 405–408

Kang LJ (1990) Rupture characteristics and the interaction between roof and chocks in gently-inclined thick coal seams using longwall top caving methods. Master's Thesis: China Coal Research Institute

Kang LJ (1995) Study on top coal stress in caving mining. Proceeding of the international mining Tech'95 symposium. Beijing, pp 18–25

Kang LJ (1996) Ground pressure characteristics in longwall top coal caving work face for gently-inclined seam coal seams. Coal Sci Technol 24(11):39–42

Kang LJ (1998) Limiting effect of strain-softening character of the top coal on loads exerted on the powered supports in a fully mechanized sub-level caving face. J China of Coal Soc 23(2):140–143

Kang LJ (1999) Mechanical coupling relationship between roof and chock in longwall top coal caving method.Coal. Min Technol 4(12):69–71

Kang LJ, Qi QX (1998) Discrete element modelling of longwall top coving process. Coal Min Technol 3(2):3–6

Kang HP, Wang JH (2007) Rock bolting theory and complete technology for coal roadways. China Coal Industry Publishing House, Beijing, pp 83–86

Kang LJ, Shi YW, Yao JG (1991) Interaction between chock and roof in longwall top coal caving for gently-inclined coal seam. Ground Press Strata Control 1:19–24

Kang LJ, Wu G, Zhang YJ (1997) Top coal movement characteristics during longwall top caving process in the Yanquansikuang coal mine. Coal Sci Technol 25(9):13–16

Kang LJ, Zheng XZ, Feng ZC (1998) Investigation on the ground pressure characteristics of the tailgate driven alongside the #10 longwall gob in the Nantun coal mine. Coal Sci Technol 26(7):41–45

Kang LJ, Liu JH, Ceng HF, Wang LX (1999) Study on the behavior and controlling method for hard roof strata of seam No. 4 in Wang ping Colliery. Coal. Min Technol 1:69–71

Kang TH, Chai ZY, Li YB, Ge YY, Zhang HB, Liu RR (2007a) Study on physical simulation of full-seam mining for a 20 m very thick and medium hard seam by sub-level caving mining with high bottom cutting Height. Chin J Rock Mech Eng 26(5):1065–1072

Kang HP, Wang JH, Lin J (2007b) High pre-tensioned stress and intensive bolting system and its application in deep roadways. J China Coal Soc 32(12):1233–1238

Kang HP, Jiang TM, Gao FQ (2008) Design for pre-tensioned rock bolting parameters. J China Coal Soc 33(7):721–726

Kong LH, Jiang FX, Liu J, Yie GX, Wang CW, Song GD (2009) High-precision microseismic monitoring system to reasonable width of segment coal pillar in extra-thick coal seam fully mechanized top-coal caving mining. J China Coal Soc 34(7):871–874

Kong LH, Jiang FX, Yang SH, Song JW, Wang CW (2010a) Movement of roof strata in extra-thick coal seams in top-coal caving mining based on a high precision micro-seismic monitoring system. J Univ Sci Technol Beijing 32(5):552–558

Kong LH, Jiang FX, Liu J, Wang CW (2010b) Relationship between support and strata in extra-thick coal seam fully-mechanized sublevel caving mining based on high precision microseismic monitoring technology. Chin J Geotech Eng 32(3):401–406

Kong LH, Jiang FX, Wang CW (2010c) Study of reasonable working resistance of support in fully-mechanized sublevel caving face in extra-thick coal seam. Chin J Rock Mech Eng 29(11):2312–2318

Li MZ, Liu KM, Zeng MS (2006) Full seam cutting and caving mining technology in and its development prospect. Coal Mining Technology 11(5):28–29, 43

Li L, Xu G (2009) Researchon supporting density of powered support for full-mechanized caving ming in Shuiliandong colliery. Coal Min Technol 14(5):86–88

Liu CY (1996) Mechanical characteristics of immediate roof and chock-roof relationship in longall mining face. Doctor's Thesis, China University of Mining & Thechnology

Liu J (2008a) Microseismic monitoring in longwall top coal caving for an extra-thick coal seam. Min Saf Environ Prot 35(1):44–46

Liu KG (2008b) Studying about mining hight fully-mechanized coal winning caving the roof activity. Coal 17(6):9–11

Liu QM (2010) Research on space effect of face length of fulimechanized caving mining with large mining height. Coal Min Technol 15(3):27–29

Liu CY, Huang BX (2009) Technical parameters of drawing and coal-gangue field movements of a fully mechanized large mining height top coal caving working face. Min Sci Technol 19(5):549–555

Liu CY, Qian MG, Cao SG, Miao XX (1997) Influencing mechanism of immediate roof on the relation between supports in the workings and surrounding rocks. J China Coal Soc 22(5):471–476

Liu CY, Wan ZJ, Cao SG (2002) The deformation destroying interaction regulation of immediate roof and top-coal in long wall sublevel caving face. Proceedings in mining science and safety technology. Science Press, Beijing, pp 105–109

Liu YD, Zhang DS, Wang HS (2006) Simulation analysis of coal mining with top-coal caving under hard-and-thick strata. J China Univ Min Technol 16(2):110–114

Lu XS, Zhao XL (2006) XU XD, Xue SJ. Kit technique research about fully-mechanized coal winning equipment. Coal 15(6):6–7

Ma QH (2005) Study of the overlying strata "O" letter type spatial structures of long wall and correlate rock pressure. Doctor's Thesis: Shandong University of Science and Technology

Ma JY, Zhang ZQ (2008) Research on coal roadway supported by bolting with mesh. J Taiyuan Univ Technol 39(5):233–235

Mao DB, Yao JG (2010) Adaptability of long wall top coal caving with high cutting height. J China Coal Soc 35(11):1837–1841

Meng QK (2007) Fully mechanized high cutting height and caving mining technology and hydraulic powered support selection in Wobei mine. Coal Sci Technol 35(8):53–54

Meng D, Shen J, Zhao Y (2010) Determination of the hydraulic supporting resistance based on a simplified structural system. Metal Mine 10:24–28

Miao SJ, Lai XP (2011) Top coal flows in an excavation disturbed zone of high section top coal caving of an extremely steep and thick seam. Min Sci Technol (China) 21(1):99–105

Nan H, Xu T, Wei ZD (2008) Application of RFPA 2D in sublevel caving mining extra-thick coal seams. Materials Sci Forum 575:1246–1251

Qian MG, Shi PW (2003) Coal mine pressure and roof control. China Coal Industry Publishing House, Beijing

Qian MG, He FL, Miao XX (1966) The system of strata control around long wall face in China. Mining Science & Technology, pp 15–18

Qian MG, Miao XX, He FL, Liu CY (1996) Coupling mechanics of chock and roof in longwall mining face. J China Coal Soc 21(1):40–44

Ren RH (2005) Surface movement regularity of super-wide mining face with top-coal caving. J China Univ Min Technol 15(1):68–71

Shi H (2005) Study and applications ability of hard and massive overlying stratain fully mechanized sub level caving face. Doctor's Thesis: Shandong University of Science and Technology

Shi YW, Chen SJ (1993) Ground control of floor in retreating mining face. China Coal Industry Industry Publishing House, Beijing

Shi H, Jiang FX (2004) Mechanical analysis of rupture regularity of hand and massive overlying strata of longwall face. Chin J Rock Mech Eng 23(18):3066–3069

Shi H, Jiang FX (2005) Study on relationship between roof stability and recovery ratio of top coal at first weight period in fully-mechanized sublevel caving face. Chin J Geotech Eng 27(4):414–417

Shi YW, Ning Y, Qi QX (1993) Ground control and optimum parameters in longwall top coal caving technologies. China University of Mining & Technology Press, Xuzhou

Shi H, Jiang FX, Wang HJ (2005) Study on relationship between roof stability and top coal's recovery ratio burning cyclic weighting in fully-mechanized sublevel caving face. Chin J Rock Mech Eng 24(23):4233–4238

Shi HM, Du JW, Li ZQ (2009) Mining height long caving mining stope pressure control technology. Coal 18(11):14–16

Song JW, Gao RP, Wang AW (2010) Science and Technology of West China 20(9):6–10

Su H (2008) Stability analysis and ground control of coal measures during longwall top coal caving process China. Coal 34(2):35–36

Wang JC (2005) Fully mechanized longwall top coal caving technology in China and discussion on issues of further development. Coal Sci Technol 33(1):14–17

Wang JH (2006a) Present status and development tendency of fully mechanized coal mining technology and equipment with high cutting height in China. Coal Sci Technol 34(1):4–7

Wang HJ (2006b) Resarch on the movement and control of "O"letter type overlying milti-strata spatull structures of face surrounded by mined areas. Doctor's Thesis: Shandong University of Science and Technology

Wang JC (2009) Theories and technologies for mining thick coal seams. Metallurgical Industry Press, Beijing

Wang GF (2010a) Top-caving powered support and fully-mechanized caving technology. China Coal Industry Publishing House, Beijing

Wang M (2010b) Study on the laws of strata behavior in the top-coal caving of fully mechanized mining. Master's Thesis: Taiyuan University of Technology

Wang JH (2011) Present status and prospects of modernized mining technology and equipment in China coal mine. Coal Sci Technol 39(1):1–5

Wang JH (2013a) Key technology of complete equipment for fully-mechanized top coal caving face with large mining height in extra-thick seam. Coal Sci Technol 41(9):1–5

Wang JH (2013b) Key technology for fully-mechanized top coal caving with large mining height in extra-thick coal seam. J China Coal Soc 38(12):2089–2098

Wang ZF, Li QS, Yang LP, Liu J (2001) Research to ensure the height of displacement caving zone in overlying strata at Xiashijie coal mine combined working face. Coal 16(1):1–3

Wang JS, Chai ZY, Kang TH, Li YB (2007) Numerical simulation of full seam mining for a 20 m very thick and medium hard seam by fully mechanized sublevel caving. J Taiyuan Univ Technol 38(2):175–178

Wang JH, Kang HP, Gao FQ (2008) Numerical simulation on load-transfer mechanisms and stress distribution characteristics of cable bolts. J China Coal Soc 33(1):1–6

Wang GF, Pang YH, Liu JF (2012) Determination and influence of cutting height of coal by top coal caving method with great mining height in extra thick coal seam. J China Coal Soc

37(11):1777–1782
Wu J (1998) Reviews on the longwall top coal caving technologies in the past 15 years. Proceedings of the 3rd Longwall Top Coal Caving Theory and Application Conference
Wu YP (2008) Research on full mechanized caving mining technology in extremely thick and complicated coal seam. Coal Min Technol 13(1):18–20
Wu J, Fu Q (1998) Current situation and prospect of top coal caving technology in China. J China Univ Min Technol 8(2):108–112
Wu J, Yan SH (1995) Basic concepts of determining resistance of face support with fully mechanized mining and top coal caving. Mine pressure and roof management 3(4):69–71
Wu J, Zhang Y (2001) The new concept of relationship between support and surrounding-rock in longwall top-coal caving faces. J China Coal Soc 26(4):350–356
Xie GX (2005) Mechanical characteristics of fully mechanized top-coal caving face and surrounding rock stress shell. J China Soc 30(3):303–313
Xie GX (2006) Influence of mining thickness on mechanical characteristics of working face and surrounding rock stress shell. J China Coal Soc 31(1):6–10
Xie GX, Luo Y (2005) Study on the Countermeasures against methane outburst of mining multiple upper protective layers in coal seams C luster. J Coal Sci Eng 11(1):31–35
Xie YS, Zhao YS (2009a) Numerical simulation of the top coal caving process using the discrete element method. Intern J Rock Mech Min Sci 46(6):983–991
Xie YS, Zhao YS (2009b) Technique of top coal caving with vibration. Procedia Earth Planet Sci 1(1):219–226
Xie GX, Chang JC, Yang K (2006a) Investigation on the stress field characteristic of top coal at FMTC faces under the influence of caving thickness. J Coal Sci Eng 13(2):123–125
Xie GX, Yang K, Liu QM (2006b) Study on distribution laws of stress in inclined coal pillar for fully-mechanized top-coal caving face. Chin J Rock Mech Eng 25(3):545–549
Xie GX, Yang K, Chang JC (2007a) Study on distribution characterics of 3D stress and thickness effects of coal-seam in unsymmetrical disposal and fully-mechanized top-coal caving. Chin J Rock Mech Eng 26(4):775–779
Xie GX, Yang K, Hua XZ (2007b) Influence of mining velocity on mechanical characteristics of surrounding rock in fully mechanized top-coal caving face. Chin J Geotech Eng 29(7):963–967
Xie GX, Chang JC, Yang K (2009) Investigations into stress shell characteristics of surrounding rock in fully mechanized top-coal caving face. Intern J Rock Mech Min Sci 46(1):172–181
Xing SJ, Wang JZ, Zhao YB (2009) Development and application of top coal caving hydraulic supports with large mining height. China Coal 35(4):81–83
Yan SH (2008) Research on side and roof falling mechanism and control approaches in caving mining with large mining height. Coal Min Technol 13(4):5–8
Yan SH (2009) Theory study on the load on support of long wall with top coal caving with great mining height in extra thick coal seam. J China Coal Soc 34(5):590–593
Yan SH, Fu Q (2003) Movement characteristics of the strata above the coal seam under longwall top coal caving. China Coal Industry Publishing House, Beijing
Yan SH, Yin XW (2008) Discussing about the main theoretical problems of long wall with top coal caving. J China Coal Soc 33(5):481–484
Yan SH, Jia GS, Liu XL (1996) Analysis of mechanism of overlying strata structure transmitting toward upper position in top coal caving. Mine Press Roof Manag 3:3–5
Yan SH, Yin XW, Xu HJ, Xu G, Liu QM, Yu L (2011) Roof structure of short cantilever-articulated rock beam and calculation of support resistance in full-mechanized face with large mining height. J China Coal Soc 36(11):1816–1820
Yang JL (2007) Stability and ground control of coal measures during longwall top coal caving process.China. Coal 33(11):37–39
Yang PJ (2009) Research on the relationship between two-leg sublevel caving shield support and surrounding rocks and adaptability. Doctor's Thesis: China University of Mining and Technology
Yasitli NE, Unver B (2005) 3D numerical modeling of long wall mining with top-coal caving. Intern J Rock Mech Min Sci 42(2):219–235
Yin XW, Yan SH, An Y (2008) Characters of the rib spalling in fully mechanized caving face with great mining height. J Min Saf Eng 25(2):222–225
Yu L, Fan ZZ, Xu G (2011) Study on the structure characteristics and ground pressure behavior of overlying strata with shallow buried coal seam. Adv Mater Res 255–260:3780–3785
Yu L, Yan SH, Liu QM (2012) Determination of support working resistance of top coal caving in extra thick coal seam. J China Coal Soc 37(5):737–742
Zhang DL (1995) Stability and ground control of coal measures in longwall top coal caving. Doctor's thesis: China University of Mining & Technology
Zhang DL (1999) Ground control in longwall top coal caving. China Coal Industry Publishing House, Beijing
Zhang HJ, Liu QM (2008) Research on the fuzzy clustering of comprehensive support working resistance. China Min Mag 17(10):93–98
Zhang CW, Yang WB (2010) Technology of supporting deep full-seam coal roadway with high stress. Coal Min Technol 15(4):63–64
Zhang BS, Kang LX, Yang SS (2006) Numerical simulation on roof separation and deformation of full seam roadway with stratified roof and large section. J Min Saf Eng 23(3):264–267
Zhang Z, Kang HP, Wang JH (2010) Pre-tensioned stress coordination function analysis of bolt-cable anchor support in coal roadway. J China Coal Soc 35(6):881–886
Zhang JG, Zhao ZQ, Gao Y (2011) Research on top coal caving technique in steep and extra-thick coal seam. Procedia Earth Planet Sci 2:145–149
Zhou ZL, Su H (2008) Support technologies for longwall top coal caving face ends. Coal 17(6):18–20
Zhu LF, Yan SH (2011) Numerical simulation of top-coal movement rule in fully-mechanized caving mining with large mining height. Coal Min Technol 16(1):11–13

6

A review on investigation of water-preserved coal mining in western China

Limin Fan[1,2] · Xiongde Ma[3]

Abstract Yushenfu mining area is located in an ecological fragile area in western China, the coal seam of which is the Jurassic Yan'an Formation. The Jurassic Yan'an Formation contains five minable coal seams, the top layer of which is thick, covered by shallow overburden and located under aquifers. Therefore, the mining induced water flowing fractured zone can easily extend to the aquifers of both the Quaternary Sarahu and Jurassic Zhiluo Formation. This would result in a series of negative hydrological and ecological effects, including groundwater leakage, groundwater lowering, furtherly causing surface vegetation withering and dying, surface water body reduction, spring drying out, and water flow of river being decreased substantially. To solve these environmental problems, several technologies have been carried out by Chinese scientists, one of which is water-preserved coal mining. This paper presents a review of the origin, definition and development of water-preserved coal mining, and its applications in Yushenfu mining area. The applicable conditions, research contents, research methodology, and technical foundation of water-preserved coal mining are addressed in this paper. The future research focuses regarding water-preserved coal mining in China are also discussed in this paper. Its results serve as a guide for selecting the methods to be preferred for mining in case the geological conditions, roof overburden structure and coal mining process are similar to Yushenfu mining area.

Keywords Water-preserved coal mining (WPCM) · Ecological water table · Ecological fragile mining areas · Western China · Water conducting fissure zone

1 Introduction

Yushenfu mining area of Jurassic coal field is located in the west of China, more specifically, the southeastern margin of the Ordos basin, which is a transition zone between the Musu desert and the Shanbei plateau. Two of 14 major coal bases of China are located in this area, Shanbei and Shendong coal bases. The Yushenfu mining area is about 8900 km^2, with proven coal reserves of more than 140 billion tons. The development of this mining area, started from 2000, are divided into 4 stages, with mining area of 873, 1098, 864 and 2430 km^2 at each stage, respectively. By the year of 2015, 149 coal mines have been constructed with a total annual production of over 1 billion tons.

At the beginning of mine development, lacking of thorough consideration of the geological environment constraints, unreasonable layout of the mine planning have resulted in a series of negative environment impacts (Fan 1992). For instance, the groundwater table dropped (Fan et al. 2016), the water quality deteriorated, surface water and wetland decreased (Ma et al. 2015), arable lands, forest and grasslands were destroyed by a large number of surface cracks (Bi et al. 2014), landslide collapsed and mining-

✉ Limin Fan
fan19880629@163.com

1 Key Laboratory of Mine Geological Hazard Mechanism and Control, Ministry of Land and Resources, Xi'an 710054, China

2 Shaanxi Institute of Geo-Environment Monitoring, Xi'an 710054, China

3 College of Environmental Science and Engineering, Chang'an University, Xi'an 710054, China

induced seismicity occurred (Ji et al. 2015). Those environment damages have been drawing a lot of social concerns.

In order to reduce the negative environment impacts of mining, Fan (2005, 2017) has proposed a method to protect environment while coal mining, namely, Water-preserved Coal Mining (WPCM). The main idea of this method is to rationally divide the mining area into different subareas and apply different mining methods to different mining subareas to achieve the goal of water protection while mining coal (Wu et al. 2017; Wang et al. 2010).

This paper is to review the concept, technical foundation and application of WPCM technology in Yushenfu mining area.

2 Research progress of WPCM

2.1 Problems

The full development of coal mining in Yushenfu mining area began in the late 1980s. As the increasing knowledge of the hydrogeology and environmental geology in this mining area and understanding of practical problems encountered during coal mining, the environmental problems associated with coal mining have been drawing more and more attentions from scientists. The problems include water loss and environmental degradation, which could in turn bring safety issues to the mine. This is evidenced by two roof fall accidents occurred in Ciyaowan coal mine of Shenbei mining area on April 28 and December 20, 1990, respectively. Due to the thickness of bedrock aquiclude is less than 100 m, the mining induced water flowing fractured zone extended to the aquifers of the Quaternary Sarahu, which caused water inrush. Although the amount of mine water inrush was small, a large amount of sand inrush into the roadway resulted in belt lane was filled by sand, continuous groundwater table dropping, drying of reservoirs, reduction of spring flow, and vegetation withering (Ma et al. 2017).

Therefore, protection of groundwater during coal mining in ecological fragile areas is important for both environmental protection and mine safety (Fan 1992). Groundwater resources should be protected during coal mining (Han et al. 1992). A new concept of water-preserved coal mining is proposed by Fan (2005), which uncovered the prelude of research on water protection while coal mining.

2.2 The concept of water-preserved coal mining

The Yushenfu mining area is located in an arid and semi-arid region of average annual precipitation about 400 mm. The water resources in this area are limited and are unevenly distributed in different areas and through different time in a year. Groundwater is the only water source for industry, agriculture, human and animals consumption, and ecological systems. However, the past coal mining activities broke the overburden of the coal seams and damaged the water conducted zone, resulting in unstable bottom structures of the aquifers. The water resource balance was then broken because the aquifers of water supply significance lost water storage capabilities. Therefore, scientists started to focuse on the maintenance of the stability of ecological valuable aquifers. It is important to implement proper ideas and methods to mine planning, mine design, and mine development to ensure the harmonious development of resources and environment.

The aim of water-preserved coal mining is to maximize the social and economic efficiency. WPCM refers to the technology to seek the best solution to balance coal mining efficiency and water carrying capacity in arid and semi-arid area. The solution is to be achieved by controlling the movement of rocks to maintain the stability of aquifers of water supply significance and ecological values or by controlling the changes of water table within a reasonable range.

The research objects of water-preserved coal mining are aquifers of water supply significance and ecological values in the western arid and semi-arid areas. Based on ground control theories and technologies, coal mining technologies can be developed to control the development of water conducting fissure zone. The main study objects are the structures and the water level of aquifers. The structure of aquifers should be stable or could still restore water for a certain time in the situation of being destroyed. The purpose of water-preserved coal mining is to balance the supply and demand between coal resources and water resources, to achieve the resource development and water protection simultaneously.

The protection aquifers of water-preserved coal mining are aquifers of Quaternary Salawusu formation (Q_{3S}) and Jurassic Yan'an formation (J_2y) (Fan et al. 2016) in the north of Ordos basin, Ordovician Karst in the south of Ordos basin (Ma et al. 2016), and Cretaceous Luohe formation in the west and southwestern margin of Ordos basin. The aquifers which have significant contributions to the base flow of rivers, aquifers which have important effects on vegetation succession, surface reservoir, and deep aquifers should also be included in the system of water-preserved coal mining.

2.3 The research category of water-preserved coal mining

The research of water-preserved coal mining focuses on four aspects, survey, analysis, and assessment of geological conditions, stability study of aquicludes, constraint conditions of ecology, and methodologies of water-preserved coal mining.

2.3.1 Geological survey

The study of geological and hydrogeological conditions is the basis of water preserved coal mining. The mining area is divided into water rich and water poor subareas according to the survey of geological conditions, geological structure, water bountifulness of the aquifers and rechargeable condition of the area. Coal should be mined by highly mechanized mining technology in water poor mining areas while the water preserved coal mining technology should be implemented in waterrich mining areas. The current geological data (Fan 1992) shows that the hydrogeological conditions of Yushenfu coal mining area can be divided into four categories, namely isolated small water rich basins, large aquifers without a clay aquifuge, aquifers with a stable clay aquifuge and SalaWuSu-burnt rock aquifers. However, due to the area variability of geological conditions and inadequate drilling holes, the accurate knowledge of geological conditions is absence within each partition. For each working face, detailed geological conditions should be explored through exploration drilling, geophysical exploration, and geochemical exploration. In 2015, the water rich burnt rock area of 5^{-2} coal seam of Hongliulin coal mine was determined through geophysical exploration, which provided a support for implementing water-preserved coal mining technology.

The normally used empirical formulas to predict the height of water conducting fissure zone originate from «Regulations of buildings, water, railway and main well lane leaving coal pillar and press coal mining» and «Mine Hydrogeology Regulations» (Hu et al. 2012). However, height of water conducting fissure zone is highly influenced by various factors such as geological conditions, roof overburden structure and coal mining process. More and more researches have focused on correction of the empirical formula by validating the height of the water conducting fissure zone, or to propose a new formula for predicting the height of water conducting fissure zone.

The multi-factor weighted evaluation model which based on geological and hydrogeological conditions is a direct method of risk prediction for water and sand inrush. This method also relies on detailed geological data with spatial analysis of GIS to predict the possibility of water and sand inrush during coal mining. This method is also able to determine the key areas of implementing water-preserved coal mining technology.

2.3.2 Structure stability of aquifuges

The aquifuges of Yushenfu coal mining area are mainly overlying bedrocks (siltstone) and clay (yellow and red clay) (Huang et al. 2010; Huang 2003). The thickness of bedrock aquifuges is normally less than 150 m and the thickness/load ratio of bedrock is no more than 1 (Huang 2002). The coal seam is shallow and bedrocks are thin, thus the working face bears large dynamic loadings which could result in step subsidence. The collapse of overlying rocks only exists in "two zones", namely water conducting fissure zone and fractured zone, and the fragmentation degree of overburden rock is large. If the clay layers are absent in overlying bedrocks, the "two zones" will directly contact with the aquifer. The structure of aquifuge will be completely destroyed, causing loss of water. If there is a clay layer in the overlying bedrocks, whether the "two zones" will penetrate into the clay layer depends on the thickness and mechanical properties of the layer.

The evolution characteristic of water-resisting property of clay during mining has been a key point in water-preserved coal mining, the study which is still insufficient. According to the water-resisting property change of clay caused by strata displacement and deformation, Liu (2016) divided the evolution of water-resisting property of clay into three stages, impermeability keeping stage, impermeability decreasing stage, and impermeability losing stage. Huang et al. (2010) stated that the breaking of bedrocks is the main factor that influence development of the "upward fissures", while the hydrological properties of clay are the main factors that influence close of "downward fissures". The relationship between "upward fissures" and "downward fissures" determined the stability of the aquifuges impermeability.

2.3.3 Constraints of groundwater level on coal mining

The constraints of groundwater level on coal mining in arid and semi-arid areas are represented by two aspects, water table and water quantity. Water table is related to ecological value and river base flow (Liu and Zhang 2006). Water quantity is related to available exploitation capacity of groundwater. Previous researches (Yang et al. 2006; Wang et al. 2008) have shown that the relationship between river baseflow and groundwater table is linear. Thus to ensure the sustainably development of regional ecological environmental, coal mining must meet the requirement of a certain water table threshold. The upper limit is to reduce evaporation amount of phreatic water and the lower limit is to sustain virtuous development of an ecological system. To achieve this threshold, the only way is to control movement and deformation of the overlying rocks and control water resources by implementing water-preserved coal mining technology (Ma 2016).

2.3.4 Water-preserved coal mining technology

Restraining the height of fissure zones is an effective way to realize water-preserved coal mining. The drilling results from YuShen mining area showed that under general conditions, the ratio of the height of the fissure zones to the mining height is between 20 and 26 (Miao and Jiang 2014). In the case that thickness of the overlying rocks is less than height of fissure zones, water-preserved coal mining technologies could be applied using the following techniques.

(1) Backfilling coal mining. The mixture of both water and sand is used to fill the mined areas to reduce the space of gob area and reduce the height of fracture zone (Xu et al. 2007).
(2) Narrow strip coal mining. This method applies for small mines. The width of the working face is reduced by reserved pillars, thus reduce the height of water conducting fissure zone (Shao et al. 2014).
(3) Slice mining. Based on the distribution of thick coal seams in Yushenfu coal mining area (Wang et al. 2015), mining height was reduced to control the height of fissure zones (Wang et al. 2014a, b).
(4) Longwall mining by speeding face advance rate. In some of the coal mines in the Shendong mining area, the height of water conducting fissure zones can be reduced through increasing face advance rate (Ma et al. 2008, 2014).

3 Engineering practice of water-preserved coal mining technology

3.1 Protection of overlying aquifers

3.1.1 Backfilling

The backfilling mining is an effective method for roof control which is beneficial to both environmental protection and coal mine safety. Backfilling is widely applied in room and pillar mines and longwall mines. The backfilling method provides a technical approach for aquifer protection and rock movement control.

The Yuyang coal mine implemented water preserved coal mining by filling aeolian sands in the mine (Lv 2013). The filling materials are mainly aeolian sands, fly ash, cement, special auxiliary materials and water of certain percentages. The percentages of water is less than 35% (paste-like fill). 50 entries of No. 2301 working face have been filled with a total volume of 52000 m^3. The strength of filling bodies after 28 days reached 5.30 MPa. The filling ratio was between 50% and 70%. After the backfilling, the surface subsidence decreased by more than 50%. Geological survey showed that the water tables decreased by about 0.45–2.76 m and the depths of water table was controlled in the range of 4–6 m.

Local backfilling and limited height mining have also been carried out in some coal mines of Shendong mining area. The main approach is to reduce mining height properly or partially fill the areas near the open-off cuts of working faces and the recovery line.

3.1.2 Narrow strip coal mining

In narrow strip coal mining, the mining system is arranged according to the original longwall mining system. The original longwall face is divided into a number of mining strips which are parallel to the original open-off cuts (Shao et al. 2009). When mining each strip, the working face is developed between the head entry and tail entry. After an exhaust ventilation system is formed, retreating mining is then conducted in the mine.

The narrow strip coal mining has been conducted in about 10 coal mines of the Yuyang District. A working face was divided into 20 strips, with 12 strips mined and 8 strips reserved in the mines. The recovery rate of narrow strip coal mining was increased by 20% or more compared to room and pillar mining method. The production of a single mine was increased to 1–3 million tons.

3.1.3 Slice longwall mining (limited mining height)

Two mining methods are normally applied in mining thick coal seams, cave mining and slice longwall mining (Wang et al. 2014a, b). Though high efficiency, overlying rocks are seriously damaged and overburden stress are increased intensely in cave mining. It has shown in practice that reducing initial mining thickness and increasing repeated mining thickness can effectively reduce the height of water conducting fissure zones and mining cost. Proper design of thickness of initial mining and repeated slice longwall mining can prevent damage of overlying aquifers.

The Yushuwan coal mine contains 5 coal seams. The top coal seam, No. 2^{-2} coal seam, is 11 m thick which could be mined by slice mining. Results of numerical simulation (Wang et al. 2014a, b) showed that the groundwater in over 45% area of the Sarahu Formation would be lost if the initial mining height exceeds more than 7 m using cave mining method. However, water preserved coal mining can be realized in most areas if the height of the upper slice is about 5 m. The drilling data (Miao and Jiang 2014) showed that the water conducting fissure zones has not reached up to the aquifer of Salawusu Formation when the mining height of the upper slice of Yushuwan coal mine was designed to be 5.5 m. This provides evidence that Yushuwan coal mine achieved water preserved coal mining.

3.1.4 Longwall mining by increasing face advance rate

The coal seam in Bulianta coal mine of Shendong coal mine area is shallow covered by overlying bedrocks. To implement water-preserved coal mining in Bulianta coal mine, three measurements were applied in 32201 working face and 32202 working face. These measurements included increasing face advance rate, limiting mining height or partially backfilling, and utilizing a reasonable support resistance in the key areas. These measurements effectively reduce the damage of roof bedrocks and avoid full development of water conducting fissure zones. Both field survey and numerical simulation (Ma et al. 2008, 2014) showed that the change of water table was small, with decline of groundwater level larger than the subsidence only at the two sides of the working face. The results indicated that the Salawusu formation aquifer was protected effectively.

3.2 Protection of aquifer for coal seam floor

The Chenghe mining area of Weibei coal field is located in western China. Under the coal seams of this area is a water rich Ordovician limestone aquifer, the water table of which is 375 m uniquely. The Ordovician limestone aquifer is an important water source for industrial and agricultural water supply in Weibei district and is the key recharge source for the wetlands in the Weibei area. However, the Ordovician limestone aquifer also brings water inrush hazards for mining to this area (Pang et al. 2014). The protection of this aquifer is of significant importance. Therefore, it is necessary to carry out WPCM in Weibei coalfield to ensure safe coal mining and maintain the water table stability.

The water inrush coefficient in Dongjiahe coal mine is 0.13–6.0 MPa/m when the mining height is 4 m. According to the "Rule of mine prevention and cure water disaster" (Wu et al. 2011), the critical threshold value of water inrush coefficient is 0.06 MPa/m in an area with tectonic damage. Since the water inrush coefficient is higher than 0.06 MPa/m in 43% of the mining areas, the Donajiahe coal mine is considered as a risk zone of mining under pressure. In order to solve the problems associated with mining under pressure, Donjiahe coal mine built a water-resisting floor grouting system to enhance the strength of coal seam floor, which is the largest in China. Using direct current method to detect the water bearing property, the floor grouting showed an excellent effect to improve the water bearing property and prevent water inrush caused by aquifers.

4 The future research in water preserved coal mining

There are four scientific problems that need to be solved in water preserved coal mining.

(1) A fluid–solid coupling model to study the evolution of compressive strength and variation mechanism of permeability in rocks needs to be built. A fluid–solid coupling equation based on stress equilibrium equations and seepage continuity equations should be established to determine the fissure flow. This equation should also be able to solve the expansion, deformation, and evolution problems for initial fissures and mining fissures under water pressures.

(2) Theories and methods to evaluate groundwater and ecology constraints in mining areas based on evapotranspiration mechanism of groundwater and unsaturated zones in arid and semiarid sandy areas should be established. The main focuses of this method are to study the water absorption characteristics of typical plants in arid environment, the optimum depth of water table to reduce the evaporation rate of stable sand, and the reduction of ineffective evaporation.

(3) Deep study should be conducted on characteristics of movement, deformation and collapse of shallow coal seams with thin overlying bedrocks. The study can build a foundation for computing the height of the water conducting fissure zones. A new empirical formula should be further built to compute the height of water conducting fissure zones according to the geological structure and rock mechanics in the areas.

(4) A low cost and high efficiency backfilling mining technology should be developed. Utilizing this technology, shallow coal seams would be efficiently mined while the aquifers are protected to ensure the environmental protection in the shallow coal seam mining areas.

5 Conclusion

The Jurassic Yan'an Formation of Yushenfu coal mining area contains five minable coal seams, the top layer of which is thick, covered by shallow overburden and located under aquifers. Therefore, the mining induced water flowing fractured zone can easily extend to the aquifers of both the Quaternary Sarahu and Jurassic Zhiluo Formation. This would result in a series of negative hydrological and ecological effects. To solve these environmental problems, several technologies have been carried out by Chinese scientists, one of which is water-preserved coal mining. The main idea of this method is to rationally divide the mining area into different subareas and apply different mining methods to different mining subareas to achieve the goal of water protection while mining coal.

The research of water-preserved coal mining focuses on four aspects, including survey, analysis, and assessment of geological conditions, stability study of aquicludes, constraint conditions of ecology, and methodologies of water-preserved coal mining. The commonly used water-preserved coal mining technology are backfilling coal mining, narrow strip coal mining, Slice longwall mining and Longwall mining by increasing face advance rate. Owing to the good performance of these methods, the review also floats a suggestion for future investigation to develop more effective methods to ensure the environmental protection in the shallow coal seam mining areas.

Acknowledgements This work was supported by the National Key Foundation (973) Research Program (No.2 013CB227901) and Science and Technology Extension Program of Shaanxi (No. 2011TG-01).

References

Bi Y, Zou H, Zhu C (2014) Dynamic monitoring of soil bulk density and infiltration rate during coal mining in sandy land with different vegetation. Int J Coal Sci Technol 2:198–206

Fan L (1992) Environmental geology in Shenmu mining area. Hydrogeol Eng Geol 6:37–40

Fan L (2005) Discussing on coal mining under water-containing condition. Coal Geol Explor 5:53–56

Fan L (2017) Scientific connotation of water-preserved mining. J China Coal Soc 1:27–35

Fan L, Xiang M, Peng J (2016) Groundwater response to intensive mining in ecologically fragile area. J China Coal Soc 11:2672–2678

Han S, Fan L, Yang B (1992) Some hydrogeological and engineering-geological problems concerning development of north Shaanxi Jurassic Coalfield. Coal Geol China 1:49–52

Hu X, Li W, Cao D, Liu M (2012) Index of multiple factors and expected height of fully mechanized water flowing fractured zone. J China Coal Soc 4:613–620

Huang Q (2002) Ground pressure behavior and definition of shallow seams. Chin J Rock Mech Eng 8:1174–1177

Huang Q (2003) Simulating test on damage regularities of thick sandy soil layer and load distribution on key roof block in shallow coal seams. Coal Geol Explor 6:22–24

Huang Q, Wei B, Zhang W (2010) Study of downward crack closing of clay aquiclude in shallow-buried coal seam. J Min Saf Eng 1:35–39

Ji R, Peng S, Fan L, Zhao Y, Li C (2015) Effect of coal exploitation on groundwater circulation in the Shen-Fu mine area: an example from middle and lower reaches of the Kuye river basin. J China Coal Soc 4:938–943

Liu S (2016) Study on mining failure law of overburden and mining instability mechanism of clay waterproof layer in thick coal seam. China University of Mining and Technology, Beijing

Liu X, Zhang Y (2006) Essence and indicators of the healthy Yellow River. J Hydraul Eng 6:649–654

Lv W (2013) The application of the backfill mining in Yu-yang coal area. Sci Technol Innov Herald 33:45–48

Ma L, Zhang D, Liu Y et al (2008) Aquifer-protective mining technology in shallow coal seam with thin bedrock. J Hunan Univ Sci Technol (Nat Sci Edi) 1:1–5

Ma L, Sun H, Wang F et al (2014) Analysis of the ground water level change of aquifer protective mining in longwall coalface for shallow seam. J Min Saf Eng 2:232–235

Ma X, Fan L, Zhang X et al (2015) Driving force analysis for water and wetlands evolution at Yushenfu mining area. J China Coal Soc 5:1126–1133

Ma X, Du F, Qi P, Li W (2016) Technology and engineering practices on water preserved coal mining inseam with pressurized water floor. Coal Sci Technol 8:61–66

Ma X, Fan L, Yan G et al (2017) Vegetation responses to groundwater level change in mining area. J China Coal Soc 1:44–49

Miao J, Jiang Z (2014) Drilling detection of the height of water conducting fissure zone in fully mechanized mining face. Shaanxi Coal 6:33–36

Pang Y, Wang G, Ding Z (2014) Mechanical model of water inrush from coal seam floor based on triaxial seepage experiments. Int J Coal Sci Technol 4:428–433

Shao X, Shi P, Wang H (2009) Study on pillars stability by keeping water in strip mining for small and medium-sized mines in northern Shanxi Province. Coal Technol 12:58–61

Shao XP, Ding ZW, Zhang J (2014) Study on long time effect and replacement mining method for coal pillar in strip mining. Coal Eng 8:57–59

Wang Y, Wang W, Qian Y, Duan L, Yang Z (2008) Change characteristics and driving forces of base flow of Yellow River Basin. J Nat Resour 3:479–486

Wang S, Huang Q, Fan L et al (2010) Study on overburden aquclude and water protection mining regionazation in the ecological fragile mining area. J China Coal Soc 1:7–14

Wang W, Sui W, Dong Q (2014a) Influence of cover stress re-establishment on the permeability evolution of the mining-induced fractures. J China Coal Soc 06:1031–1038

Wang Y, Xia Y, Du R (2014b) Discussion on maximum mining height of coal mining under water-containing condition in one mine field of northern Shaanxi province. J Min Saf Eng 4:558–563

Wang J, Yu B, Kang H et al (2015) Key technologies and equipment for a fully mechanized top-coal caving operation with a large mining height at ultra-thick coal seams. Int J Coal Sci Technol 2:97–162

Wu Q, Zhao S, Li J et al (2011) The preparation background and the main points of Rule of mine prevention and cure water disaster. J China Coal Soc 1:70–74

Wu Q, Shen J, Wang Y (2017) Mining techniques and engineering application for "Coal-Water" dual-resources mine. J China Coal Soc 1:8–16

Xu J, You Q, Zhu W, Li X, Lai W (2007) Theorectical study of strip-filling to control mining subsidence. J China Coal Soc 2:119–122

Yang Z, Wang W et al (2006) Huang Jinting Research on buried depth of eco-safety about groundwater table in the blown-sand region of the Northern Shaanxi province. J Northwest Sci Tech Univ Agric For 8:67–74

7

Theory and practice of integrated coal production and gas extraction

Liang Yuan[1]

Abstract The integrated extraction of coal and gas combines coal mining with gas capture. Taking into account the gas deposition and flow conditions in the Chinese coal basins, this paper describes the status of the theory and key technologies of this integrated extraction system, and presents its application and practice in the Shaqu, Zhongxing, Fenghuangshan and Pingmei mines. Areas for further improvements in future studies are discussed, focusing in particular on the fundamentals of the extraction system to make it greener, more scientific, and more advanced in both the exploitation and utilization of coal and the gas in coal.

Keywords Integrated coal production · Gas extraction · De-stress mining · Pillarless mining · Circular overlying zone

1 Introduction

Most coal mines in China extract coal from seams with low permeability and high gas content. Gas control in such mining seams is a worldwide problem and the associated techniques are not completely optimized. Gas outburst accidents frequently occur during mining of these seams, therefore the productivity is lower, and it is very difficult to achieve safe and efficient mining. The Huainan mining area is a typical example of these conditions, having low permeability, high gas content, and complex geological conditions. Since 2004, the mining depth in Huainan coal mines has reached 700–1000 m, all 15 mines are classified as being outburst-prone, and the coal seams are characterized by low permeability (0.001 mD), high gas content (12–36 m^3/t), and high gas pressures (up to 6 MPa). Mining of these seams by conventional methods generates many problems, including coal pillars left in the panels, high abutment stress in the protected seams, the risk of outbursts, high gas emissions from the first-mined seam, gas accumulation at the upper corner of the longwall working face, a large amount of roadway excavation in rocks and at great depths, and the high cost of gas control.

Integrated coal and gas production combines coal mining with gas capture. This is a typical model for the development of dual resources (coal and gas in coal mine). Mining activity leads to desorption and flow of gas in the coal and creates favorable conditions for gas drainage. Furthermore, coal mining and its associated ventilation systems provide operating space for gas extraction. Efficient drainage of gas reduces the gas concentration and amount at the working face, which can effectively prevent gas explosions and eliminate coal and gas outbursts, thereby promoting safer and more efficient coal mining (Fields 1973; Garcia and Cervik 1985; Hargraves and Martin 1993; Creedy et al. 2001; Cheng et al. 2004, 2006). In recent years, integrated coal production and gas extraction has been developed and promoted rapidly in China, Australia, and other coal-mining countries, and has been demonstrated to be an effective method for safe and efficient mining of coal seams with high gas content and low permeability (Xie et al. 2006; Yuan 2008a).

✉ Liang Yuan
yuanl-1960@sina.com

[1] State Key Laboratory of Deep Coal Mining & Environment Protection, Huainan Mining Group Co., Ltd., Huainan 232001, China

This paper summarizes the fundamental theory and key technologies of integrated coal production and gas extraction in an effort to promote its wider application. Applications to four coal mines in China are also presented to showcase its applicability and effectiveness.

2 Research status

Integrated coal production and gas extraction must be considered as a system which involves the exploration and measurement of stress field redistribution, deformation and failure of coal and rock, gas pressure and permeability, and gas desorption and migration during the process of coal mining.

2.1 Fundamental theory

2.1.1 Deformation and failure of overlying roof strata and overburden pressure

Theories of a natural equilibrium arch, pressure arch, cantilever beam, prefabricated fissures, articulated rock, elastic foundation beam, and many other well-known hypotheses have been proposed. In China, the masonry-beam structure model, passing-stress rock beam structure model, and key-layer theory have been proposed (Qian and Liu 1984; Song 1988; Qian and Shi 2003).

2.1.2 Crack distribution in the overlying strata

To optimize the design of the bore for drainage of pressure-relieved gas from a mining area, detailed fracture distribution behaviors generated by mining have been developed by Chinese scholars. For example, the O-ring fracture (Qian and Xu 1998), the elliptical parabolic band (Li et al. 1999), and the annular ring fracture distributions (Yuan 2008b) have been investigated and proven.

2.1.3 Gas relief caused by mining flow patterns

Based on empirical data, two models of relieved gas emissions have been established in Germany (Noack 1998), as shown in Fig. 1. One model the relationship between the extent of gas emission and different roof and floor heights, while the other relates gas pressure to different heights of the roof and floor. Both models have been widely applied to forecast gas emission volumes from adjacent layers and evaluate gas drainage technologies. In addition, empirical models of gas emissions from adjacent layers in the typical high-gas mining areas of Yangquan and Huainan coal mines have been established (Cheng et al. 2004, 2006).

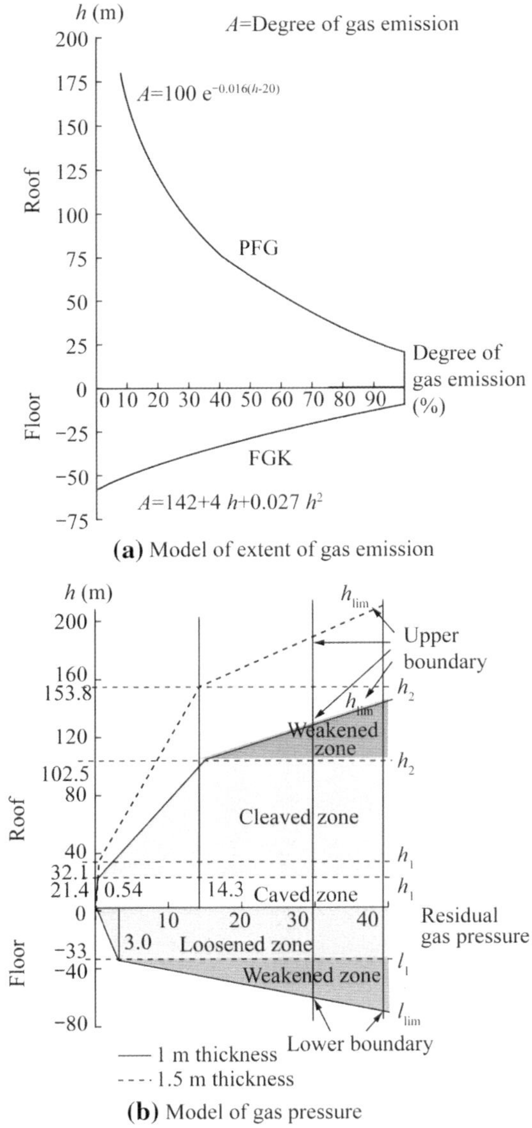

Fig. 1 Model of mining gas desorption emissions (Noack 1998). *PFG* curve of degree of roof gas emissions; *FGK* curve of degree of bottom gas emissions

2.1.4 Variation of permeability and flow of relieved gas

The relationship between stress distribution and permeability in the front of the working face and the surrounding goaf rock was analyzed. The permeability coefficient reduces as a function of the abutment pressure of the working face, while the permeability sharply increases in the gob with the transfer of mining-induced stress and then reduces again as the mining stress is removed. However, the permeability in the surrounding rock after compaction is still much larger than the original permeability. Variation of rock permeability was also studied by Palchik (2005), Unver and Yasitli (2005), and Usanov et al. (2014). Permeability distribution of the overlying relief gob seam along the advancing direction of the working face was analyzed by dividing it into three stages: initially increasing permeability and flow by pressure relief, followed by high permeability and flow by continued relief, and then reducing permeability and flow by pressure recovery (Yu 1984).

2.2 Technical practice

Gas extraction can be carried out before, during, and after mining. In countries such as the US or Australia, coal seams have higher permeability and occur at shallower depths (Fields 1973; Hargraves and Martin 1993; Steven et al. 2007). The main methods of gas extraction include pre-extraction by vertical wells (Garcia and Cervik 1985) and long ground directional drilling or bedding drilling. In recent years, coal bed methane development has gradually achieved a certain economy of scale. Gas extraction during and after mining is generally applied to low permeability coal seams by methods such as drilling through the coal seam, horizontal roof drilling, gob area wells, and special gas extraction tunnels (Noack et al. 1995; USEPA 1998, 1999). Drilling through coal seams and horizontal roof drilling are widely used in China and other countries, and gob area wells are widely used in countries with shallower coal depths, such as the US and Australia (Creedy et al. 2001; Michael 2007). In recent years, long ground directional drilling has become a new trend, and has been successfully applied in the US and Australia (Fig. 2). Coal mine methane extraction by directional drilling from the ground into the coal seam has been applied in Australia, and advanced navigation systems for directional drilling have been developed. In China, techniques for gas extraction have been developed for conditions when pressure relief has been extended at a coal mine, such as in the Huainan, Yangquan, and Huaibei groups. Simultaneous extraction coal and gas by roadway retention and borehole drilling has been successfully implemented at Huainan coal mines.

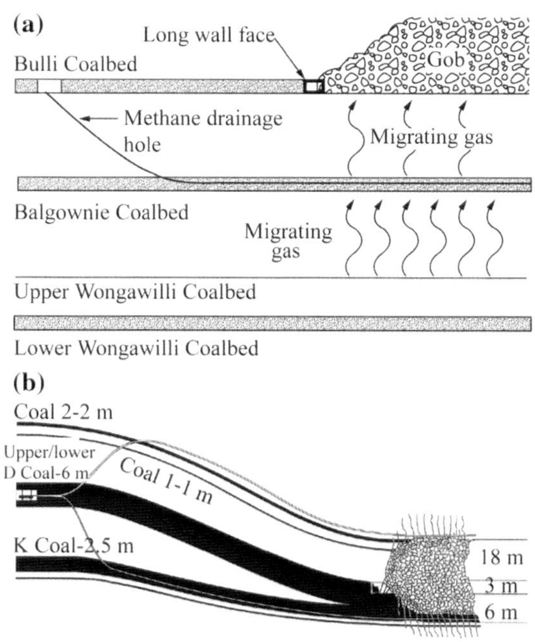

Fig. 2 Gas extraction by long ground directional drilling in the US and Australia

3 Technique of integrated coal production and gas extraction

3.1 Principles

The principle of integrated coal production and gas extraction follows a sequence of phenomena. Following mining-induced failure and the consequent movement of surrounding rocks, the gas desorbs from the broken coal seams and becomes free-flowing, and is thereby able to move into fractures and accumulate in the coal and rock strata. This accumulated gas is extracted by gas drainage boreholes and/or tunnels concurrently with the mining of the coal seams. Figure 3 illustrates this principle.

By integrating coal production with gas extraction, coal seam gas becomes a useful resource. Advantage is taken of the effect of strata movement on gas pressure-relief and flow during the coal mining process to establish a single production system for simultaneous coal and gas extraction. By including gas drainage as part of the regular coal mining process, the traditional single coal resource mining pattern is transformed into an integrated production facility for both coal and gas.

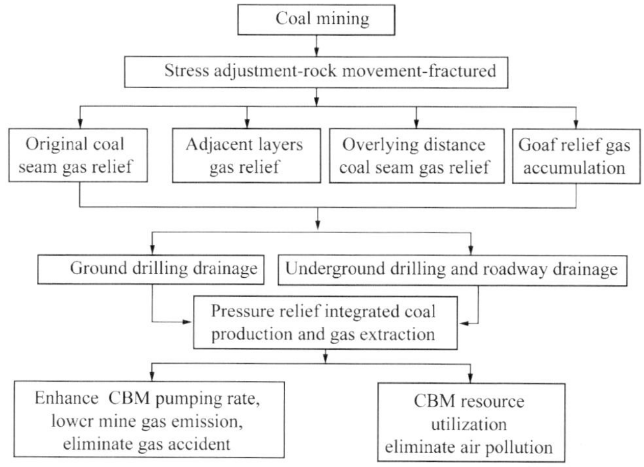

Fig. 3 Principle of integrated coal production and gas extraction

3.2 Theoretical foundations

3.2.1 The annular-shaped overlying zone theory

The "annular-shaped overlying zone" theory has been proposed. The stress-relief coefficient of the overlying strata is obtained from numerical simulation results (Yuan et al. 2011a). The overlying zone can be considered an "annular-shaped fractured zone", which is located at a certain height above the goaf and has a particular width and extends upward at a particular angle. The zone has the following characteristics: (i) it can be divided into three sections along the strike, i.e., front, center, and rear. The front section is adjacent to the working face and runs through the working face in the direction of dip, and is the main region where gas desorption and emission occur. The central section connects the front and rear sections along two mining boundaries, and forms channels between these sections to allow large-scale gas flow. The rear section is near the cut and is smaller than the front section in size; (ii) the overlying annular-shaped fracture zone extends upward at a certain angle in the dip section, and the extension angles of the outward and inward boundaries are different; (iii) the overlying annular-shaped fracture zone is located in the strike section above the goaf at particular intervals and extends upward at a particular angle. The front range is larger than the rear range.

3.2.2 Three-band structure in stress-relief layer

The distribution and evolution of three bands in the stress-relief layers due to mining of the first seam have been clearly identified. These provide for the occurrence of desorption and formation of flow channels of gas in gas-enriched regions. In the process of mining a working face without a pillar, the front abutment stress region is located 0–30 m ahead of the working face and the concentration factor is about 2–3. The stable stress-relief region is located 300–500 m ahead of the working face in the advancing direction, the fracture development area is 0–50 m behind the working face of the first-mined seam, the active strata movement region is some 50–500 m behind the working face, and the attenuation region is more than 500 m behind the working face developed toward the goaf in the form of a wedge (Yuan 2008a).

3.2.3 Gas drainage by stress relief

Technical principles for enhancing gas drainage by stress relief have been proposed. Mining-induced stress, gas pressure and permeability, strata movement, and fracture evolution behavior have been investigated. A linear relationship between gas pressure and stress and a negative exponential relationship between permeability and stress have been demonstrated. Additionally, mining of the first seam can induce expansionary deformation, stress relief, and permeability enhancement of neighboring coal seams (Yuan 2008b).

3.3 Key Technologies

3.3.1 Excavation of a special roadway for gas drainage

A special roadway should be excavated in the de-stressed and fractured region formed in the first or protective seam for gas drainage. This technology has been developed to target the gas-rich zones in this region. This gas-rich region, which has a width of 0–30 m and height of 8–25 m, is located in the annular-shaped fractured zone above the two roadways of the working face (Yuan 2008c).

This technology has been further developed to extract desorbed gas from overlying de-stressed coal seams that are spaced some distance from the first-mined seam. Once the first (protective) seam is mined, the range of the de-stressed region develops upward at a strike fracture angle of 80°–85° and the de-stressed layer is located 10–150 m above the protective seam (Yuan 2009a).

This technology has also been applied to extract desorbed gas from the underlying de-stressed protected coal seams. After the first seam is mined, the de-stressed range develops downward at strike and dip fracture angles of 99°–100° and 102°–110°, respectively, and the de-stressed layer is located 15–100 m below the protective seam (Yuan 2009b).

Ground well technology for gas drainage from de-stressed regions has been developed. This includes well construction and operating procedures. Technologies for controlling deformation of the surrounding roadway rocks in soft coal seams have also been developed. These include bolt-grouting support of soft rock roadways with high stress and controlled pre-stressing of fractured roofs with obviously inclined separation.

3.3.2 Roadway controlling technologies in pillarless mining

Technologies for roadways include support with strengthened bolting, meshing, and grout injection, a self-moving shield for reinforced support in mining-perturbed roadways, and building of packing walls for retaining gob-side roadways. Other key technologies include the development of materials capable of withstanding high loading for the packing walls, a modular shield for constructing the packing walls, and pre-fracturing of the working face roof for de-stressing. The initial and periodic caving intervals can be reduced by weakening the roof strength using advanced deep-hole pre-splitting blasting. An initiative for support roofing with high stiffness has been proposed, which emphasizes the "three high" bolting system, i.e., high strength, high pretension, and high stiffness of bolts (Yuan et al. 2011a, b).

3.3.3 Gas drainage by boreholes drilled in the retained roadway

A new concept for safe and efficient gas extraction has been proposed based on pillarless mining with a Y-shaped ventilation system. Boreholes in the retained roadways are drilled into the de-stressed and gas-rich zones in the overlying and underlying regions to extract gas (Yuan 2010).

4 Main achievements

4.1 Overlying annular fractured zone

Three-dimensional computational fluid dynamics modeling of a typical mining panel has revealed the existence of an overlying annular-shaped fractured zone in which gas-rich sections are located. The basic geometric features for highly efficient gas drainage are fully understood. A zone with a certain width locates at a particular height above the goaf and extends along the vertical direction at a particular angle, as shown in Fig. 4.

The basic criteria for identifying this zone in multiple coal seams with low permeability are as follows:

(1) High gas desorption rate: Based on specific coal seam conditions, the effective stress reduction factors can be determined when gas desorbs significantly, and the zones in which high gas desorption occurs are identified according to the overlying strata stress-relief model.

(2) High permeability: High permeability is used to identify zones where mining-induced fractures are highly developed and compaction starts to occur. Regions with high permeability are determined using the overlying permeability distribution model. The upper boundary of the range for effective gas drainage does not exceed the upper boundary of the horizontal high permeability zone.

(3) High concentration of gas extracted: The minimum purity requirement of the gas extracted is determined and the range of gas purities is determined by applying the basic laws of gas dynamic flow.

Based on the above, the regions of highly efficient gas extraction from multiple coal seams can be determined.

4.2 De-stressed extraction of coal and gas

Gas drainage is implemented by placing roadways and/or boreholes in the de-stressed and/or fractured zones formed by mining of the first (protective) seam in multiple coal seams. A sketch of the gas drainage system is shown in Fig. 5.

4.3 Pillarless extraction of coal and gas

With the use of a Y-type ventilation system, the coal in front of the working face in the first seam is mined, while gas from both the protective and protected seams is extracted through boreholes drilled in the retained roadways. In this way, integrated coal production and gas extraction is achieved. The extracted gas is separated into high- and low-purity components, and transported to the surface for use in electric power plants and domestic cooking and heating, respectively.

Depending on the location of the gas-rich zones, boreholes in the retained roadway can be drilled up- and/or

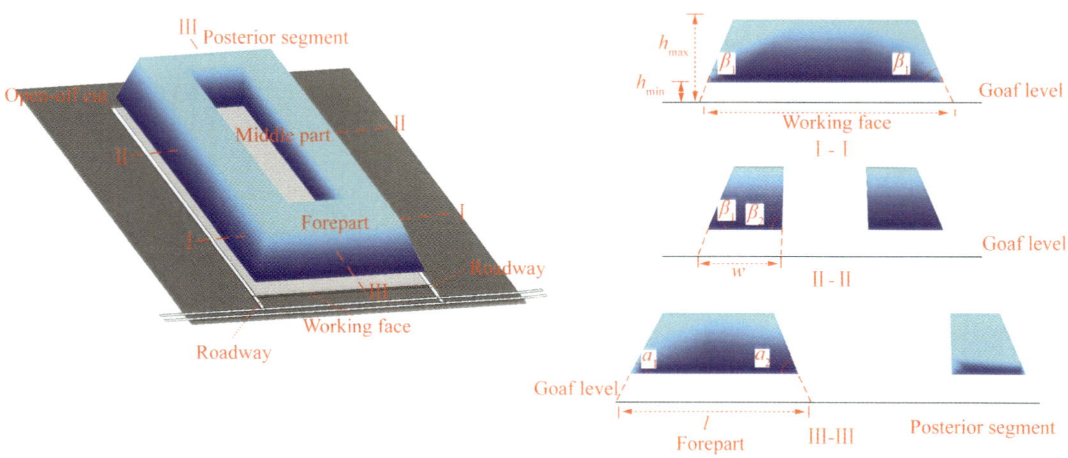

Fig. 4 Annular-shaped overlying zone

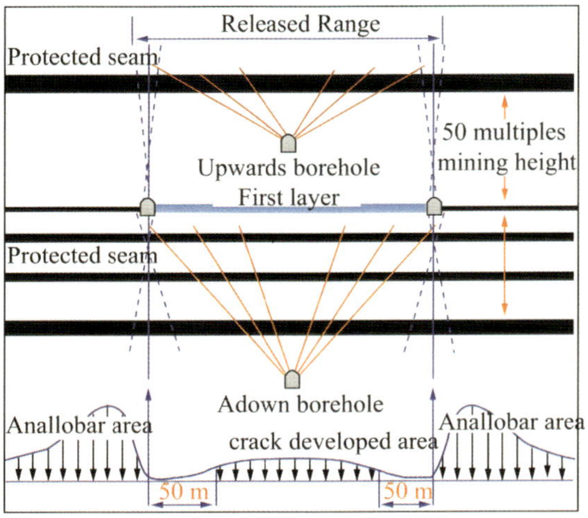

Fig. 5 De-stressed gas drainage system

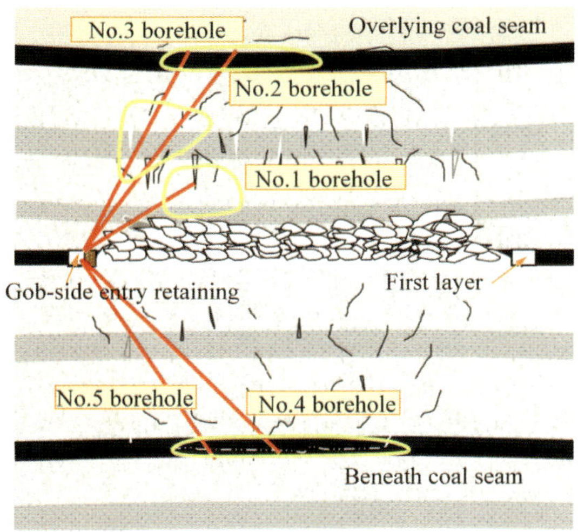

Fig. 6 Pillarless extraction of coal and gas

downward to target the gas-rich zones in both overlying and underlying strata. The pillarless extraction of coal and gas is schematically shown in Fig. 6.

4.4 Innovative research methods

A large-scale physical model has been developed for fundamental research into integrated coal production and gas extraction, as shown in Fig. 7. The main components of the model include an outburst simulation instrument, an excavation system, a data monitoring and recording system, a dynamic data-analyzing system, and a virtual early-warning system. The geometric dimensions of the model are 3000 mm × 3500 mm × 2000 mm (length × width × height) and it can be easily transported. Large-scale geomechanical tests can be implemented with advanced monitoring systems to study integrated coal and gas production in seams with low permeability. Development

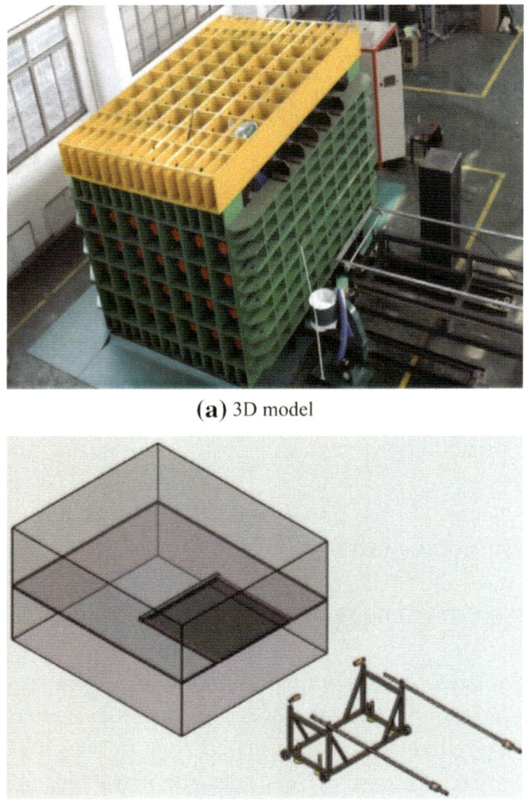

(a) 3D model

(b) Excavation system

Fig. 7 Large-scale physical model for simulation of coal and gas extraction

and evolution of the stress and fracture zones in a retained roadway and in overlying or underlying seams with pillarless mining can be investigated under various stress, geological, and excavation conditions.

5 Operating practice

The core ideas relating to the integrated extraction of coal and gas are consistent with scientific, efficient, and practical engineering principles. Coal resources in China differ vastly in the nature of their deposition; the theoretical and technological innovations with respect to integrated extraction of coal and gas take different deposition conditions and different mining methods into account.

Integrated extraction of coal and gas originated in the Huainan coal mining group. Huainan was once one of the areas with the most serious gas outburst and explosion hazards in the Chinese coal industry. Because of the high gas content and low permeability of coal seams in Huainan, innovative methods were developed to enhance seam permeability to resolve these issues. Coal permeability is improved by mining-induced de-stressing, thereby enabling the integrated extraction of coal and gas. These technological developments have completely prevented the reoccurrence of gas explosion accidents and greatly improved coal productivity and gas extraction efficiency. All coal mines in Huainan now simultaneously extract coal and gas. Annual gas extraction from the mines is about 500 Mm^3.

The successful demonstration of this technology in Huainan played an important role in resolving gas control in other gas-containing Chinese coal mines. It has been applied in Shanxi, Henan, Shanxi, Heilongjiang, Hebei, Xinjiang, and other major coal-producing provinces. Coal production from mines using this technology is some 21 billion tons.

As a first example, the geological conditions of the Shaqu coal mine in Shanxi province are very complex and it is one of the highest gas-emitting mines in China. The gas content is 29 m^3/t, the pressure is up to 2.4 MPa, and the absolute gas gushing quantity is 473 m^3/min. In 2008, the gas concentration exceeded the statutory limit in 3635 incidents, in each incident exceeding a concentration of 2.5 % (Yuan 2008c). In the 6 years prior to 2012, coal production was about 200 Mt (design capacity was 300 Mt), and production capacity was seriously restricted by this gas problem. Integrated extraction of coal and gas was implemented in 2011, with comprehensive gas-control measures, such as protective mining, progressive drainage, Y-type ventilation, expansion of the gas extraction system, underground gas drainage from retained roadways, surface well drainage, and gas use. By the end of 2012, the number of gas incidents had dropped to five, and none occurred in 2013.

The second example is the successful implementation of integrated coal production and gas extraction with pillarless mining in the 1205 working face of the Zhongxing coal mine, Fenxi Coal Mining Group. This mine comprises typical "three soft" multiple coal seams with close intervals and composite fragmented thick roof conditions. Field observations indicate that the retained roadway (cross-sectional area of 12 m^2) is well maintained and its cross-sectional shrinkage is only 11.11 %. The average gas extraction from the working face is 43.17 % and the average gas concentration at the tail entry is 0.33 %, indicating of the effective gas extraction of the mine.

A third example is the successful application of integrated coal production and gas extraction at the working face 154,307 (with thick and hard roof conditions) of the Fenghuangshan coal mine, Jincheng Coal Group. The overlying seam of the mine was outburst-prone and the accumulated gas needed to be drained. Gas drainage is carried out through boreholes in the retained roadway of the working face, arranged in the overlying seam and supplemented by surface drainage wells. Field observations show that the number of gas incidents dropped by 88 % and the risk of outbursts in the overlying seam was eliminated.

As a final example, the mining depth of the working face in the Henan Pingdingshan Coal Group was 1100 m and the retained roadway was under conditions of high stress with a thick, hard roof. The roadway floor seriously heaved, and the abutment stress resulting from sudden breaking of a large area of overhanging roof in the working face was very high. Various measures were taken to resolve these issues. Roof stress was relieved by pre-splitting explosion with roof long boreholes and the "three high and one strong" support technique was used to control roadway stability. In addition, a Y-shaped ventilation system, gas drainage boreholes arranged in the de-stressed seams, development of a new high-strength material for the packing wall, and the use of a mechanized filling technique to build the packing wall were also implemented.

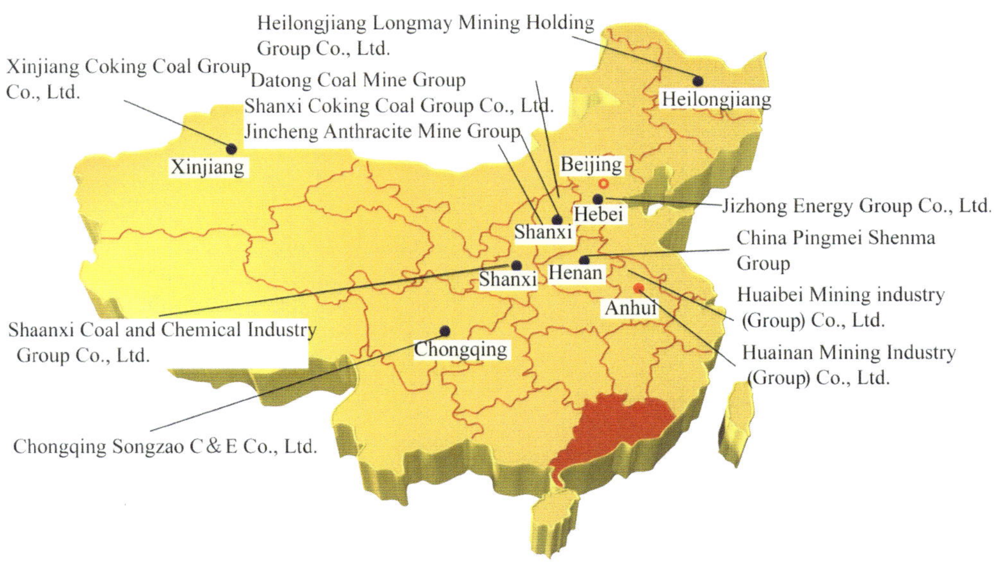

Fig. 8 Chinese coal mines practicing integrated extraction of coal and gas

These achievements provide robust technical support for the extraction of deep coal and gas resources. Figure 8 shows the Chinese coal mines in which integrated extraction of coal and gas is applied.

6 Conclusions

China has rich coal resources. With increasing energy consumption and a greater demand for a cleaner environment, integrated coal production and gas extraction will play an increasingly important role. As demonstrated over the last 10 years in Huainan and other mines with similar conditions, integrated production of coal and gas is probably the only way to make coal mining greener and more scientific. The main theoretical and technological achievements for this technology are summarized as follows:

(1) Dynamic evolution and distribution of mining-induced stress, fracture, and gas flow fields in mining multiple seams of high gas content and low permeability;
(2) Identification of a circular overlying zone for efficient gas extraction;
(3) Excavation of overlying and underlying gas drainage roadways in de-stressed fractured zones induced by mining the first (protective) seam;
(4) Pillar mining with a retained roadway and Y-type ventilation;
(5) Gas drainage via boreholes drilled in the retained roadway into gas-rich zones in overlying and underlying strata.

Implementation of integrated coal and gas production can not only secure energy supply for sustainable economic development in China, but also enhances the scientific and technological level of China's coal mining industry. Further theoretical and technological research is required to make coal mining greener, more scientific, and more efficient.

References

Cheng YP, Yu QX, Yuan L (2004) Experimental research on coal and gas safely and efficiently remote relief exploitation. J China Univ Mining Technol 33(2):132–136
Cheng YP, Zhou DY, Yu QX (2006) Research on protection layer relief gas drainage and emission laws. J Mining Saf Eng 23(1):12–18
Creedy DP, Garner K, Holloway S et al (2001) A review of the worldwide status of coalbed methane extraction and utilisation. Report No. COAL R210, DTI/Pub URN 01/1040
Fields HH (1973) Degasfication through vertical shafts. In: Methane Control in Eastern U.S. oal ines. United States Department of Interior & Bureau of Mines Information Circular IC 8621, 25–38
Garcia F, Cervik J (1985) Methane control on longwalls with corss-measure boreholes. Report of Investigations 8985, United States Department of the Interior
Hargraves AJ, Martin CH (1993) Australasian coal mining practice. The Australasian Institute of Mining and Metallurgy
Li SG, Shi PW, Qian MG (1999) Research on overburden elliptical parabolic crack with dynamic distribution. Ground Press Strata Control 3–4:44–46
Michael B (2007) High performance longwall mining in deep and methane rich coal deposits in Germany taking into account methods and technologies for controlling the estimated gas liberation. In: Proceeding of 2007 China (Huainan) international symposium on coal gas control technology, Huainan, China, 57–75
Noack K (1998) Control of gas emissions in underground coal mines. Int J Coal Geol 35:57–82
Noack K, Koppe U, Kunz E (1995) Control of high gas emissions in underground coal mines. In: International symposium workshop on Management and Control of High Gas Emissions and Outbursts in Underground Coal Mines. 20–24 March, Wollongong, Australia, 16–24
Palchik V (2005) Localization of mining-induced horizontal fractures along rock layer interfaces in overburden: field measurements and prediction. Environ Geol 48:68–80
Qian MG, Liu XC (1984) Mine pressure and control. China Coal Industry Publishing House, Beijing
Qian MG, Shi PW (2003) Mining pressure and strata control. China University of Mining & Technology Press, Xuzhou
Qian MG, Xu JL (1998) Research on overburden fracture distribution "O" ring features. J China Coal Soc 23(5):466–469
Song ZQ (1988) Practical mine pressure. China University of Mining & Technology Press, Xuzhou
Steven S, Bruce A, Crosdale P et al (2007) Coal petrology and coal seam gas contents of the Walloon Subgroup-Surat Basin, Queensland, Australia. Int J Coal Geol 70:209–222
Unver B, Yasitli EE (2005) Modelling of strata movement with a special reference to caving mechanism in thick seam coal mining. Int J Coal Geol 66(4):227–252
Usanov SV, Mel'nik VV, Zamyatin AL (2014) Monitoring rock mass transformation under induced movements. J Min Sci 49(6):913–918
USEPA (1998) Coalbed methane recovery and electric power generation project-opportunity at the Wesola Mine Myslowice, Poland
USEPA (1999) White paper: guidebook on coalbed methane drainage for underground coal mines
Xie HP, Peng S, He M (2006) Deep mining basic theory and engineering practice. Science Press, Beijing
Yu QX (1984) Gas prevention in coalmine. China University of Mining & Technology Press, Xuzhou
Yuan L (2008a) Key technique of safe mining in low permeability and methane-rich seam group. Chin J Rock Mech Eng 27(7):1370–1379
Yuan L (2008b) Theory and practice of integrated pillarless coal production and methane extraction in multi-seams of low permeability. China Coal Industry Publishing House, Beijing
Yuan L (2008c) Gas distribution of the mined-out side and extraction technology of first mined key seam relief-mining in gassy multiseams of low permeability. J China Coal Soc 33(12):1362–1367

Yuan L (2009a) Theory of pressure-relieved gas extraction and technique system of integrated coal production and gas extraction. J China Coal Soc 34(1):1–8

Yuan L (2009b) Theory and practice of integrated pillarless coal production and methane extraction in multiseams of low permeability. Eng Sci 11(5):72–80

Yuan L (2010) Concept of gas control and simultaneous extraction of coal and gas. China Coal 6:5–12

Yuan L, Guo H, Shen BT, Qu QB, Xue JH (2011a) Circular overlying zone at longwall panel for efficient methane capture of mutiple coal seams with low permeability. J China Coal Soc 36(3):357–365

Yuan L, Xue JH, Liu QS, Liu B (2011b) Surrounding rock stability control theory and support technique in deep rock roadway for coal mine. J China Coal Soc 36(4):535–543

8

Characteristics and applications of gas desorption with excavation disturbances in coal mining

Jiachen Wang[1,2] · Renlun Wu[1,2] · Peng Zhang[1,2]

Abstract According to the deficiency of experiment system for gas adsorption and desorption in coal mass, a large scale experiment system is developed independently by researchers. This experiment system is composed of primary and auxiliary boxes, power transmission system, mining system, loading system, gas charging system, data monitoring and intelligent acquisition system. The maximum experiment coal consumption is 1200 kg, the mining system is developed to conduct experiment for gas desorption under excavating disturbance, and the plane-charging cribriform ventilation device is developed to realize uniform ventilation for experiment coal sample, which is accord with the actual gas source situation of coal bed. The desorption characteristics of gas in coal are experimentally studied under the conditions of nature and mining using the experiment system. The results show that, compare with nature condition, the permeability of coal and the velocity of gas desorption could significantly increase under the influence of coal pressure relief and destruction caused by mining, and the degree of gas desorption could somewhat increase too. Finally, pressure relief gas extraction of current seam and adjacent seams after mining in a certain coal mine of Yangquan mining area are introduced, and the gas desorption experiment results is verified by analyzing the effect of gas extraction.

Keywords Gas adsorption and desorption experimental system · Mining · Gas desorption · Simultaneous extraction of coal and gas

1 Introduction

Coal seam gas content increases with greater depth and mining intensity in underground coal mines, and is a primary factor in mining safety and efficiency. Coal is a complex porous medium that consists of primary pores and many fissures that resulted from tectonic movement. It therefore has large free space and pore surfaces. Coal seam gas exists in the adsorbed and free state. Only free gas can flow to the working face or be extracted. Therefore, an increase in coal seam gas desorption, an improvement in free gas content and the efficient extraction of gas is key to controlling gas in Chinese mines that contain large gas volumes. The low gas pressure, poor coal seam permeability and high adsorption capacity (adsorption gas accounts for about 80 %–90 % of the total gas in coal seams) are general characteristics in China's coal seam (Ye et al. 1999; Miao et al. 2004). Barrer believes that gas adsorption and desorption is a reversible process under certain conditions (Barrer 1951). Coal pore structure, gas pressure, temperature, coal metamorphic degree, moisture, effective stress, electromagnetic field, acoustic field, physical field, mining activities and other factors are related closely to coal gas desorption characteristics. Therefore, research on coal gas desorption characteristics as a function of different factors has important significance to improve preventive

✉ Jiachen Wang
wangjiachen@vip.sina.com

1 College of Resources and Safety Engineering, China University of Mining & Technology (Beijing), Beijing 100083, China

2 State Key Laboratory of Coal Resources and Safe Mining, China University of Mining & Technology (Beijing), Beijing 100083, China

measures for gas accidents and on the effect on gas extraction.

In recent years, many researchers have developed a series of relevant experimental systems and undertaken some valuable experiments on the influence of pore pressure (Lü et al. 2010), temperature (He et al. 2010; Li 2011; Bai et al. 2014), coal metamorphic degree (Wang 2001), moisture (Wang et al. 2010; Chen et al. 2013), effective stress (Tang et al. 2006), electromagnetic field (He and Zhang 2000), acoustic field (Jiang et al. 2008, 2010), vibration (Li et al. 2010) and other factors affecting the characteristics of coal gas desorption. However, improving the degree of gas desorption and the effect of gas extraction by changing the above factors artificially is of great difficulty and high cost. Practices in coal mines show that coal seam permeability is improved in mining, which creates conditions for gas desorption, gas migration and gas extraction (Wang and Fan 2008; Wu 2013).

Almost all existing gas desorption experimental systems feed gas into static coal samples and simulate gas desorption, adsorption and migration without key influential mining conditions. Relatively large differences therefore result in practical production (Zhang 2012). In this paper, we independently developed experimental systems for gas adsorption and desorption simulations in coal samples at a large scale to research coal gas adsorption and desorption characteristics during mining. The experimental system optimizes coal gas adsorption and desorption experimental methods, and the research results provide guidance for coal gas extraction.

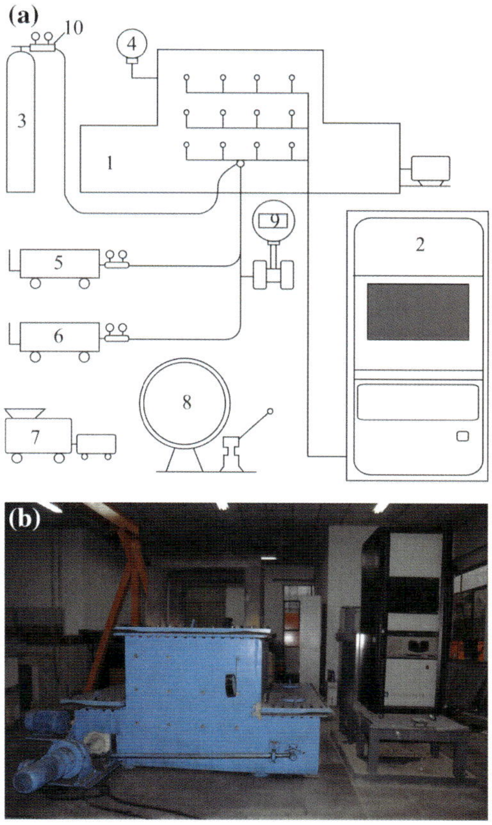

Fig. 1 Experimental system for gas adsorption and desorption simulation of large-scale coal sample. **a** Diagram of experiment system. *1* Experimental platform for gas adsorption and desorption simulation in coal sample with large scale; *2* 32-channel data acquisition system; *3* 40 L high pressure methane cylinder; *4* Hydrogen precision pressure gauge: *5* Air pump; *6* Vacuum pump; *7* Crusher; *8* Press machine; *9* Vortex flowmeter; *10* Hydrogen decrement gauge. **b** Real picture of experiment system

2 Experimental device and working procedures

2.1 Development of experimental device

As shown in Fig. 1, the experimental system for gas adsorption and desorption simulations in large-scale coal samples has been developed independently by China University of Mining and Technology (Beijing). The experimental system consists of four parts, which includes an experimental platform, 32-channel data acquisition system, high pressure methane cylinder and auxiliary system (Zhang 2012). The experimental platform consists of primary and auxiliary boxes, a power transmission system and a mining system, which are shown in Fig. 2.

(1) Primary and auxiliary boxes. As shown in Fig. 2a, the primary box is used to contain the experimental coal sample. The walls of the box body are 40-mm-thick steel plate welded into a tight complete structure, with a cover plate thickness of 60 mm, and the design pressure is 6 MPa. The primary box's length, width and height are 1600, 1200 and 1200 mm, respectively. The maximum experimental coal sample mass is about 1200 kg. The front and rear surfaces of the main box body contain 24 holes with different height distributions in three layers; 23 of these have a 14 mm diameter for conveying sensor cables and the remaining holes with 12 mm diameters are standard screw drills for gas injection. A special high-pressure hose is connected to the external side, and the hole's internal side is connected with a cribriform ventilation device at the bottom of the primary box (Fig. 2b). The cribriform ventilation device is designed as a cribriform pectinate strip. The height of the pectinate strip is 20 mm, and the upper surface of the strip contains well-distributed ventilation holes. In this way, plane-charging can be fulfilled to simulate natural gas sources. A thread-cutting coal drill pipe exists inside the box body, which is used to simulate the mining work face. Two auxiliary boxes are located on both sides of the

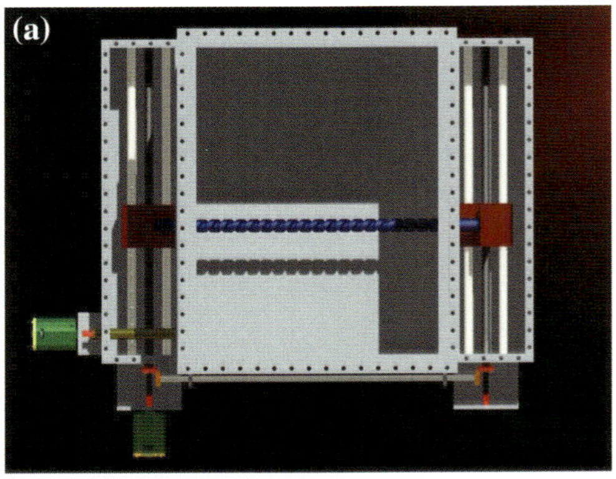

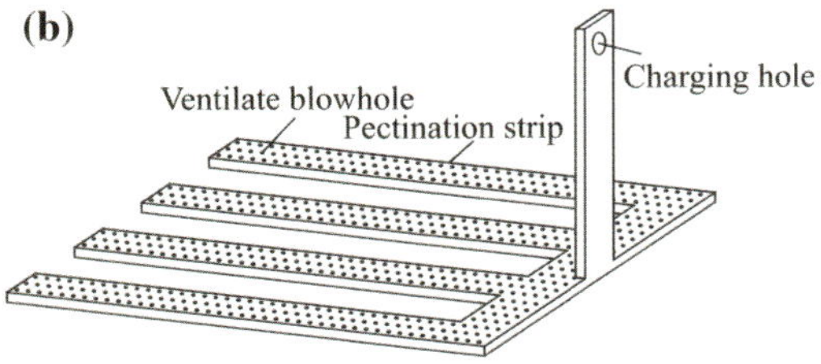

Fig. 2 Primary and auxiliary boxes of experimental platform. **a** Vertical view of experiment platform. **b** Cribriform ventillation device

Fig. 3 Real picture of mining system. **a** Rotational device of coal cutting shaft. **b** Stepping device of coal cutting shaft

primary box, where the coal cutting power transmission shaft is located. The side boxes are also used as a container for drilled coal. Two strip holes lie on both sides of the overlap of the primary and the auxiliary boxes, which provides a moving pathway for the coal drill pipe. To observe the coal drill pipe movement, a window is set at the sealing cover of one of the auxiliary boxes.

(2) The power transmission system platform is composed of two perpendicular electric motors. Each motor is used to control the rotation and parallel motion of the coal-cutting pipe.

(3) Mining system. The smaller gear shaft fitted in the auxiliary box is driven by the motor parallel to the coal-cutting shaft. The larger gear shaft welded with a coal-cutting shaft is driven by the smaller gear shaft via a chain (Fig. 3a). The coal-cutting shaft can rotate to simulate mining, and the unloaded drill pipe design speed is 300 r/min. Thread pipes on two sides of the primary box are driven by a motor arranged perpendicularly to the coal-cutting shaft, and two sliders fixed with a coal drill pipe are driven by a thread pipe. Thus the coal drill pipe can move forwards and backwards (Fig. 3b). To ensure that the two sliders are located at the same horizontal position, five slick fixing rods are fitted on the same lever with thread pipe in two auxiliary boxes. The coal thrill pipe can move on an absolute horizontal with the help of seven rods in total.

The 32-channel data acquisition system is composed of four 8-channel data acquisition boards, an 8-port hub, a 7.5 V direct current switching power supply, three kinds of sensors and a computer processing system. The sensors are used to collect stress data at different locations under simulated mining conditions, gas concentrations and temperatures. The gas concentration is evaluated by gas pressure. The auxiliary system includes a crusher, portal jib crane, air pump, vacuum pump, vacuum gauge, hydrogen precision pressure gauge, vortex flowmeter and pressure machine.

Characteristics of the experimental system are as follows:

(1) Large-scale experimental coal corresponds with the geometric similarity theories (Yang 2005).

(2) The cribriform ventilation device is developed to feed well-distributed gas into the coal sample, which agrees with the in situ gas source in coal seams.

(3) The mining system is developed to simulate the excavation process to reveal gas adsorption and desorption performance under mining conditions.

(4) The loading system was also developed. Four hydraulic pressure load holes exist, which provide loads to the experimental coal sample by grade or step.

(5) Intelligence and centralization of the data monitoring and information collection. All data acquisition work over a series of experimental periods is undertaken by a computer able to conduct preliminary statistical

analysis. The experimental system can be used to study the performance of gas movement under different gas adsorption equilibrium pressure, moisture content and mining conditions.

2.2 Experimental procedure

(1) Preparing coal sample. Raw coal is crushed to a diameter of 0.2–0.5 mm before being mixed with water to a certain proportion. During the coal preparation, moisture determination of the crushed raw coal was achieved by determining the total moisture in the coal (GB/T 211—2007, 2007). The crushed coal was well mixed with a certain amount of water, which was calculated from the difference between the required moisture content in the coal and the moisture determination. Further moisture determination was conducted to ascertain whether the required coal moisture had been achieved. It is necessary to wet or dry coal during this period until the coal sample moisture meets the experimental requirements.

(2) Coal charging and seal. Certain 70-mesh stainless steel wire meshes were placed under the cribriform ventilation device at the bottom of the experimental platform primary box. Coal granules cannot cross the steel wire meshes, so the cribriform ventilation device under the steel wire meshes would not be blocked by coal granules. With the steel wire meshes in place, a layer of stones (10 cm thick; 20, 10 and 5 mm diameter) was laid to enable even gas to charge. After the stone paving had been laid, several 70-mesh steel wire meshes were placed on the stone layer. The stable stone pad forms the structure of the gas-charging base. The prepared coal is placed into a primary box and tamped. Pressure, stress and temperature sensors were placed in the coal sample at different layers and locations. The top shell was set in place and sealed. Leak detection was conducted to ensure that the experimental box was airtight.

(3) Gas charging. When leak detection had been completed, the remaining air in the experimental box was extracted by vacuum pump. Gas was then charged into the experimental box through the pressure reducing valve. The valve was not closed until the pressure sensor indicated a specified adsorption equilibrium pressure.

(4) Cyclic gas charging. The valve was closed, the gas pressure in the experimental box was reduced to a stable value and the gas adsorption reached equilibrium because of gas adsorption. Gas was charged repeatedly until the gas pressure reached a specified adsorption equilibrium value.

(5) Gas desorption. A vortex flowmeter was installed and gas desorption was performed with the data acquisition system working. The mining system was enabled to simulate the mining work.

3 Performance of gas desorption experiment while mining

3.1 Experimental schemes

The scheme used to determine the gas desorption performance is shown in Table 1. Methane (99.9 %) was substituted for coal gas.

3.2 Experimental and analytical results

The isothermal adsorption curve is used to research the gas adsorption and desorption characteristics in the coal sample. Based on prior research, the isothermal adsorption curve follows the Langmuir equation (Yu 1992; Tsang 1999; Minkoff et al. 2003).

$$X = \frac{abP}{1 + bP} \tag{1}$$

where X is the gas adsorption per unit mass of coal at a certain temperature and equilibrium pressure P (m^3/t), P is the gas adsorption equilibrium pressure (MPa), a is the adsorption constant that denotes the maximum gas adsorption quantity per unit mass of coal at a certain temperature and infinite gas pressure (m^3/t) and b is the adsorption constant (MPa^{-1}).

The gas adsorption quantity per unit mass of coal can be expressed as:

$$X_i = \frac{\Delta Q_i}{G_r} \tag{2}$$

where X_i is the gas adsorption quantity per unit mass of coal in the ith experiment (m^3/kg), ΔQ_i is the total gas

Table 1 Experimental scheme

Experiment scheme	Scheme 1	Scheme 2
Coal thickness (cm)	45	45
Dry coal sample (kg)	575	575
Moisture content (%)	5	5
Total mass (kg)	605	605
Adsorption equilibrium pressure (MPa)	0.1	0.1
Temperature (°C)	17	17
Mining condition	No mining	Mining

adsorption quantity at adsorption equilibrium and G_r is the mass of experimental coal (kg).

$$\Delta Q_i = Q_{ci} - Q_{di} \tag{3}$$

$$Q_{ci} = \left(\frac{P_{1i}}{Z_{1i}} - \frac{P_{2i}}{Z_{2i}}\right)\frac{273.2V_0}{(273.2+T)\times 0.101325} \tag{4}$$

$$Q_{di} = \frac{273.2V_d P_i}{Z_i(273.2+T)\times 0.101325} \tag{5}$$

where Q_{ci} is the standard gas volume charged into the experimental platform (m^3); Q_{di} is the quantity of remaining free gas (m^3); P_{1i} and P_{2i} are the gas cylinder pressures before and after charging, respectively (MPa); Z_{1i} and Z_{2i} are the coefficients of compressibility at P_{1i} and P_{2i} respectively; V_0 is the volume of the high-pressure gas cylinder, $n \times 0.04$, $n = 1, 2, 3$ (m^3); V_d is the total remaining volume in the experimental platform (1.6 m^3); P_i is the gas pressure at adsorption equilibrium (MPa); T is the average temperature during the experiment ($T = 17$ °C) and Z_i is the coefficient of compressibility at pressure (P_i) and temperature (T) (Yu 1992).

The gas cylinder pressure before and after charging was recorded. The quantity of adsorbed and free gas in the experimental coal was determined from Eqs. (1)–(5). The results are given in Table 2.

After the gas adsorption in the experimental coal reached its ultimate value, the vortex flowmeter was installed and the gas valve was switched on. The pressure of the experimental box was established at atmospheric pressure within a minute. The desorption experiment was performed according to the two schemes mentioned above, and data were collected by the vortex flowmeter to upload to the data processing system. The relationship between the quantity of gas desorbed and the time was obtained as shown in Figs. 4 and 5.

By analyzing Figs. 4 and 5, characteristics of this experiment are presented as follows.

(1) Under nature conditions, gas desorption increases with time in an approximately parabolic shape. The desorption velocity decreases gradually. After 2200 min, the quantity of gas adsorbed reaches a stable value of 1.0 m^3, and the desorption velocity tends to zero.

(2) Under mining conditions, the coal sample undergoes unloading and collapse, and the coal sample permeability increases significantly. Thus the velocity and quantity of desorption increases simultaneously. After 120 min, the coal sample under mining will aid desorption. The quantity of gas desorption continues increasing, but the gas desorption velocity declines. After 420 min, the quantity of gas desorption stabilizes at 1.1 m^3, and the velocity of desorption is zero.

(3) The velocity of gas desorption is significantly higher than that under natural conditions. The degree of gas desorption is also increased.

Table 2 Parameters of adsorption equilibrium

No.	Parameters	Values
1	Initial pressure of gas cylinder (P_{1i}) (MPa)	9.25
2	Terminal pressure of gas cylinder (P_{2i}) (MPa)	4.11
3	Total quantity of gas charging (Q_{ci}) (m^3)	2.698
4	Total quantity of free gas (Q_{di}) (m^3)	1.574
5	Total quantity of adsorption gas (ΔQ_i) (m^3)	1.124
6	Adsorption equilibrium pressure (P_i) (MPa)	0.1
7	Adsorption quantity (X_i) (m^3/t)	1.955

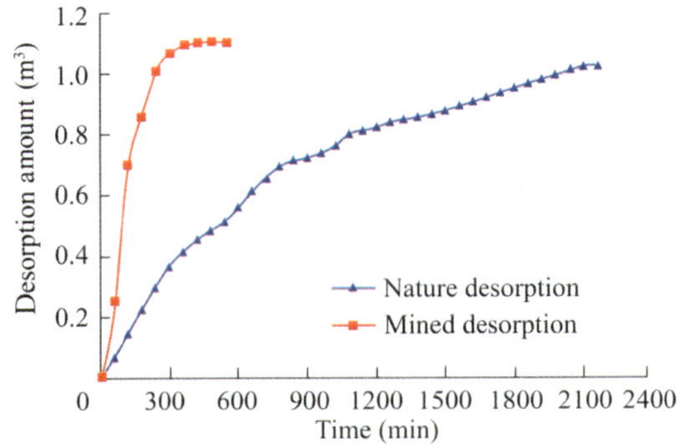

Fig. 4 Relationship between quantity of gas desorption and time under natural and mining conditions

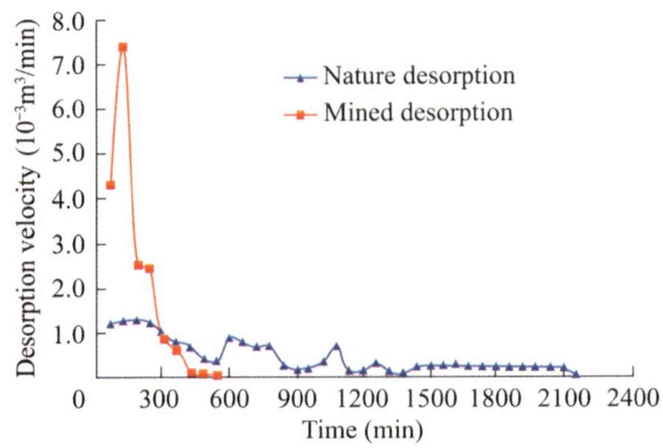

Fig. 5 Relationship between velocity of gas desorption and time under natural and mining conditions

4 Application

According to the characteristics of the desorption experiment mentioned above, the degree and velocity of gas desorption in the coal sample increase significantly. This is the most important law for improving the effect of gas extraction.

4.1 Gas extraction in current coal seam

The buried depth of fully mechanized caving face 15101 in a coal mine in the Yangquan mining area is 344–337 m. The tend strike length and the inclination length of the working face are 500 and 121 m, respectively. The thickness of the coal seam is 4.5–5.3 m, and the average thickness is 4.9 m. The dip angle of the seam is 8°–12°, and some parts of the seam are as large as 20°. The gas content of the working face is 18.48 m^3/t, the gas pressure is 1.25 MPa and the permeability coefficient of the coal seam is 0.175 $m^2/(MPa^2 \cdot d)$. The coefficient of gas flow attenuation per 100 m of coal seam bore is 0.0417/d. To extract gas in the current coal seam, extraction boreholes are located perpendicularly to the tunnel, parallel to the coal seam and 1.5 m from the coal seam floor. Gas extraction boreholes have a 60 mm diameter and 90 m length, and the interval between the boreholes is 5 m.

The current coal seam gas extraction boreholes arrangement of working face 15101 is shown in Fig. 6 (Wu et al. 2010).

The characteristic gas extraction of the No. 7 current seam extraction borehole is shown in Fig. 7, and the statistics of gas extraction of the current coal seam extraction borehole are shown in Table 3. According to Fig. 7 and Table 3, the coal mass permeability and gas desorption velocity are increased significantly because of the lead abutment pressure, and the gas extraction velocity

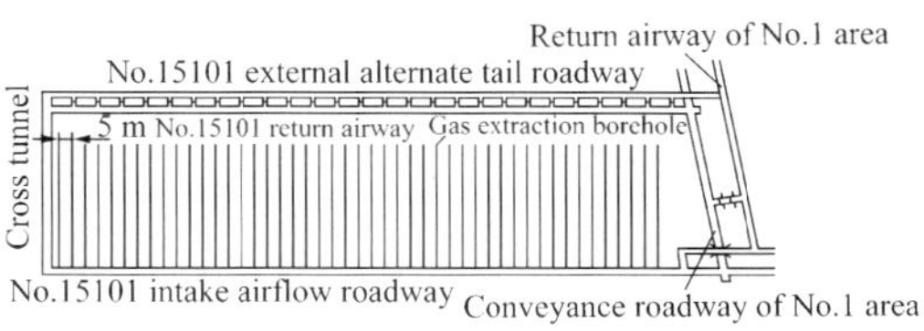

Fig. 6 Diagram of current seam gas extraction boreholes arrangement of working face 15101

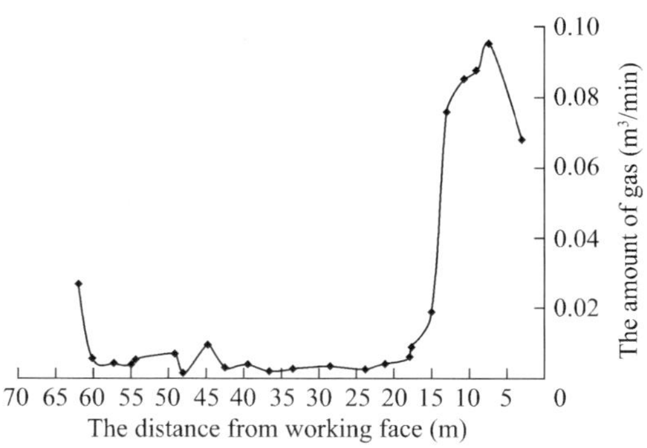

Fig. 7 Characteristics of gas extraction of current seam extraction borehole of working face 15101

Table 3 Statistics of gas extraction of current seam extraction borehole of working face 15101

Zones	Average distance from working face (m)	Average quantity of gas extraction (m^3/min)
Natural zone	>20	0.031
Mining influenced zone	≤20	0.212

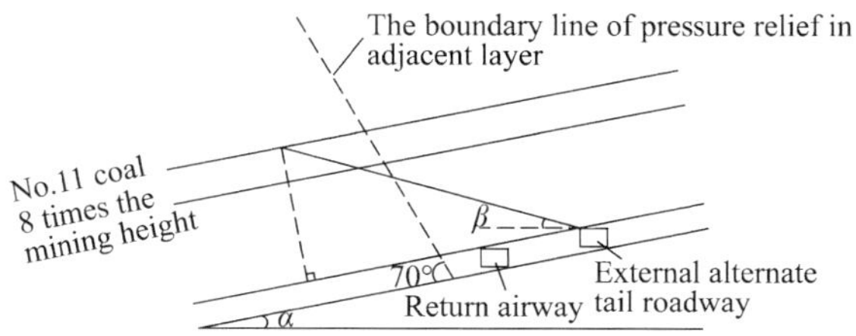

Fig. 8 Diagram of arrangement of adjacent seams gas extraction boreholes of working face 15101

20 m in front of the working face is increased 5.82 times that of the borehole in farther places. The coal mass gas pressure decreases from the inner coal seam to the working face because of the existence of gas extraction boreholes. Gas desorption and migration is therefore encouraged. The area influenced by mining, which exists in front of the working face, is a key area for gas extraction, especially in coal seams with low permeability.

4.2 Gas extraction in adjacent seams

The gas emission from adjacent seams in working face 15101 is 25.65 m^3/min. External alternate tail roadways are arranged for gas drainage and boreholes are drilled through strata in the roadway for preventing adjacent seams gas emission accident. The elevation angle of the boreholes is 46.5°, and they are drilled towards the gob. The angle between the boreholes and the coal side is 60°. The interval and borehole depth are 15 and 58.3 m, respectively. The ends of the boreholes are located at the fractured zone over the caving zone. Adjacent seams gas extraction boreholes of working face 15101 are arranged as shown in Fig. 8.

The relationship between the quantity of gas extraction and the distance of the No. 5 adjacent seams extraction borehole is shown in Fig. 9. The statistics of gas extraction of the adjacent seams extraction borehole are shown in Table 4. The zone of improved gas extraction appears 25 m away from the working face, and the effect of gas extraction further away remains at a low level. The overlying strata would undergo alternate collapse and caving with an increase in surrounding strata stress. The degree of

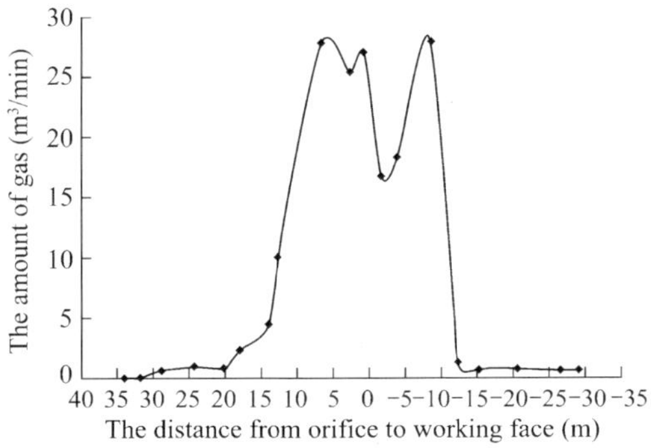

Fig. 9 Characteristics of gas extraction of adjacent seams extraction borehole of working face 15101

Table 4 Statistics of gas extraction of adjacent seams extraction borehole of working face 15101

Zones	Average distance from working face (m)	Average quantity of gas extraction (m^3/min)
Natural zone	>25	0.03
Mining influenced zone	25–−20	16.40
Extraction boreholes broken zone	≤−20	2.25

gas desorption increases. Consequently, the quantity of gas extraction increases significantly to 545.6 times the value under undisturbed conditions. The gas extraction borehole remains behind with the advance of the working face. Boreholes may be blocked because of the moving strata, and the quantity of gas extraction may decrease immediately. Therefore, effective measures should be taken to extract the adjacent seams gas. The area of influence of mining in the overlying strata and a reasonable arrangement of gas extraction boreholes parameters should be recognized.

5 Conclusions

(1) A new experimental system was developed to research characteristics of gas adsorption and desorption for mining large-scale coal samples, with a maximum experimental coal sample mass of 1200 kg. The experimental system could simulate panel-charging, which agrees with the gas source under natural conditions. Mining systems were also developed to simulate mining conditions. A system was established to achieve gradual stepwise loading.

(2) The research was conducted using the experimental system to establish the gas desorption characteristics under mining. The statistics from a series of experiments indicate that the velocity and degree of gas desorption increased significantly compared with natural conditions. The coal seam permeability improved because the pressure relief and failure of the overlying strata were influenced by the mining.

(3) Based on the characteristics of gas desorption with mining, the area influenced by mining was determined and a reasonable arrangement of the parameters of gas extraction boreholes could improve the effects of gas extraction, and realize the simultaneous extraction of coal and gas.

Acknowledgments This work is supported by the National Key Basic Research Program of China (2013CB227903) and the National Natural Science Foundation of China (U1361209).

References

Bai JP, Zhang DK, Yang JQ, Zhang H (2014) Thermodynamic characteristics of adsorption-desorption of methane in coal seam 3 at Sihe Coal Mine. J China Coal Soc 39(9):1812–1819

Barrer RM (1951) Difussion in and through solid. Cambridge University Press, Cambridge

Chen XG, Cheng YP, Wang L (2013) Experimental study on the inhibition of injection water to the gas desorption of coal. J Min Saf Eng 30(2):296–301

GB/T 211—2007 (2007) The determination methods of total moisture in coal

He XQ, Zhang L (2000) Study on influence and mechanism of gas adsorption and desorption in electromagnetic fields exerted. J China Coal Soc 25(6):614–618

He MC, Wang CG, Li DJ, Liu J, Zhang XH (2010) Desorption characteristics of adsorbed gas in coal samples under coupling temperature and uniaxial compression. Chin J Rock Mech Eng 29(5):865–872

Jiang YD, Xian XF, Yi J, Liu ZF, Guo CY (2008) Experimental and mechanical on the features of ultrasonic vibration stimulating the desorption of methane in coal. J China Coal Soc 33(6):675–680

Jiang YD, Xiong L, Yang XY, Xian XF (2010) Mechanism research on sound field promoting the coal bed methane desorption. J China Coal Soc 35(10):1649–1653

Li H (2011) Experiment research on the influence of ambient temperature on gas desorption laws of particle coal. Master's Thesis, Henan Polytechnic University

Li SG, Zhao Y, Zhang TJ (2010) Test systems of the coal sample adsorption/desorption characteristics based on low-frequency vibration. J China Coal Soc 35(7):1142–1146

Lü XF, Pan YS, Liu JJ, Tang JP, Di JZ (2010) Experimental study of effect of pore pressure on desorption deformation of coal matrix. Rock Soil Mech 31(11):3447–3451

Miao XX, Liu WQ, Chen ZQ (2004) Seepage theory of mining rock mass. Science Press, Beijing

Minkoff S, Stone C, Bryant S, Peszynska M, Wheeler M (2003) Coupled fluid flow and geomechanical deformation modeling. J Pet Sci Eng 38:37–56

Tang JP, Pan YS, Li CQ, Shi Q, Dong ZX (2006) Experimental study in effect of effective stress on desorption and seepage of coalbed methane. Chin J Rock Mech Eng 25(8):1563–1568

Tsang CF (1999) Linking thermal, hydrological, and mechanical processes in fractured rocks. Annu Rev Earth Planet Sci 27:359–384

Wang ZF (2001) Gas desorption laws and application of coal in air, water and mud medium. Doctor's Thesis, China University of Mining and Technology

Wang JC, Fan ZZ (2008) Key issues of coal and gas co-mining in thick seam. Coal Sci Technol 36(2):1–5

Wang ZF, Li XH, Qi LL, Sun R, Liu XQ (2010) The study of the moisture effect on gas desorption speed of Yangquan anthracite. Saf Coal Mines 7:1–3

Wu RL (2013) Effects of key stratum on the scope of the "three zones" of gas pressure relief and migration in coal seam group mining. J China Coal Soc 38(6):924–929

Wu RL, Xu JL, Qin W, Wang ZW, Zhang HP, Tian LX (2010) Experiment on advance pressure relief gas extraction in coal seams being mined in Xindadi coal mine. China Coal 36(4):83–85

Yang JJ (2005) Similarity theory and the structural model test. Wuhan University of Technology Press, Wuhan

Ye J, Shi BS, Zhang CC (1999) Coal reservoir permeability and its controlled factors in China. J China Coal Soc 24(2):118–122

Yu QX (1992) Methane prevention in coal mine. China University of Mining and Technology Press, Xuzhou

Zhang P (2012) Experimental and theoretical studies on longwall mining impact to the gas movement in solid coal. Doctor's Thesis, China University of Mining & Technology (Beijing), Beijing

9

Updating the coal quality parameters in multiple production benches based on combined material measurement

Cansın Yüksel[1] **· Jörg Benndorf**[2] **· Matthias Lindig**[3] **· Oliver Lohsträter**[3]

Abstract An efficient resource model updating framework concept was proposed aiming for the improvement of raw material quality control and process efficiency in any type of mining operation. The concept integrates sensor data measured online on the production line into the resource or grade/quality control model and continuously provides locally more accurate estimates. The concept has been applied in a lignite field with the aim of identifying local impurities in a coal seam and to improve the prediction of coal quality attributes in neighbouring blocks. A significant improvement was demonstrated which led to better coal quality management. So far, the proposed concept and the application in coal mining was limited to a case where online measurements were unambiguously trackable due to a single extraction face being the point of origin for the material. This contribution presents an extension to the case, where characteristics from blended material, originating from two or three simultaneously operating extraction faces, are measured. The challenge tackled in this contribution is the updating of local coal quality estimates in different production benches based on measurements of a blended material stream. For a practical application of the updating concept, which is based on the Ensemble Kalman Filter, a simple method for generating prior ensemble members based on block geometries defined in the short-term model and the variogram, is discussed. This method allows for a fast, semi-automated and rather simple generation of prior models instead of generating a fully simulated deposit model using conditional simulation in geostatistics. It should foster operational implementation in an industrial environment. The main purpose of this article is to investigate the applicability of the developed framework with a simplified prior resource model. In addition to this any model improvements due to the integration of sensor data obtained by observing a blend of coal from multiple extraction faces is investigated.

Keywords Resource model updating · Online sensor · Ensemble Kalman filter · Multiple excavator · Lignite · Real application

✉ Cansın Yüksel
C.Yuksel@tudelft.nl

[1] Resource Engineering Section, Department of Geoscience and Engineering, Delft University of Technology, Stevinweg 1, 2628 CN Delft, The Netherlands

[2] Institute for Mine Surveying and Geodesy, Faculty of Geosciences, Geoengineering and Mining, University of Technology Bergakademie Freiberg, Reiche Zeche, Fuchsmühlenweg 9, 09599 Freiberg, Germany

[3] Mitteldeutsche Braunkohlengesellschaft mbH, Glück-Auf-Straße 1, 06711 Zeitz, Germany

1 Introduction

One of the main challenges in mining is the control of product quality, which is impacted by impurities in the deposit, such as waste intrusions in coal seams. In lignite operations, these impurities can lead to high ash values (e.g. more than 15% ash) and cannot be localized completely by exploration data and captured in the predicted deposit models.

Utilizing online-sensor techniques for coal quality characterization in combination with rapid resource model updating, a faster reaction to the unexpected deviations can be implemented during operations, leading to increased production efficiency. This concept was first proposed as a closed loop framework by Benndorf et al. (2015). The developed framework is based on the concept of data assimilation, in particular to the Ensemble Kalman filter (EnKF) (Evensen 1994; Evensen and Van Leeuwen 1996; Evensen 1997a, b; Burgers et al. 1998; Evensen and Van Leeuwen 2000). It integrates online-sensor data obtained during the extraction process, e.g. measured on a belt conveyer, into the resource model, as soon as they are obtained.

The first investigation (Benndorf 2015) has proven the approach to work well within a synthetic case study under a variation of several control parameters (number of excavators, precision of the sensor, update interval, measurement interval, extraction mode/production rate). Wambeke and Benndorf (2016) extended the framework for practical application, including the handling of attributes and measurements showing a non-Gaussian distribution, dealing with localization and inbreeding issues, avoiding the spurious correlations and increasing the computational efficiency. The third investigation (Yüksel et al. 2016) implemented the framework in a full case study by adapting implementation details for coal quality attributes in a continues mining environment. The applicability of the framework for a full scale lignite production environment was validated and significant improvements were demonstrated. These results have been achieved by a test case where one sensor has been placed on the excavator. This sensor observes the produced material from that excavator and the data produced by this sensor is being used for updating the neighbourhood blocks around the mined blocks. Thereafter, Yüksel and Benndorf (2016) investigated the performance of the resource model updating framework with respect to main parameters, which are the ensemble size, localization and neighbourhood strategies and the sensor precision.

In many mining operations material quality control measurements are taken at central locations in the downstream process, such as, on a central conveyor belt or from the trains that are loaded after the coal blending yard. In this case the measurements represent a blend or a combination of material originating from multiple extraction faces. The measurement of one sample cannot be tracked back to one origin of the material. However, a collection of multiple measurements over time would allow to solve this unambiguity. In this contribution the updating framework is applied while multiple excavators are producing at different benches. This is done in order to understand the updating performance when feeding the blended coal observations back to multiple excavator locations from where the production originates.

A second aspect discussed here is the practical implementation in an operational environment. The resource model updating concept is based on EnKF, which requires ensemble members (or realisations). These can be obtained by conditional simulation (Benndorf 2013; Pardo-Igúzquiza et al. 2013; Srivastava 2013; Tercan and Sohrabian 2013), which can be a time consuming effort requiring some expert knowledge. For operational implementations, the process should be rather simple and robust. Therefore, the aim is to investigate whether realizations of a prior model can be obtained rather simple without loss of updating performance.

This study aims to present a new application of the framework in a full scale lignite production, where the initial resource model generation is automated based on a short-term model. This would immediately increase the production efficiency in a real mining environment, by simply giving the opportunity to react on the changes of the resource model with newly gained information. Moreover, using the real-time updating framework would also decrease the frequency of material misallocation. Having an improved resource model helps to have a smaller amount of actual lignite being incorrectly allocated to the waste dump. And similarly, a smaller amount of actual waste send to the stockpile.

In the following paper a real lignite mining case study is presented in order to compare the updating frameworks' performance with a prior model based on conditional simulation and a prior model based on a short-term model. The comparison experiments are performed for different time intervals and different number of excavators. Results of this investigation should help the automation process when the framework is being used in a real mining environment.

2 Updating coal attributes in a resource model based on online-sensor data

For rapid updating of the resource model, sequentially observed data have to be integrated with prediction models in an efficient way. This is done by using sequential data assimilation methods, namely the EnKF based methods.

Let $\boldsymbol{Z}(\boldsymbol{x})$ be the state of a stochastic process modelling the spatial distribution, where $\boldsymbol{Z}$ refers the local ash content at excavation locations $\boldsymbol{x}$. The developed framework uses geostatistical simulation technique (e.g. conditional simulation) in order to create the ensemble of realizations, also called prior ensemble $\boldsymbol{Z}_0(\boldsymbol{x})^e$, where $e = 1, \ldots, N$ is the number of realizations. Then the updated resource model ensembles, $\boldsymbol{Z}^*(\boldsymbol{x})^e$, is calculated by the following equation:

$$\boldsymbol{Z}^*(\boldsymbol{x})^e = \boldsymbol{Z}_0(\boldsymbol{x})^e + \boldsymbol{K}^e[\boldsymbol{l}^e - \boldsymbol{A}\boldsymbol{Z}_0(\boldsymbol{x})^e] \quad (1)$$

where $\boldsymbol{Z}(\boldsymbol{x})^{e}$ and $\boldsymbol{l}^{e}$ respectively consist of an ensemble of realizations and the sensor based measurements; $\boldsymbol{A}$ represents a forward simulator of the production sequence, so the term $\boldsymbol{A}\boldsymbol{Z}_{0}(\boldsymbol{x})^{e}$ represents the predicted measurements based on the prior block model. In Eq. (2), $\boldsymbol{C}_{zz}^{*e}$ refers to the updated error covariance of the resource model, where the overbar denotes the expected values of the ensembles.

$$\boldsymbol{C}_{zz}^{*e} = \overline{\left[\boldsymbol{Z}(\boldsymbol{x})^{e} - \overline{\boldsymbol{Z}(\boldsymbol{x})^{e}}\right]\left[\boldsymbol{Z}(\boldsymbol{x})^{e} - - \overline{\boldsymbol{Z}(\boldsymbol{x})^{e}}\right]^{\mathrm{T}}} \quad (2)$$

The Kalman gain, $\boldsymbol{K}$, calculates a weighting factor that indicates the reliability of the measurements, to decide "how much to change the prior model by a given measurement". The covariance matrices represent the whole ensemble and the Kalman gain $\boldsymbol{K}^{\mathrm{e}}$ is derived from these.

$$\boldsymbol{K}^{e} = \left(\boldsymbol{A}^{\mathrm{T}}\boldsymbol{C}_{zz}^{e}\boldsymbol{A} + \boldsymbol{C}_{ll}^{e}\right)^{-1}\boldsymbol{A}^{\mathrm{T}}\boldsymbol{C}_{zz}^{e} \quad (3)$$

Two measures are implemented by (Wambeke and Benndorf 2016) to reduce computational time of Kalman gain calculations. The first measure is related to the neighbourhood. The size of the $\boldsymbol{C}_{zz}$ matrix is in the order of the size of the blocks that are in the defined updating neighbourhood. The second measure is a Cholesky decomposition which is implemented to avoid an explicit computation of the inverse in Eq. (3). This results in significant computational speed ups.

Blended measurements and differences in the scale of support are dealt through the empirical calculation of the covariances. These covariances ($\boldsymbol{A}^{\mathrm{T}}\boldsymbol{C}_{zz}^{e}$) mathematically describe the relations between the blended measurements and individual source locations.

With the goal of a continuously updatable coal quality attributes in a resource model, a framework based on the normal-score ensemble Kalman filter (NS-EnKF) (Zhou et al. 2011) approach was tailored for large scaled mining applications. The NS-EnKF is chosen to deal with the non-gaussianity of the data by applying a normal-score transformation to each variable for all locations and all time steps, prior to performing the updating step in EnKF.

Figure 1 gives general overview of the operations which are performed to apply the resource model updating framework. The concept initially starts with resource modelling, traditionally by using conditional simulation. This is the first required data set consisting of ensemble members to be updated. The second data set consists of the production data and their related actual sensor measurements. The production data provides the excavated block information, e.g. names and quantities. The actual online-sensor measurement values are collected during the lignite production and they represent the excavated material. The third data set consists of a collection of actual and predicted sensor measurements. The predicted measurements are obtained by applying the production sequence as a forward predictor to prior resource model realizations. Once all of the input data are provided, the updated posterior resource model will be obtained. This process will continue as long as new online-sensor measurement data is received.

3 A simplified prior model

As mentioned earlier, the first data set required to apply the resource model updating framework is a collection of the resource model realizations, also called the prior model. Traditionally, this is done by sophisticated geostatistical methods, such as conditional simulation. However, this requires some expert knowledge and adds an additional step prior to using the updating framework. Moreover, generating a new resource model might create disarray between geology and mine planning departments in the company, since they already have a resource model created by their own team. For these reasons, in order to apply the updating framework in real mining environment, a more practical and simplified application of the framework is required. The proposed simplification obtains the required prior model realizations by adding fluctuations around the company's short-term mining plan. This short-term model is created by the mining engineers, based on applying the defined block geometries (Fig. 2) on the company's estimated block model. In this way, each block will have an estimated ash value. Figure 3 compares both of the prior model generation processes.

In order to create the quality model based on short-term model, the following strategy is employed:

(1) Short-term block model values are generally available for each block and they deliver the prior estimation of block attributes (E-type estimate).
(2) A conditional simulation is applied to production blocks that were in the short-term block model. For this application, the previously calculated block scaled variogram model is used. Drill hole locations with zero ash content are used as the reference point while running the simulations.
(3) After this, simulated data on the production blocks refer to the uncertainty and they will be added on prior estimations of block attributes.
(4) The short-term model based on the simulations is now ready to be imported into the algorithm as the first main component (prior model).
(5) The updated resource model (posterior model) will be split in a mean part, which will be written back to update the short-term block model. The uncertainty related part will be written back in the ensemble part.

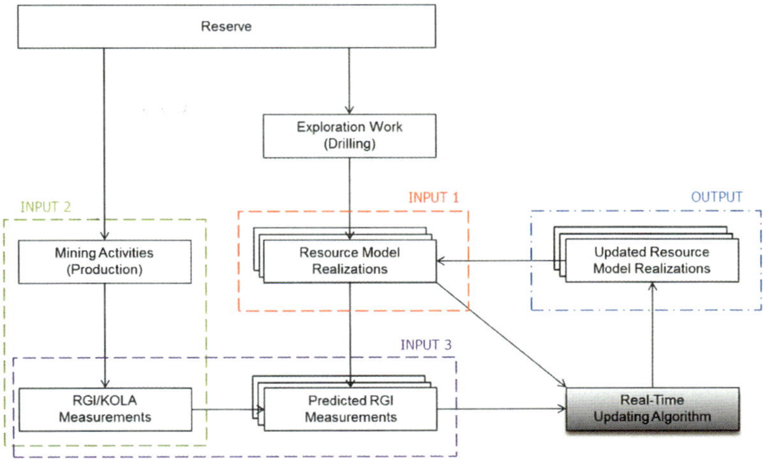

Fig. 1 Configuration of the real-time resource model updating framework, modified from (Wambeke and Benndorf 2015)

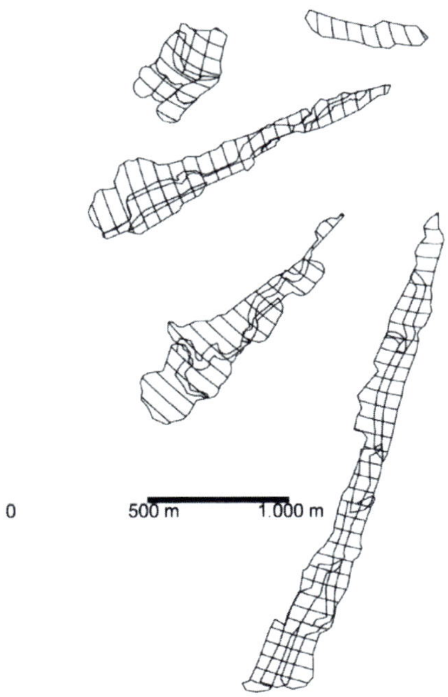

Fig. 2 Planned block geometries in the production benches

As long as new measurement data is obtained, these steps need to be applied recursively. The process can easily be automated by using a previously calculated variogram model and some interfaces. In this way, there will be no requirement for an additional complex process of creating conditional simulations since they are not part of the daily work flow. Moreover, there will be no disarray between a company's short-term model and the input prior model of the updating approach; the integration will be smooth. Additional to that, no expert knowledge will be required when applying the framework due to the automated process, contrary to conditional simulation application. All of these simplifications on application are very significant since it is important to benefit from the framework in a real mining environment.

4 Application in a full scale lignite production using multiple excavators

This case study aims to discuss two different aspects. First, it aims to test the performance of the resource model updating framework while the sensor is observing a blend of coal resulting from multiple excavators. Second, it aims to simplify and semi-automate the framework for easier application in a real mining environment.

4.1 Case description

The case study is performed on a lignite mining operation in Germany, where the geology of the field is complex, including multiple split seams with strongly varying seam geometry and coal quality distribution (Fig. 4). In this case study, the challenge originates from the complicated geology that leads to geological uncertainty associated with the detailed knowledge about the coal deposit. This uncertainty causes deviations from expected process performance and affects the sustainable supply of lignite to the customers. The aim is to improve the knowledge over the coal deposit and increase the process performance by applying a resource model updating framework.

For the case study, the target area is defined as an already mined out area of 25 km^2, where there are about 3000 drill holes. Mining operations are executed by six excavators, each working on a different bench. Among these six excavators, only five of them are continuously working on a lignite seam. Generally, the maximum number of excavators that are working at the same time is

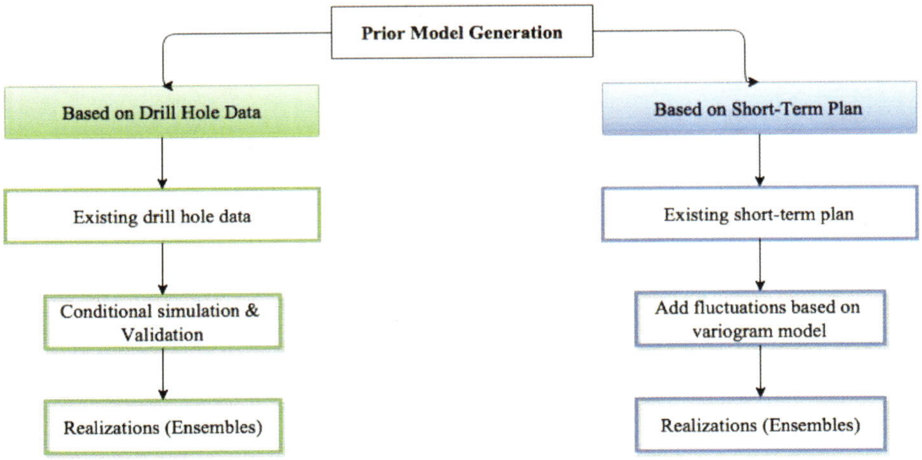

Fig. 3 Flow chart of prior model generation

Fig. 4 Complicated geology in the lignite mine

three. For this reason, the case study will apply cases where either only one excavator is working, or two excavators are working or three excavators are working at the same time.

The produced materials are being transported through conveyor belts. All conveyor belts merge at a central conveyor belt leading to the coal stock and blending yard, which is further connected to a train load. Figure 5 presents the mentioned six benches in black lined blocks, conveyor belts in blue lines, drill holes as green points and the online measuring system as an orange point (Fig. 6).

A radiometric sensor measurement (RGI) system is installed on the central conveyor belt just before the coal stock and blending yard. This system allows an online determination of the ash content of the blended mass flow directly on the conveyor belt, without requiring any sampling or sample processing. For demonstration purposes, this case study assumes the RGI values to be accurate.

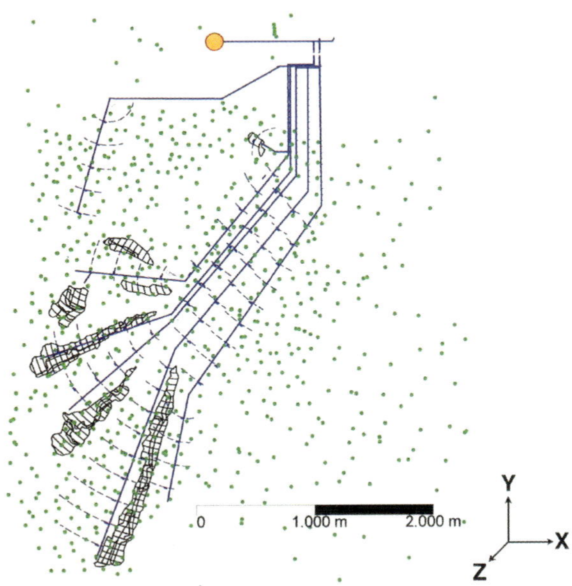

Fig. 5 Production benches, belt system and drill holes on the study area

4.2 Data preparation

To apply the resource model updating framework, preparation of input data is required (Fig. 1). The first data set is the prior model, which contains a collection of the resource model realizations. For the case study, two different prior models are prepared based on different approaches.

Fig. 6 Radiometric sensor measurement device, installed on the conveyor belt, measuring blend of coal resulting from multiple excavators, just before the stock pile

4.2.1 Prior model: based on drill hole data

A prior model based on drill hole data refers to generation of prior realizations by conditional simulation based on the given drill hole data. First, the geological model of the defined coal seam is created on a 25 m × 25 m × 1 m dimensioned block model based on the roof and floor information of the lignite seam. Second, a 25 m × 25 m × 1 m dimensioned quality model capturing the wet ash content in percentages is generated by 25 simulations based on the provided drill hole data. The simulated ash values are then merged with the previously defined coal seam. After this, the block model realizations are ready to be imported into the algorithm as the first input.

4.2.2 Prior model: based on short-term model

A prior model based on the short-term model refers generation of prior realizations by adding fluctuations on short-term mining model of the company. A detailed explanation of this application is introduced in the previous section.

The updating experiments are performed both for drill hole based prior model realizations and short-term model based prior model realizations. This is done in order to compare the performance of the updating framework while updating differently generated prior models. The aim of this performance comparison is to investigate: “If the updating framework uses a non-geostatistical set of simulations as a prior model, would the updated models still be improved?”

The second data set consists of the production data and their related actual sensor measurements. The material travelling time from each production location (excavator & bench location) to RGI location is calculated. In order to determine the location of the received RGI measurement data, in other words: “to track back where the measured material comes from”, the production data is linked with the RGI data based on the given timecodes (material travel delays are taken into account). The second input file for the algorithm is written to a file containing the following information: timecode, actual sensor measurement (RGI data), excavated block_1 id, excavated block_1 mass, excavated block_2 id, excavated block_2 mass,..., excavated block_n id, the excavated_n block mass; where $b = 1, \ldots, n$ is the excavated block number in the given time span.

The third data set consists of a collection of actual and predicted sensor measurements. An ensemble of predicted values is obtained by the forward simulator applying the digging location and the material transport model to each realization. The third input file for the algorithm is written to a file containing the following information: the block ID, central block location (X, Y, Z coordinates), a series of real measurements and predicted measurements.

4.3 Experimental set-up

The experiments that are performed both with drill hole based and short term plan based prior model realizations, fall into two different categories. The first category involves a different time span based experiments, where updating of the prior model is performed every 2, 1 h, 30, 15 and 10 min. For these experiments, the related RGI and production data are linked to each other (for every minute) and averaged for each indicated time span.

The second category involves experiments that are based on the number of excavators producing coal at a given time period. It investigates the capability of the updating framework when updating multiple benches based on a blend of material measurements. For these experiments, the data set that is prepared for every 2 h of updating is taken as the base data and divided into three different data sets. This division is done based on the number of excavators that are producing coal at a given time span such as; 1 excavator, 2 excavators and 3 excavators.

For each criteria introduced above, an experiment is performed. Each experiment initially updated the prior model for a four day time period. Based on this resulting posterior model, forward simulator is used to generate predicted posterior model values for the future mining operations (for next two days). These predictions are then compared with the related RGI data. Chosen time spans are representative for any time span that might be chosen in the future.

The updating neighbourhood size is chosen as 900 m × 900 m × 10 m in *X*, *Y*, *Z* directions based on the variogram model range, which was calculated during geological modelling. A 225, 225, 5 m sized localization is applied for each mentioned experiment in order to prevent long range spurious correlations based on a previously performed sensitivity analysis (Yüksel and Benndorf 2016).

5 Results and discussion

5.1 Results

This section presents representative results of the previously defined experiments. The following graphs provide representative information where the *X*-axis refers to the mentioned time spans, $i = 0, 1, 2, \ldots, n$. Instead of writing the full date and time information, the authors decided to use time span codes for simplicity. For example for the case where the updating is every 2 h; if $i = 0$ refers on 01.01.2014 at 00:00 o'clock, $i = 10$ refers 20 h later, which is 01.01.2014 at 20:00 o'clock. *Y*-axis refers to the ash %. The presented graphs consist of the following information:

(1) Posterior Model Box Plots: Box plot representation of posterior model simulations which are updated based on a given criteria (e.g. updating every 2, 1 h, 30, 15 or 10 min; or updating while 1 excavator, 2 excavators or 3 excavators are producing).
(2) Posterior Mean: Represents the mean of the updated models in the learning period. Essentially, it is the mean of the posterior model that is updated based on a given criteria.
(3) Predicted Mean: Represents the mean of the predictions in the prediction period. Basically, it is the prediction of future mining blocks, based on the four-day-long-updated model.
(4) Prior Mean: Mean of the prior model that is created based on either the drill hole data or short-term model. It is mined through different operation files based on a given criteria (e.g. updating every 2, 1 h, 30, 15 or 10 min; or updating while 1 excavator, 2 excavators or 3 excavators are producing).
(5) RGI: The averaged RGI data for a given time span.
(6) White area: Represents the learning period, where posterior models are produced as a result of updating the prior model, by using the RGI data.
(7) Green area: Represents the prediction period, where the mining operations are executed on the four-day-long-updated model.

In these graphs the prior model is updated for four days. Based on this updated prior model, the posterior model, further mining operations are performed for the next two days. The operation file mines through the posterior model and highlights the area as green.

5.1.1 Using a prior model that is based on conditional simulation

The achieved improvements are numerically evaluated using an absolute error measure. The absolute error is defined as the absolute difference between the measured value of a quantity and its actual value. In our case, absolute error refers to the absolute difference between the measured RGI value $\boldsymbol{l}$ of produced coal at a given time span and its prior $\mathbf{Z}_0(\boldsymbol{x})$ (or posterior $\mathbf{Z}^*(\boldsymbol{x})$) value, calculated by the following equation:

$$AE = \frac{1}{n}\sum_{i=0}^{n}\left|l_i - \mathbf{Z}^*(\boldsymbol{x})_i\right| \tag{4}$$

The absolute error values are calculated for each experiment iteration at a given time span $i = 0, \ldots, n$ and eventually averaged when the update of the block model is completed for the defined study case.

Table 1 provides the calculated absolute errors for prior models and predictions that are illustrated in the green area of the graphs. Additional to that, it indicates the improvement (IMPROV) in percentages when comparing prior's and predictions' absolute errors. Improvements indicate the decrease of the absolute errors and it can be calculated as;

$$\text{IMPROV}\ (\%) = \frac{\text{PriorModel}_{\text{AbsoluteError}} - \text{PredictedModel}_{\text{AbsoluteError}}}{\text{PriorModel}_{\text{AbsoluteError}}} \tag{5}$$

Moreover, Fig. 10 presents the calculated absolute error for the following two days after updating the prior model every 2 h for four days. Red dots illustrate the calculated absolute errors for each time span.

5.1.2 Using a prior model that is based on short-term model

The achieved improvements are presented using "absolute error" as previously introduced. Table 2 provides the calculated absolute errors for prior models and predictions. The improvement percentages (IMPROV) are calculated by comparing the absolute errors of prior and prediction models.

Table 1 Calculated absolute errors for predictions—Prior model is based on drill hole data

Time	Prior model	Predictions originated from posterior model	
	Absolute error	Absolute error	IMPROV (%)
2 h	2.25	0.82	64
1 h	2.82	1.03	43
30 min	1.20	1.14	5
15 min	2.59	2.08	20
10 min	2.57	2.39	7
1 Exc	2.22	1.87	16
2 Exc	1.72	0.96	44
3 Exc	1.12	0.92	18

Table 2 Calculated absolute errors for predictions—Prior model is based on short-term model

Time	Prior model	Predictions originated from posterior model	
	Absolute error	Absolute error	IMPROV (%)
2 h	1.09	0.70	36
1 h	0.99	0.84	14
30 min	1.49	1.27	15
15 min	4.37	1.17	73
10 min	4.24	1.59	63
1 Exc	2.10	1.36	35
2 Exc	1.13	0.91	19
3 Exc	1.18	0.89	25

Moreover, Fig. 14 presents the calculated absolute error for the following two days after updating the prior model every 2 h for four days. Red dots illustrate the calculated absolute errors for each time span.

6 Discussion

6.1 Improvements of predictions

Figures 7, 8 and 9 illustrate the improvement of the ash % predictions in the posterior model, where updating of the prior model (developed from drill hole data) is applied based on RGI data for four days. The ensuing posterior model is mined through based on production data. The predictions of the posterior model while mining the neighbourhood blocks are then compared with the actual ash % (in this case RGI measurements are assumed as reality) and prior model. Representative graphs are provided (Fig. 10).

Figure 7 presents the case where the prior model (based on drill hole data) is being updated for every 2 h for four days. The following green area represents a period of two days, where orange points represent the averaged prediction behaviour of the posterior model which is updated for four days. In this time period, it can be observed that posterior model predictions are mostly following the trend of the RGI data (red lines). Moreover, when comparing the posterior model predictions with the prior model (blue square points), significant improvements are observed in the posterior predictions. Based on Table 1, the averaged absolute error for those predictions is 0.82, while it is 2.25 for prior model. This indicates a 64% improvement.

Figure 8 presents the case where the prior model (based on drill hole data) is being updated every 2 h when only 1 excavator is operating for four days. Similarly, in the green area, orange points represent the averaged prediction behaviour. Between the 50th–52nd and the 55th–58th timecodes, posterior model predictions are remaining stable due to production of the same mining block at each time. This stable prediction averages around the reality (RGI data). Moreover, uncertainty of the predictions (box plot whiskers) covers the reality (RGI data) better than the prior model. After the 63rd timecode, posterior model predictions follow a similar trend as the prior model due to spatial variability of the lignite seam. Furthermore, since this experiment focused on a case with only one excavator

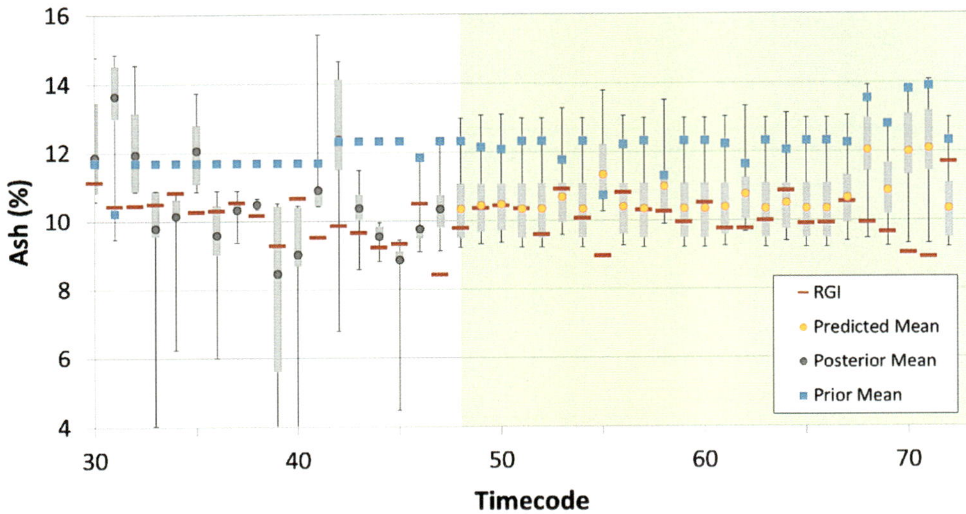

Fig. 7 Results based on conditional simulation: updating every 2 h for 4 days. The *green area* represents the prediction period. The *white area* represents the learning period

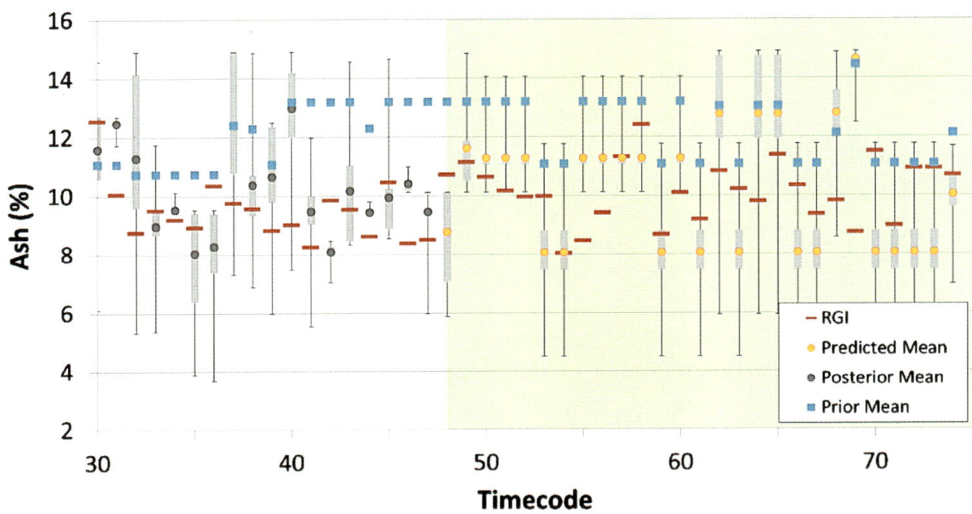

Fig. 8 Results based on conditional simulation: updating every 2 h for 4 days, 1 excavator producing. The *green area* represents the prediction period. The *white area* represents the learning period

producing, the application was not limited only for one bench. As a result, after updating four days in three benches, using only the times where one excavator is working, could only improve the future predictions for a limited time. The authors believe that for this case, the quality and the lifetime of the predictions can be improved by extending the learning phase (more than four days).

Figure 9 presents the case where the prior model (based on drill hole data) is being updated every 2 h when 2 excavators are operating for four days. By using 2 excavators at the same time, already more information becomes available about the lignite seams that are being worked on and this leads a longer time of good quality improvements. This can be seen by comparing Figs. 8 and 9.

Figures 11, 12 and 13 apply the same experiments as above, however in these figures the prior model is based on the short-term model. With these experiments similar results as before were achieved. In Fig. 11, where the updating of the prior model is every 2 h, predicted ash % values are almost always aligned with the reality (RGI data). Figure 12 presented a case with 1 excavator and Fig. 13 presented a case with 2 excavators. As above, when looking to those two graphs, better predictions are observed when using 2 excavators.

For both cases, Figs. 10 and 14 are provided in order to investigate the behaviour of the absolute error values obtained from predictions. The absolute error values are initially very low, but after approximately a day period

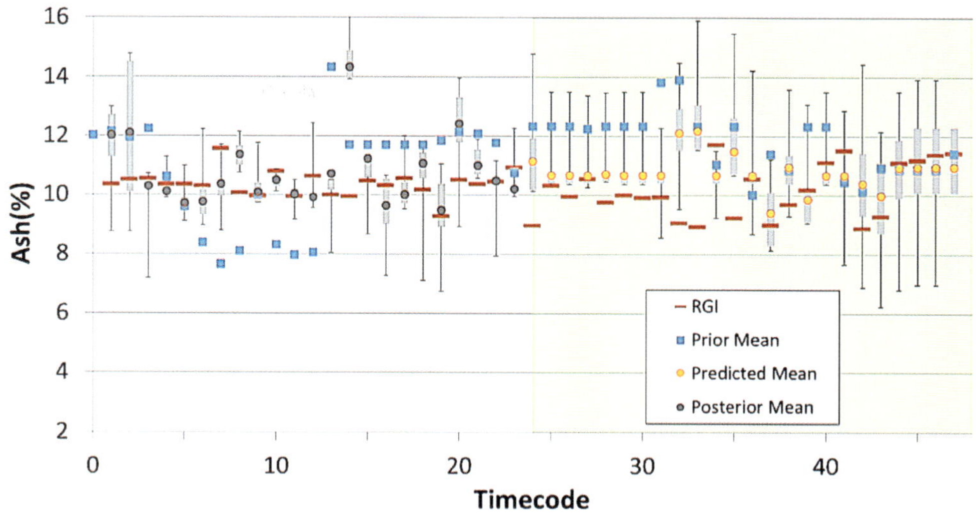

Fig. 9 Results based on conditional simulation: updating every 2 h for 4 days, 2 excavators producing. The *green area* represents the prediction period. The *white area* represents the learning period

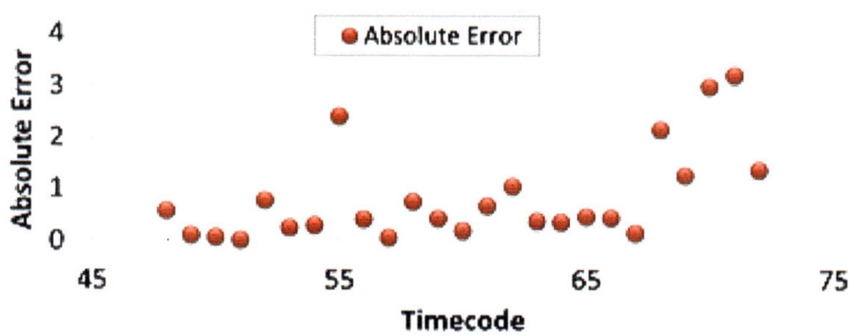

Fig. 10 Absolute error predictions (for the next 2 days) of after updating every 2 h for 4 days

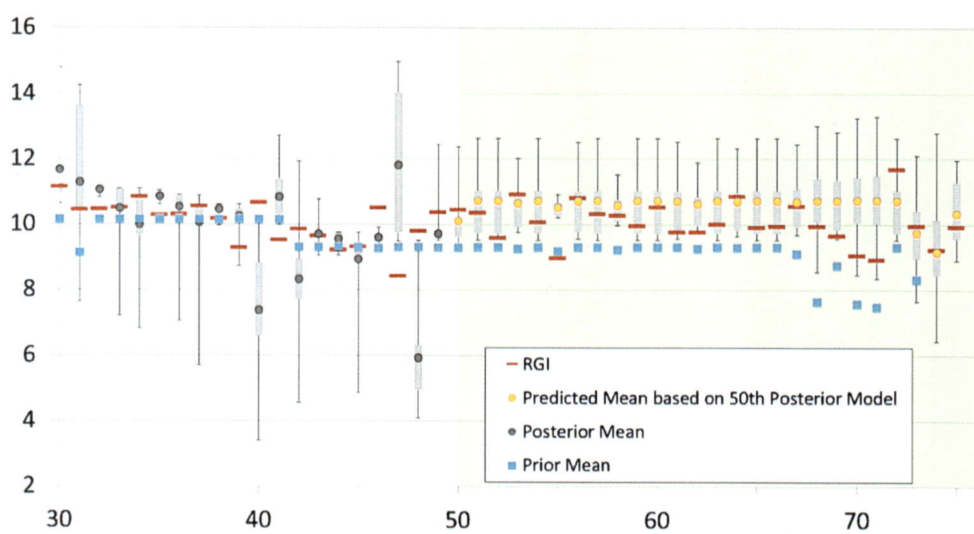

Fig. 11 Results based on short-term model: updating every 2 h for 4 days. The *green area* represents the prediction period. The *white area* represents the learning period

they indicate an increase over time. When the distance between the mined block and the neighbourhood blocks increases, it is expected to see less improvement for the neighbourhood blocks. This occurs due to the lower spatial correlation. Moreover, when predicting the neighbourhood blocks there might be some blocks that are not updated in the learning period. This causes not only an increase in the absolute error over the time, but also outliers in the early

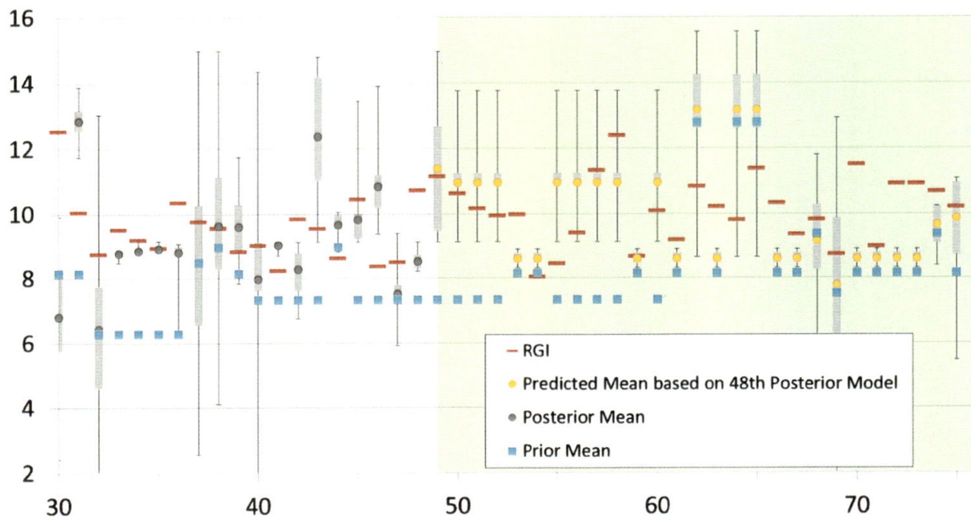

Fig. 12 Results based on short-term model: updating every 2 h for 4 days, 1 excavator producing. The *green area* represents the prediction period. The *white area* represents the learning period

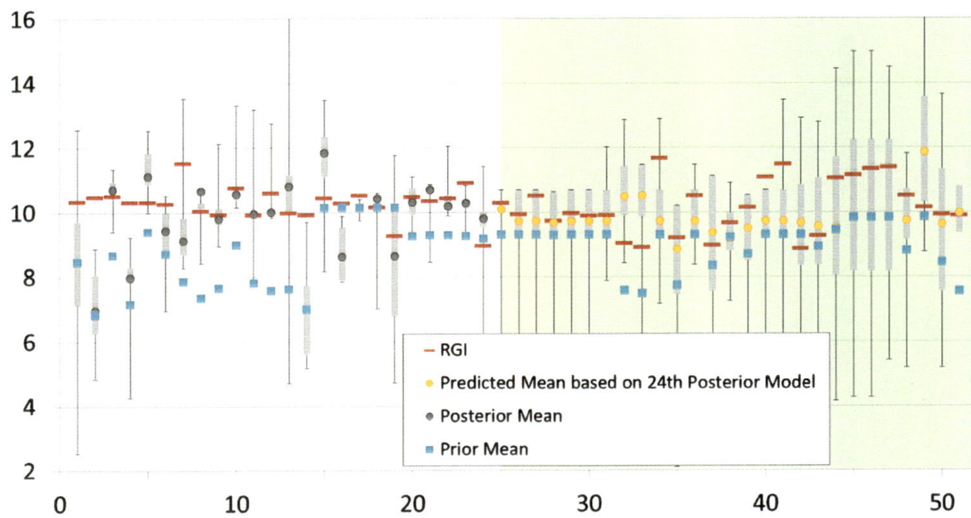

Fig. 13 Results based on short-term model: updating every 2 h for 4 days, 2 excavators producing. The *green area* represents the prediction period. The *white area* represents the learning period

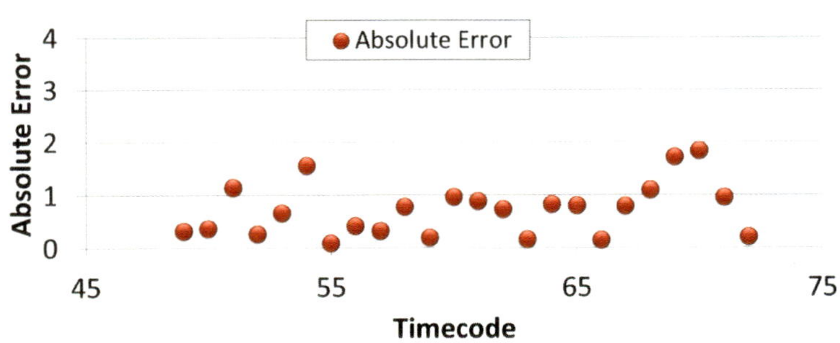

Fig. 14 Absolute error predictions (for the next 2 days) after updating every 2 h for 4 days

phases of the prediction period. For example, see timecode 55 in Fig. 10 or see timecode 51 and 54 in Fig. 14. These outliers can be observed and the reason that they occur can be explained as follows: at each individual timestamp there are different blocks being mined from different benches. If a block gets mined in the prediction period and it has not been mined in the learning period or if it has not been in the neighbourhood of any other mined blocks, it has never been updated. Thus, it still has the prior model's value assigned on it. This results in a prior biased prediction and an increase of the absolute error.

6.2 Time based experiments

Different time span based experiments are performed (every 2, 1 h, 30, 15 and 10 min) for updating the prior model based on drill hole data. In overall, significant improvements (up to 64%) are obtained while updating the prior model with measured RGI values and predicting neighbourhood blocks' qualities (Table 1). A comparison among the performed experiments between the most frequent update (every 10 min) and the least frequent update (every 2 h), shows that highest improvements are achieved by the least frequent updates of this case (every 1 and 2 h updating cases).

Similar to the above experiments, different time span based experiments are also performed for updating the prior model based on the short-term model. All of the experiments show satisfactory improvements (up to 73%) (Table 2). In this case, the highest improvements are achieved by the most frequent updates (every 15 and 30 min).

However, calculating these absolute errors does not necessarily indicate the best parameters to use. It only validates the applicability of the method for the given parameters. It should not be forgotten that the calculated absolute errors for predictions can vary depending on the quality of the posterior model that is chosen as the base of the predictions. For each case this paper has chosen the posterior models that are obtained after four days of updating the prior model. Other experiments are also applied to test this issue and they all recorded significant but varying amounts of improvements.

6.3 Excavator number based experiments

Experiments based on a different number of working excavators are performed in order to investigate the capability of the updating framework. The previous case study (Yüksel and Benndorf 2016), in which the study area was limited to one bench and one producing excavator, produced successful results. The RGI online-sensor was positioned on the producing excavator, so that the measured material was the produced material from that excavator. However, in this case study, there are three different benches and three different producing excavators (one excavator for each bench). The online RGI sensor is positioned on one of the conveyor belts just before the stock yard. Therefore, the RGI sensor measures blended material produced from different benches. The aim of performing the mentioned experiments in this section is to test the performance of the updating algorithm in when the observations are measured from a blended flow.

By looking at Table 1, a range of 16%–44% improvement is observed when using a varying amount of excavators in the updating experiments with a drill hole based prior model. This shows that the algorithm can handle a situation where the blended measurement data is fed into different benches where the material is originally produced.

By looking at Table 2, a range of 19%–35% improvement is observed when using a varying amount of excavators in the updating experiments with a short-term model based prior model. The obtained improvements are significant considering the benefits of automation while using a short-term model based prior model. Once again, the results indicate that the algorithm can handle a situation where the blended measurement data is fed into different benches where the material is originally produced.

7 Conclusions

This study provides a full-scale case study on the application of an Ensemble Kalman based resource model updating framework, with the aim of simplifying the application process.

To offer an easy application of the updating framework in a real mining environment, a simplified application method is created. This simplified application method involves creating the prior realizations based on the company's short-term model. Improvement percentages, in average, were not significantly different when the case study results are compared with the results obtained from a case study where the prior realizations are generated with geostatistical simulations. This paper validates that the automation of the developed framework during real applications can be done based on a short-term model without any additional process being required in order to prepare the prior model.

Moreover, significant improvements are observed while using blended material measurement data in order to update different production locations in different benches. This provides great flexibility for future applications.

The authors would like to point out that this method can be applied to any bulk mining operation, without changing the core method and improvements, if a material tracking system, grade or quality control model and online-sensors measurement system are in place.

Future studies will focus on the value of introducing additional information in the short-term model during the production phase.

Acknowledgements This research is a minor part of the Real-Time Reconciliation and Optimization in large open pit coal mines (RTRO-

Coal) project and it is supported by Research Fund for Coal and Steel of European Union. RTRO Coal, Grant Agreement No. RFCR-CT-2013-00003.

References

Benndorf J (2013) Application of efficient methods of conditional simulation for optimising coal blending strategies in large continuous open pit mining operations. Int J Coal Geol 112:141–153

Benndorf J (2015) Making use of online production data: sequential updating of mineral resource models. Math Geosci 47:547–563

Benndorf J et al (2015) RTRO–coal: real-time resource-reconciliation and optimization for exploitation of coal deposits. Minerals 5:546–569

Burgers G, Jan van Leeuwen P, Evensen G (1998) Analysis scheme in the ensemble Kalman filter. Mon Weather Rev 126:1719–1724

Evensen G (1994) Sequential data assimilation with a nonlinear quasi-geostrophic model using Monte Carlo methods to forecast error statistics. J Geophys Res: Oceans 99:10143–10162

Evensen G (1997a) Advanced data assimilation for strongly nonlinear dynamics. Mon Weather Rev 125:1342–1354

Evensen G (1997b) Application of ensemble integrations for predictability studies and data assimilation. In: Monte Carlo simulations in oceanography proceedings' Aha Huliko'a Hawaiian Winter Workshop, University of Hawaii at Manoa, Citeseer, pp 14–17

Evensen G, Van Leeuwen PJ (1996) Assimilation of Geosat altimeter data for the Agulhas current using the ensemble Kalman filter with a quasigeostrophic model. Mon Weather Rev 124:85–96

Evensen G, Van Leeuwen PJ (2000) An ensemble Kalman smoother for nonlinear dynamics. Mon Weather Rev 128:1852–1867

Pardo-Igúzquiza E, Dowd PA, Baltuille JM, Chica-Olmo M (2013) Geostatistical modelling of a coal seam for resource risk assessment. Int J Coal Geol 112:134–140. doi:10.1016/j.coal.2012.11.004

Srivastava RM (2013) Geostatistics: a toolkit for data analysis, spatial prediction and risk management in the coal industry. Int J Coal Geol 112:2–13. doi:10.1016/j.coal.2013.01.011

Tercan AE, Sohrabian B (2013) Multivariate geostatistical simulation of coal quality data by independent components. Int J Coal Geol 112:53–66. doi:10.1016/j.coal.2012.10.007

Wambeke T, Benndorf J (2015) Data assimilation of sensor measurements to improve production forecasts in resource extraction. Paper presented at the IAMG, Freiberg (Saxony) Germany

Wambeke T, Benndorf J (2016) A simulation-based geostatistical approach to real-time reconciliation of the grade control model. Math Geosci. doi:10.1007/s11004-016-9658-6

Yüksel C, Benndorf J (2016) Performance analysis of continuous resource model updating in lignite production. Paper presented at the Geostats 2016, Valencia

Yüksel C, Thielemann T, Wambeke T, Benndorf J (2016) Real-time resource model updating for improved coal quality control using online data. Int J Coal Geol. doi:10.1016/j.coal.2016.05.014

Zhou H, Gómez-Hernández JJ, Hendricks Franssen H-J, Li L (2011) An approach to handling non-Gaussianity of parameters and state variables in ensemble Kalman filtering. Adv Water Resour 34:844–864

10

An improved influence function method for predicting subsidence caused by longwall mining operations in inclined coal seams

Yi Luo[1]

Abstract Prediction of surface subsidence caused by longwall mining operation in inclined coal seams is often very challenging. The existing empirical prediction methods are inflexible for varying geological and mining conditions. An improved influence function method has been developed to take the advantage of its fundamentally sound nature and flexibility. In developing this method, the original Knothe function has been transformed to produce a continuous and asymmetrical subsidence influence function. The empirical equations for final subsidence parameters derived from collected longwall subsidence data have been incorporated into the mathematical models to improve the prediction accuracy. A number of demonstration cases for longwall mining operations in coal seams with varying inclination angles, depths and panel widths have been used to verify the applicability of the new subsidence prediction model.

Keywords Subsidence prediction · Influence function method · Inclined coal seam · Longwall mining

1 Introduction

Longwall mining in inclined coal seams has been a common practice in many major coal mining countries. The characteristics of the subsidence basin induced by longwall operations in inclined coal seam are different from that caused by mining in a level seam. Accurately predicting final surface subsidence over such mining operations has been a challenge in the field of subsidence research. Most of the existing subsidence prediction methods for inclined seam are empirical type such as graphical and profile function methods. They are site specific and inflexible in dealing with the variations in geological and mining conditions. As a trend in developing subsidence prediction tools, the more versatile influence function methods (IFM) is considered to be more suitable in developing subsidence prediction method for mining operations conducted in inclined coal seams. Kratzsch (1983) briefly listed some methods developed based on the IFM concept in predicting final subsidence. The more practical one, the zone area method, requires significant manual calculations and results in insufficient accuracy. The author has previously proposed one analytical method based on the IFM concept (Luo and Cheng 2009). A piece-wise asymmetric influence function depending on the seam inclination angle has been proposed. Based on the influence function, a mathematical model to evaluate the final surface movements and deformations has been developed. However, the piece-wise influence function can result in abnormal discontinuities in the resulting final movement and deformations at some points and the computation process is also cumbersome.

As an improvement to the previous mathematical model, a new influence function is proposed using an approach to mathematically project the original influence function for flat coal seam onto the inclined coal seam. The resulting influence function meets all essential requirements for its definition and produces continuous final surface movements and deformations.

In this paper, the development process of the new influence function has been detailed. For the completeness

✉ Yi Luo
yi.luo@mail.wvu.edu

1 Department of Mining Engineering, West Virginia University, Morgantown, WV, USA

of the method, the essential mathematical derivations in the previous paper are modified and presented. The mathematical formulae for evaluating the final surface movements and deformations are given. The validity of the new model is demonstrated with some examples.

2 Proposed influence function

The two fundamental steps of employing influence function methods in subsidence prediction are: (1) definition of the influence function that describes the distribution of subsidence influence on the ground surface caused by the extraction of one element of the coal seam, and (2) integration of the influence function over the "mine area". A proper form of influence function should be carefully defined so that it can well represent the mechanism involved in the subsidence process.

In this paper, the original form of the influence function for subsidence caused by the extraction in flat coal seam of the Knothe's Theory [Eq. (1)] is modified to represent the asymmetrical subsidence influence along the seam dipping direction. In the equation, x' is the horizontal distance in a local coordinate system to be explained later. The S_{max} and R are the maximum possible subsidence and radius of major influence, respectively. The definitions of these two important subsidence parameters have been given in other publications (Luo 1989; Peng and Luo 1992; Peng et al. 1995).

$$f_s(x') = \frac{S_{max}}{R} e^{-(x'/R)^2} \quad (1)$$

Figure 1 shows the proposed scheme of applying influence function method in predicting final surface subsidence caused by longwall mining in an inclined coal seam. A convention is followed to establish the global coordinate system (O-X) in which the origin (O) is located directly above the left (upper) panel edge and its positive direction points to the right (lower) panel edge side. The surface point (P) where the final surface subsidence is to be predicted is located x_p distance from the origin of the global coordinate system. A number of important parameters involved in defining the influence function are:

(1) *The limit angles on the lower and upper sides of the panel* (γ_H and γ_L), *respectively*. When they are plotted upwards from the lower and upper edges of the panel, they specify the edges of the final subsidence basin, respectively. When they are plotted from a surface point downward, they delineate the zone in which the extraction in the coal seam would influence the surface point to subside. These two lines are the lower and upper influence boundary lines. Generally, the limit angles depend on the angle of the seam inclination, α. Rom (1964) developed a graph to determine γ_H and γ_L based on German subsidence experience. Chinese also derived their own empirical formulae for the limit angles for a number of mining districts (Fengfeng Subsidence Research Group 1982).

(2) Nadir angle (μ) shows the spatial relation between a specified surface point (point P in Fig. 1) and the extraction point in the coal seam that influences surface point P to subside the most amount (point Z). It should be noted that the P-Z line equally divides the angle formed between the lower and upper influence boundary lines drawn downwards from the surface point P. The nadir angle is determined by

$$\mu = \frac{1}{2}(\gamma_L - \gamma_H), \text{deg.} \quad (2)$$

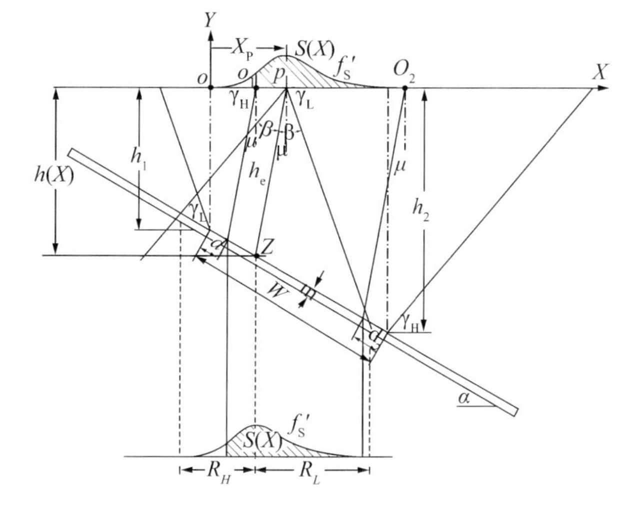

Fig. 1 Relations for using influence function method to predict final surface subsidence over a longwall gob in inclined coal seam

The effective radii of influence function (R_L and R_H) on lower and upper sides of the point of maximum extraction influence (Z) are shown in Fig. 1. They can be determined by finding the intersection points between the inclined coal seam and the lower and upper influence boundary lines, respectively. The line equations for the lower and upper major influence boundaries (γ_L and γ_H) that intersect the prediction point at x_p are shown in Eqs. (3) and (4), respectively.

$$y = (x_p - x)\tan\gamma_L \quad x > x_p \quad (3)$$

$$y = (x - x_p)\tan\gamma_H \quad x \le x_p \quad (4)$$

The line equation for the inclined coal seam is defined by Eq. (5) where h_1 is the overburden depth at upper panel edge.

$$y = -h_1 - x\tan\alpha \quad (5)$$

The coordinates of the intersection point between the lower influence boundary and the coal seam are determined from Eqs. (3) and (5) as:

$$x_L = \frac{x_p \tan\gamma_L + h_1}{\tan\gamma_L - \tan\alpha} \tag{6}$$

$$y_L = (x_p - x_L)\tan\gamma_L \tag{7}$$

Similarly, the coordinates of the intersection point between upper influence boundary and the coal seam are determined as:

$$x_H = \frac{x_p \tan\gamma_H - h_1}{\tan\alpha + \tan\gamma_H} \tag{8}$$

$$y_H = (x_H - x_p)\tan\gamma_H \tag{9}$$

The location of the maximum extraction influence point (*Z*) is determined by finding the intersection point between the line equation of *P-Z* and that of the inclined coal seam. The coordinates of the maximum extraction influence point are:

$$x_Z = \frac{\frac{x_p}{\tan\mu} - h_1}{\frac{1}{\tan\mu} + \tan\alpha} \tag{10}$$

$$y_Z = -h_1 - x_Z \tan\alpha \tag{11}$$

The effective radii of major influence on the lower and upper sides can be simply determined using Eqs. (12) and (13), respectively. These radii of major influence are dependent on x_p, α, γ_L, γ_H and overburden depth.

$$R_L = x_L - x_Z \tag{12}$$

$$R_H = x_Z - x_H \tag{13}$$

The other parameter to define the influence function is the maximum possible subsidence, S_{max}. In general, S_{max} decreases as the overburden depth increases. Therefore, S_{max} varies with x_p. It is reasonable to assume that for an inclined coal seam S_{max} at a prediction point is a function of the effective distance (h_e) instead of the true depth. The effective distance is simply calculated from the coordinates of the prediction point (x_p, 0) and the point of maximum extraction influence (x_Z, y_Z).

$$h_e = \sqrt{(x_Z - x_p)^2 + y_Z^2} \tag{14}$$

The true subsidence factor can be estimated by substituting h_e into the empirical equation derived previously by the author from the collected US and Australian longwall subsidence data (Luo and Peng 2000). It should be noted that the unit of the effective distance h_e in the following empirical equation is ft and one foot is 0.3048 m.

$$a = 1.9381(h_e + 23.4185)^{-0.1884} \tag{15}$$

The maximum possible subsidence at surface point x_p for an inclined coal seam is determined by Eq. (16).

$$S_{max} = a\,m\cos\alpha \tag{16}$$

The influence function for subsidence in Knothe's theory (Knothe 1957) proven to be accurate for mining in flat seam [Eq. (1)] should be modified to represent the distribution of subsidence influence due to the extraction in an inclined coal seam according to a special transformation scheme. The transformation involves a mathematical projection of the symmetric influence function for flat seam, f_s, onto the inclined seam to form the new asymmetric influence function of f'_s as shown in Fig. 2. It should be noted that local coordinate systems are used in the transformation of the influence functions. For the flat seam, the local coordinate system is o'–x' while that for the inclined seam is o'–ξ'.

The projection starts from point (*P*) on ground surface. The first step is to project the influence boundaries for the upper and lower sides to the flat and inclined seams, respectively. When they intersect with the flat seam, the resulting radii of influence on both sides of the origin of the local coordinate system *o'* are the same as *R*. However, the resulting radii of major influence of the intersections with the inclined seam are different. On the upper side, the radius is R_H which is smaller than *R* while that on the lower side R_L is larger than *R*.

To confirm with influence function definition, the first two essential requirements for the mathematical projection should be met: (1) the total area under the influence function f_s is equal to that under f'_s; (2) the area under each

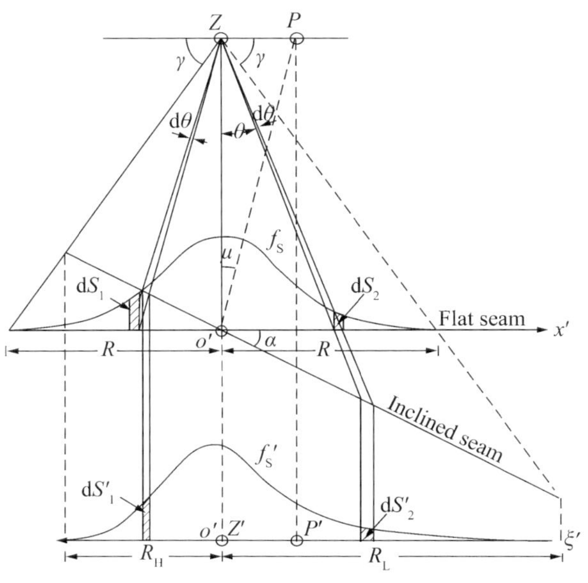

Fig. 2 Schematic of transforming influence function from flat seam to inclined seam

half of the influence functions on the left side of the origin o' should be equal to that on the right side.

In the second step of defining the influence function, an incremental angle $d\theta$ is drawn from the projection angle θ as shown in Fig. 2. The left and right boundary lines of $d\theta$ intersect the flat and inclined seams at different depths. The resulting widths along the horizontal direction for the intersections are different on the flat and inclined seams. Apparently, on the upper side the origin o', the width of the intersection on inclined seam is smaller than that on the flat seam while on the lower side where a reversed phenomenon is observed. The change in width can be more precisely described by the transformation of the horizontal coordinate in the respective local coordinate systems from x' to ξ' in Eq. (17). In the equation, h is the actual overburden depth and $\theta = \tan^{-1}(x'/h)$. Apparently, the transformed local coordinate ξ' in the inclined seam depends on the seam inclination angle α, overburden depth h and the local coordinate x'.

$$\xi = \frac{x'h}{\sqrt{x'^2 + h^2}} \frac{\cos\alpha}{\sin\left[180^\circ - \alpha - \tan^{-1}\left(\frac{x'}{h}\right)\right]} \tag{17}$$

The transformation of the influence function must ensure that the elemental influence for the flat seam, dS_2 as shown in Fig. 2, is equal to that for the inclined seam, dS'_2—the third essential requirement. To meet this requirement, adjustments must be made on the height or value of f'_s ccording to the change in intersection widths. This adjustment is accomplished through the dividing the original influence function $f_s(x')$ in Eq. (1) by the first derivative of ξ' [Eq. (18)] with respect to x'. The derivative term is shown in Eq. (19).

$$f'_s(\xi') = \frac{f_s(x')}{\frac{d\xi'}{dx'}} \tag{18}$$

$$\frac{d\xi'}{dx'} = \frac{h^2[1 + \cos(2\alpha)]}{2\left[\alpha + \tan^{-1}\left(\frac{x'}{h}\right)\right](h^2 + x'^2)} \tag{19}$$

The new subsidence influence function is then defined using Eqs. (18) and (19). Both Eqs. (1) and (19) are continuous and Eq. (19) has no zero point in the range from $-\infty$ to ∞. The features ensure the new subsidence influence function to be continuous through the same range—an important requirement to result in continuous final surface movements and deformations.

Based on the proposed transformation, the new influence functions for a number of seam inclination angles (i.e., $\alpha = 0°, 15°, 25°, 35°$ and $45°$) for an overburden depth of $h = 800$ ft (244 m) are plotted in Fig. 3. The continuous nature for each of the new influence function is apparent. The asymmetry of each influence function, skewing toward the upper panel edge side, becomes more severe as the angle of seam inclination increases. The peak of the influence function moves toward the upper panel edge side and the magnitude increases slightly also with the inclination angle. Therefore, the transformed influence function meets all the essential requirements as stated previously.

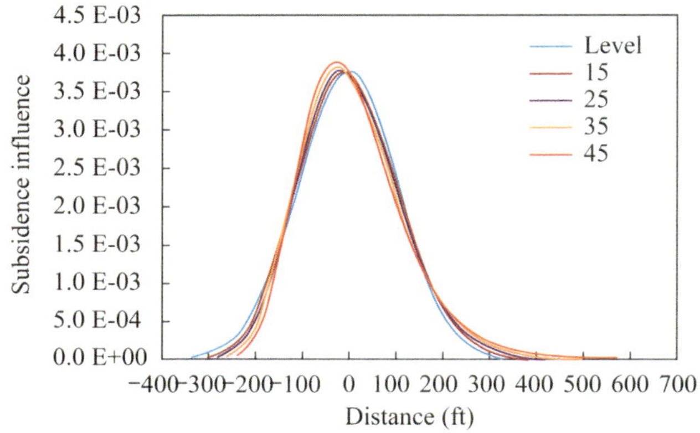

Fig. 3 Influence functions for various seam inclinations ($h = 800$ ft or 244 m)

3 Final surface movements and deformations

The final surface movements are obtained by integrating the influence functions. The final deformations are the derivatives of the final surface movements.

3.1 Determination of final subsidence

Based on the concept of the influence function method, the final subsidence at a surface point is the summation of all influences received at this point caused by the extraction in the coal seam within the influence boundaries. Mathematically, it is the integral of the subsidence influence function over the "mined" area. After considering the overhanging overburden strata over the panel edges and an equivalent transformation of the coordinate system, the final surface subsidence at the prediction point can be determined by integrating the transformed influence function between the left and right inflection points, O_1 and O_2 (Fig. 1). The offsets of the inflection points at the coal seam level (d_1, and d_2) are first determined using the empirical formula [Eq. (20)] derived from the collected US subsidence data (Peng et al. 1995). The actual overburden depths on the left and right edges of the panel (h_1 and h_2) are used in Eq. (20) to obtain the offsets of inflection points on the left and right sides of the panel, respectively. It should be noted that the units for the depth and offset distance in the empirical equation are ft.

$$d_{1,2} = h_{1,2}\left(0.382075 \times 0.999253^{h_{1,2}}\right) \tag{20}$$

It is reasonable to project the two inflection points from the coal seam to the surface with the nadir angle (μ) since this angle signifies the line of the maximum influence on the ground surface caused by the underground extraction (Fig. 1). Through such projection, the coordinates of the left and right inflection points in the global coordinate system, x_1 and x_2, are determined as:

$$\begin{cases} x_1 = d_1 \cos\alpha + (h_1 + d_1 \sin\alpha)\tan\mu \\ x_2 = (W - d_2)\cos\alpha + (h_2 - d_2 \sin\alpha)\tan\mu \end{cases} \tag{21}$$

In performing the integration of the influence function between the inflection points, the inflection points should be expressed in local coordinates.

$$\begin{cases} \xi_1' = x_1 - x_{\mathrm{p}} \\ \xi_2' = x_2 - x_{\mathrm{p}} \end{cases} \tag{22}$$

The Final subsidence at prediction point, x_{p}, is shown in Fig. 1 as the shaded area and is determined as

$$S(x_{\mathrm{p}}) = \int_{\xi_1'}^{\xi_2'} f_s'(\xi')\mathrm{d}\xi' \tag{23}$$

3.2 Determinations of final movement and deformations

The other final surface movement (i.e., horizontal displacement) and deformations (i.e., slope, strain and curvature) along a major cross-section in the dipping direction of the inclined seam are directly related to or as the derivatives of the final surface subsidence. Since the depth at a given prediction point h is dependent on the location x, obtaining the analytical expressions for the derivatives will be very cumbersome if it is possible. Numerical differentiation techniques are used to evaluate the final horizontal displacement, slope, strain and curvature. The final surface slope is defined as:

$$i(x_{\mathrm{p}}) = \frac{\mathrm{d}S(x_{\mathrm{p}})}{\mathrm{d}x_{\mathrm{p}}} \tag{24}$$

Based on the traditional subsidence theories, the final horizontal displacement is proportional to the final slope. For a flat coal seam, the proportionality coefficient is defined as R^2/h where R is the radius of major influence and h is the overburden depth (Luo 1989). It has been found that R is normally one third of h worldwide and the proportionality coefficient is $h/9$. For mining in an inclined coal seam, the effective distance h_{e} replaces h for determining the proportionality coefficient. Therefore, the horizontal displacement at the prediction point is defined as:

$$U(x_{\mathrm{p}}) = \frac{h_{\mathrm{e}}(x_{\mathrm{p}})}{9} i(x_{\mathrm{p}}) \tag{25}$$

The final surface strain and curvature at the prediction point are the first derivatives of the final horizontal displacement and slope as shown in Eqs. (26) and (27), respectively.

$$\varepsilon(x_p) = \frac{\mathrm{d}}{\mathrm{d}x_{\mathrm{p}}} U(x_{\mathrm{p}}) \tag{26}$$

$$K(x_{\mathrm{p}}) = \frac{\mathrm{d}}{\mathrm{d}x_{\mathrm{p}}} i(x_{\mathrm{p}}) \tag{27}$$

3.3 Verification of the new model

In order to verify the proposed mathematical model, a computer program developed in MathCad is used to carry out the required computations of final surface subsidence prediction for longwall mining operations conducted in inclined coal seams. Figure 4 shows the predicted final subsidence profiles of a 1000 ft (305 m) wide longwall panel in a 5.5 ft (1.68 m) thick coal seam with different inclination angles (i.e., 0°, 15°, 30° and 45°). The overburden depth at the upper panel edge is 600 ft (183 m). For mining conducted in level coal seam (0° inclination angle), the final subsidence shows a super-critical final subsidence basin symmetric about the panel center. As the seam inclination increases, the asymmetric nature becomes severer with the shape of the subsidence basin on the lower half being gentler than that on the upper side. The maximum possible subsidence, $S_{\max}$, decreases as the inclination angle increases as a result of Eq. (16). The subsidence profiles also shift toward the lower side and more and more significant subsidence occurs beyond the lower panel edge. Different from the other three subsidence profiles, the one

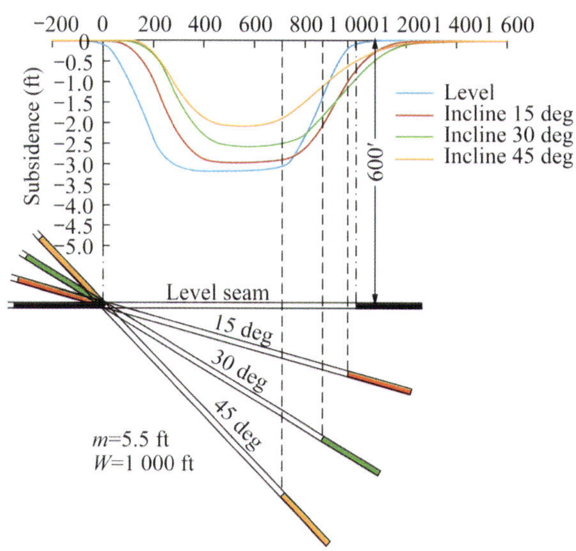

Fig. 4 Predicted final subsidence profiles over 1000 ft (305 m) wide panel in 5.5 ft coal seam with different inclination angles

for the 45° inclined seam shows a subcritical subsidence basin.

Figure 5 shows three predicted final subsidence profiles over a 30° inclined coal seam when the widths of the longwall panel are 800, 1000 and 1500 ft (244, 305 and 457 m), respectively. The mining height and the overburden depth at the upper panel edge are 7.0 and 700 ft (2.13 and 213 m), respectively. The locations of the lower panel edges are also plotted in the figure. In addition to producing asymmetric subsidence profiles, the ability for the influence function method to adapt the change in panel width automatically from sub-critical (w = 800 ft or 244 m), critical and super-critical (w = 1500 ft or 457 m) conditions is evident. This is the most important advantage of the influence function methods over the empirical methods. In the supper-critical condition, a large "flat" subsidence basin bottom appears in the central portion of the basin. However, the maximum possible subsidence in the "flat" bottom is no longer an uniform amount as expected for flat coal seam but decreases slightly toward the lower side of the panel.

The most important improvement for introducing this new continuous influence function is to ensure the continuity of the resulting final horizontal displacement, slope, strain and curvature. Figure 6 shows the predicted final strain profiles for the same examples shown in Fig. 5. The three profiles show smooth and continuous nature throughout their respective entire ranges. The peak values of the maximum tensile and compressive strains decrease as the overburden increases.

The application examples in this section show the desirable attributes of surface final subsidence caused by longwall mining operations in inclined coal seam. The magnitudes and distributions of predicted final surface movements and deformations are reasonable. Therefore, the improved model based on the concept of influence function method is a good tool for subsidence prediction caused by longwall mining operations in inclined coal seam.

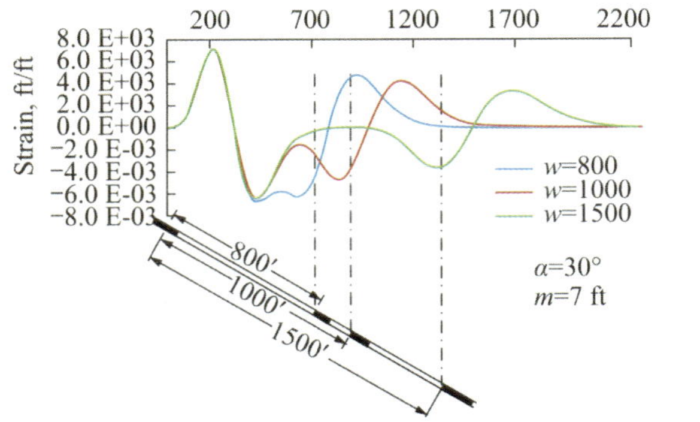

Fig. 6 Predicted final strain profiles for different panel widths in 30° coal seam

4 Case demonstrations

In order to further demonstrate the capabilities of the improved mathematical model, two Chinese cases with moderate and steep coal seam inclinations are predicted using the developed model and presented in this section.

Figure 7 shows the predicted final surface and deformation profiles for a typical subsidence case in Jixi mining district. In this case, the coal seam dips toward the right at an angle of 18°. The depth on the upper panel edge is 215 ft (65 m). The mining height is 4.5 ft (1.37 m). The panel is 600 ft (183 m). Due to the relatively small depth, the resulting subsidence basin is a supercritical one. However, the flat bottom portion of a supercritical subsidence basin is not necessarily flat but the maximum subsidence there decreases as the depth increases. The subsidence profile on the deeper side is much gentler than that on the shallow side. Another observation is that the maximum horizontal displacement, tensile and

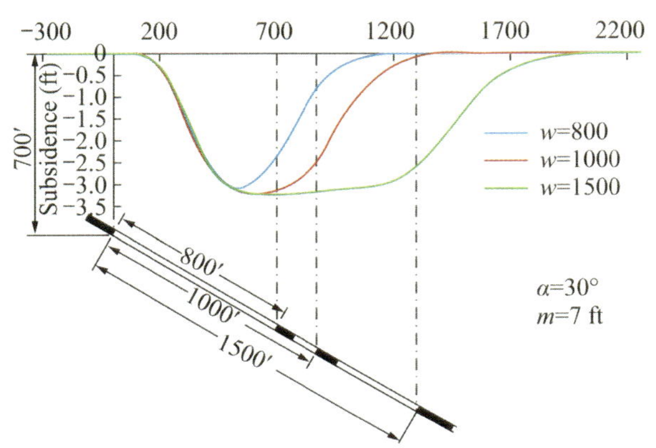

Fig. 5 Predicted final subsidence profiles for different panel widths in 30° coal seam

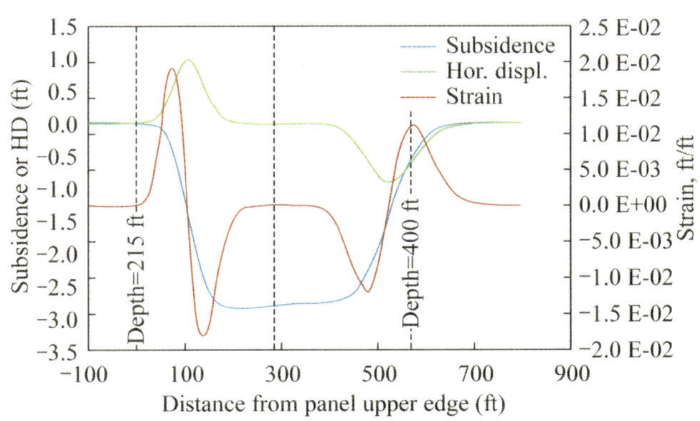

Fig. 7 Predicted final subsidence, horizontal displacement and strain profiles for Jixi mining district

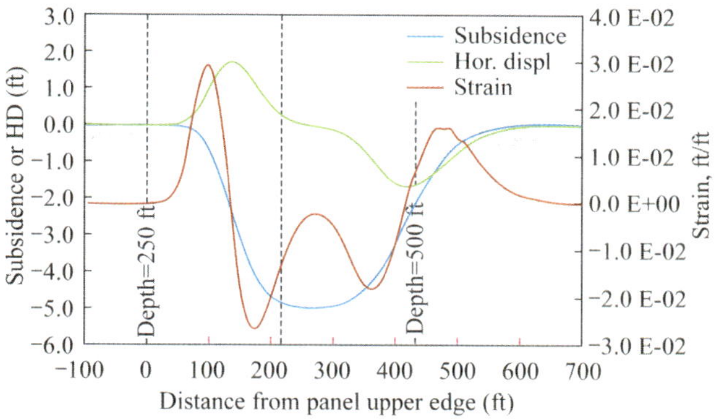

Fig. 8 Predicted final subsidence, horizontal displacement and strain profiles for Huainan mining district

compressive strains on the deeper side are smaller than those on the shallower side as expected.

Figure 8 shows the predicted final surface movement and deformation profiles typical for Huainan mining district. The geological and mining information includes: $\alpha = 30°$, $h_1 = 250$ ft (76.2 m), $W = 500$ ft (152 m) and $m = 9.0$ ft (2.74 m). Due to the much larger mining height, the maximum movements and deformations are larger than those in the previous cases.

Compared to the previous model (Luo and Cheng 2009), the discontinuity problems at some prediction points are totally avoided in the predicted strain profile (similarly in slope, horizontal displacement and curvature profiles) using the new prediction method.

5 Conclusions

The mathematical model to employ influence function method for predicting final surface subsidence over a longwall mining operation in an inclined coal seam has been improved. The main improvement is to define a continuous asymmetric subsidence influence function to represent the influence to surface subsidence induce by extracting an element in an inclined coal seam. The degree of asymmetry of the influence function is dependent on the angle of the seam inclination. The determination of the final subsidence at a prediction point is performed through integrating the influence function between the inflection points. Once the final surface subsidence is determined, the other movement and deformations are obtained through numerical differentiations.

Based on the case demonstrations, the improved influence function method represents the characteristics of the final subsidence basin formed over longwall gob in an inclined coal seam well. Its flexibility to deal with varying seam inclination and degree of critical extraction makes the influence function method a much better tool in subsidence prediction than the existing methods. It provides a more accurate and faster tool to perform the complicated calculations for the final surface movements and deformations.

References

Fengfeng Subsidence Research Group (1982) Research findings on subsidence characteristics in Fengfeng Mining District. Fengfeng Coal Sci Technol 3 (in Chinese)

Knothe S (1957) Observation of surface movements under influence of mining and their theoretical interpretation. In: Proceedings of the European congress of ground movements, Leeds, pp 210–218

Kratzsch H (1983) Mining subsidence engineering. Springer-Verlag, New York, p 543

Luo Y (1989) Integrated computer model for predicting surface subsidence due to underground coal mining-CISPM. PhD Dissertation, Morgantown: West Virginia University, 168. (UMI order No. 9020385)

Luo Y, Cheng JW (2009) An influence function method based subsidence prediction program for longwall mining operations in inclined coal seams. J Min Sci Technol 19(9):592–598

Luo Y, Peng SS (2000) Prediction of subsurface subsidence for longwall mining operations. In: Proceedings of the 19th international conference on ground control in mining, pp 163–170

Peng SS, Luo Y (1992) Comprehensive and integrated subsidence prediction model-CISPM (V2.0). In: Proceedings of the 3rd workshop on surface subsidence due to underground mining, Mor-gantown, pp 22–31

Peng SS, Luo Y, Zhang ZM (1995) Subsidence parameters - their definitions and determination. AIME-SME Trans 300:60–65

Rom H (1964) A limit angle system. Mitt Markscheidew 71:197–199

11

Design of experiments to optimize the extraction parameters of a power grade Indian coal

Heena Dhawan[1] · Sreedevi Upadhyayula[2] · D. K. Sharma[1]

Abstract Low grade high ash coals with difficult physical washability characteristics require chemical cleaning techniques such as solvent extraction to obtain super clean coal with lower ash contents. Kedla, an Indian low grade coal has ash content around 44%–50% and there is a need to maximize its solvent extraction yield so as to make this power grade coal industrially useful. This paper highlights the use of definitive screening design and response surface methodology to extract the maximum organic matter from Kedla coal using ethylenediammine (EDA) and *N*-methyl pyrollidone (NMP) as solvents under mild atmospheric pressure conditions. The coal was pre-treated with acetic acid before extractions were carried out. The parameters such as the acid pre-treatment time, acid concentration, particle size, the co-solvent concentration and the solvent concentration were optimized in this paper. With this design of experiments technique about 40% of the organic matter with minimal ash (less than 2%) content was rendered extractable unlike the already postulated designer solvent system—e, N system (i.e. NMP containing small amount of EDA) that uses coal to solvent ratio as 1:17 (wt/vol) and coal to co-solvent ratio as 1:1(wt/vol) thus only giving 19% extraction yield. Therefore, the present design of experiments technique helps in conserving the number of experiments for process intensification.

Keywords Coal · Extraction yield · Acid pre-treatment · Definitive screening design · Response surface methodology

1 Introduction

Coal is one of the most widely used fuels. It is used both as a fuel for generation of electricity and as a non- fuel for the production of value added chemicals and products (Song and Schobert 1993, 1996; Nyathi et al. 2013). Coal is also the biggest contributor to the pollution in the world. With the limitations being imposed on the CO_2, NO_2 and other harmful gases emissions into the atmosphere, it becomes all the more important to have cleaner coal combustion and utilization. Coal is basically a condensed macro-molecular polymeric structure (Krevelen 1993; Larsen et al. 1985; Hall and Larsen 1993; Chen et al. 1998). Most of its structural constituents are aromatics, paraffinics, heteroaromatics and hydroaromatics which are held by various covalent and non-covalent interactions. These interactions are vanderwaal interactions, aromatic-aromatic, charge transfer and donor acceptor interactions between the aromatics and heterocyclic units (Haenel 1992; Nishioka and Larsen 1990; Larsen and Mohammadi 1990; Otake and Suuberg 1997; Painter et al. 1990; Rahman et al. 2017).

Indian coals are of the drift origin and the mineral matter is finely dispersed in the organic matter of these coals (Kumar and Patel 2008; Sharma and Giri 2016). High ash contents also lead to the increase in the transportation cost of coals, reduced calorific values and also results in slagging and fouling of the boilers, turbines, gasifiers etc. Therefore, it becomes important both economically and environmentally to reduce the ash contents in these coals.

✉ Heena Dhawan
heena.6sep@gmail.com

1 Centre for Energy Studies, Indian Institute of Technology, New Delhi 110016, India

2 Department of Chemical Engineering, Indian Institute of Technology, New Delhi 110016, India

Ash content varies from 20% to 55% in Indian coals but their sulphur contents are comparatively lower (< 0.6%) (Mishra 2009). There are various methods to reduce the ash present in coals- physical, chemical and biological. However, physical techniques use less expensive resources such as water, air pulsations etc. Chemical treatments such as organo-refining and chemical leaching give better extractability of coals with much reduced ash contents (Balakrishnan et al. 2015; Meshram et al. 2015). Biological treatments use microbes to digest the mineral matter in coals (Jain and Sharma 2004). The effectiveness of each method depends upon the degree of dispersion of ash in the organic matter of coals.

Coal consists of two components—organic matter (carbonaceous matrix) and inorganic matter (mineral matter). Organic matter can be extracted using solvent extraction which leaves behind solvent refined clean coal product which has negligible ash contents (Kihong et al. 2016). The role of the solvent is to break the coal–coal interactions and replace it with the coal solvent interactions. Coal being a heterogeneous polymer shows a tendency to swell in organic solvents (Takanohashi and Iino 1995; Hernández et al. 2012; Hu et al. 1998; Mathews et al. 2015). Some solvents that have the characteristic property of donating a pair of electrons show good swelling. Mostly N and O donors are considered to be good solvents for coal extraction. Swelling gives a good idea of the extent of coal interaction with the solvent that may not necessarily always account for good extraction e.g. ethylenediammine (EDA) (Krevelen 1993). Early studies on coal extraction include the use of some basic solvents such as benzene, $CHCl_3$, alcohol, ether, carbon disulphide as reported by Van Krevalan (1993). Coal derived solvents are also used for extraction. Coal tar and coal tar pitch were used by Honda et al. Higher boiling fractions like anthracene oils gave better extraction yields (Mishra and Sharma 1990; Elliott 1981). About 95% extraction yield has been reported by Orchin et al. with the use of boiling phenanthrene as a solvent (Krevelen 1993).

N-methyl pyrollidone (NMP) is considered to be a good solvent for coal extraction (Renganathan and Zondlo 1993; Takanohashi et al. 1996; Kirk and Othmer 2006; Dryden 1948). It is an H-bond acceptor and has good interacting ability with the aromatic structures. It is used as a solvent in oil refineries for the separation of olefins and aromatics. Renganathan and Zondlo have reported as high as 50% extraction yields with NMP under atmospheric pressure conditions (Renganathan and Zondlo 1993; Renganathan et al. 1988). EDA has the ability to interact with the coal structure most powerfully by disrupting the –OH bonds. It is considered to be a good swelling solvent (Shibaoka et al. 1979; Suuberg et al. 1994; Kwak 2005).

Response surface methodology (RSM) is an analytical technique used for solving several scientific problems with minimal experimentation (Chandaliya et al. 2012). The purpose behind using this technique is to maximize the dependent output variable (response) based on the different independent input variables. A number of parameters rule a particular reaction and this methodology gives an optimum outcome of response surface corresponding to a certain reaction for maximum desirability of a particular product (Li et al. 2004; Kincl et al. 2005; Aslan and Cebeci 2007; Gunaraj and Murugan 1999; Azimi et al. 2013; Segurola et al. 1999). If y is the response corresponding to the x set of variables that govern the reaction, the relationship could be expressed as shown in Eq. (1),

$$Y = f(x_1 \times x_2 \times x_3 \times \ldots x_m) \qquad (1)$$

The parameters or the factors for a reaction could be continuous, i.e. the ones that continuously affect the course of a reaction or categorical, that do not continuously affect the course of a reaction and generally this expression is expressed in the form of a second order equation as shown in Eq. (2),

$$Y = \beta_0 + \sum \beta_i x_i + \sum \beta_{ii} x_i^2 + \sum \sum \beta_{ij} x_i x_j + £ \qquad (2)$$

where x stands for the variables that are the input parameters of the reaction and the β for the unknown coefficients and £ is the error. The β values are calculated using the second order model. The software using these equations gives a minimum number of experiments to be performed in order to get the desired output.

The present paper highlights the use of definitive screening design (DSD) and response surface methodology (RSM) in order to maximize the extraction yield of an Indian Coal, Kedla. It involves a two step process, pre-treatment with acetic-acid followed by extraction with NMP containing a small amount of EDA (co-solvent). Being a high ash coal, pre-treatment with acid enhances the extraction yield as the acid loosens the coal structure and breaks important linkages in coal making the inroads of the solvent easier inside the intermolecular spaces in coal (Masaki et al. 2005; Li et al. 2004). The work on the development of designer solvent e, N where coal: NMP ratio is taken as 1:17 (wt/vol) and coal:EDA ratio taken as 1:1 (wt/vol) has already been studied before in author's laboratory (Pande and Sharma 2002; Iino et al. 1988). However, presently the optimization of acid pre-treatment conditions followed by solvent extraction using design of experiments has not been reported before. Parameters taken for the study in the present paper are coal particle size, pre-treatment of coal with acetic acid by varying the coal to acid ratio, acid pre-treatment time, coal to solvent ratio, % of co-solvent in the solvent. The solvent extraction time has been optimized separately.

2 Materials and methods

2.1 Coal sample and the solvents

Kedla coal, an Indian non-coking power grade coal was procured from NTPC (National Thermal Power Corporation), New Delhi. The proximate and ultimate analysis was performed according to IS 1350 standards. The ultimate and proximate analysis of the coal sample and the properties of the solvents used are presented in Tables 1 and 2. The acetic acid from Fischer Scientific, *N*-methyl pyrrolidone (NMP) from SRL and ethylenediamine (EDA) from LOBA chemicals were used for present extraction studies.

Equations used for the calculations:

On clean coal product basis:

$$\text{Extraction yield (\% daf basis)} : \frac{weight(clean\ coal)}{1 - \frac{ash\%}{100} \times weight(clean\ coal)} \times 100\%$$

On loss in weight basis:

$$\text{Extraction yield (\% daf basis)} : \frac{initial\ weight(raw\ coal) - weight(clean\ coal)(100 - ash\%)}{1 - \frac{ash\%}{100} \times weight(clean\ coal)} \times 100\%$$

Daf: dry ash free basis

2.2 Design of experiment (DOE)

The experiment was designed based on three levels—low, medium and high level factorial design with the variation of five main factors—the coal particle size, pre-treatment time of coal with acid, coal/acid (wt/vol) ratio, co-solvent concentration and the solvent concentration (Tables 3, 4).

Table 1 Proximate and ultimate analysis (wt%) (on air dried basis) of Kedla coal

Proximate analysis			Ultimate analysis (daf)					Calorific value (cal/g)
M_{ad}	A_d	VM_{daf}	C	H	N	S	O[a]	
0.99	44.86	17.18	52.9	3.3	1.1	0.5	42.2	4209

[a]By difference

Then based upon the results for those five parameters, a three level factorial design using RSM methodology was used for further optimization (Tables 5, 6).

Particle size is one of the most important parameters for coal extraction studies as coal extraction largely depends on the lump size that is fed into the reactors. Lesser the particle size better is the interaction of coal with the solvents which ultimately affects the extraction yields. Therefore, the upper limit for particle size was chosen as + 250 μm and lower limit as − 125 μm for the present study. The coal to acid (wt/vol) ratio was taken as 1:10–1:20. This range was of coal amount was found to be suitable for the effective interaction and pre-treatment with acetic acid. Moreover, the acid recovered after filtration can be reused. The time of acid pre-treatment was taken from 30 min (low level) to 120 min (high level) as after the pre-treatment with the acetic acid, extraction for 2 h with the solvents (NMP and EDA) was performed. The coal to solvent ratio was kept between 1:10 and 1:20 as in between these limits, the fluidity of the coal–solvent mixture is maintained. Also with excessive use of solvents, their recovery becomes difficult and the process becomes more uneconomical. 10%–15% of co-solvent EDA of total solvent concentration (vol/vol) was found enough to effectively break important linkages such as H-bonds in coal in order to make inroads for the main solvent NMP into the condensed coal structure (Pande and Sharma 2002).

Another important parameter that was optimized separately was the time of the solvent (NMP–EDA) extraction. It was kept for 2 h for each run and was optimized separately, which has been explained later in the text. Based on the responses of the definitive screening design, a three level continuous factor based RSM experimental design was developed as presented in Table 6.

2.3 Experimental procedure

For each run of the DSD experiment, Kedla coal (2 g) was taken in a 250 mL round bottomed flask along with the acetic acid and refluxed for the desired time as shown in Tables 3 and 4. The solution was then filtered and the coal dried overnight in oven at 105°C. This coal was used for refluxing in a 250 mL round bottomed flask under milder ambient pressure conditions with the given concentration

Table 2 Properties of the two main solvents, NMP and EDA

Solvent	T_b (°C)	ρ (20 °C)	η (N/m^2)	μ (D)	DN	δ (jcm^{-3})$^{1/2}$
N-methyl pyrrolidone	202	1.028	1.67	4.09	27.3	22.9
Ethylenediamine	118	0.903	1.54	1.99	55.0	25.3

T_b, boiling point; ρ, specific gravity; η, viscosity; μ, dipole moment; DN, donor number; δ, solubility parameter

Table 3 Three level factorial design for optimization of five parameters

Levels	Particle size (μm)	Time of acid pre-treatment (min)	Coal/acid ratio(wt/vol)	% Co-solvent (EDA) in solvent (vol%)	Coal/solvent (NMP) ratio (wt/vol)
Low level (− 1)	− 125	30	1:10	10	1:10
Average (0)	125–250	60	1:15	12.5	1:15
High level (1)	+ 250	120	1:20	15	1:20

Table 4 13 Run definitive screening design (DSD) for optimization of five parameters

Run no.	Particle size (μm)	Time of pre-treatment (min)	Coal/acid ratio (wt/vol)	% Co-solvent (EDA) in solvent (NMP + EDA) (vol%)	Coal/solvent (NMP) ratio (wt/vol)	Response (Y) (extraction yield %)
1	1	0	− 1	1	1	23.18
2	1	− 1	0	− 1	1	19.01
3	− 1	− 1	− 1	1	0	27.28
4	− 1	− 1	1	0	1	37.95
5	1	1	1	− 1	0	22.74
6	− 1	1	− 1	− 1	1	30.56
7	1	− 1	1	1	− 1	18.25
8	0	1	1	1	1	27.63
9	− 1	1	0	1	− 1	18.35
10	1	1	− 1	0	− 1	23.66
11	− 1	0	1	− 1	− 1	34.73
12	0	0	0	0	0	12.54
13	0	− 1	− 1	− 1	− 1	28.02

Table 5 Three levels for the factorial design (for RSM methodology—further optimization of extraction parameters)

Item	Particle size (μm)	% Co-solvent (EDA) in solvent (NMP + EDA) (vol%)	Coal/solvent (NMP) ratio (wt/vol)
Low level (− 1)	− 125	10	1:10
Average (0)	125–250	12.5	1:15
High level (1)	+ 250	15	1:20

in Table 4 of the solvent—NMP and co-solvent EDA for 2 h. The coal solvent mixture was cooled and then filtered. The filtrate was distilled-off to remove the excess solvent and the clean coal product was precipitated using water as the anti-solvent. The recovered solvent can be reused. The residual coal and the clean coal product were washed with 2%–5% aq. HCl solution and water. Both were then dried overnight in an oven at 100 °C and weighed to obtain the extraction yield.

For each run of the RSM experiment, the coal sample (2 g) of the particle size, as mentioned in Table 6, was taken in a round bottomed flask and refluxed for about an hour with 30 mL (Based on DSD responses) of acetic acid (Li et al. 2004). The coal–acid mixture was cooled and filtered. The acid treated coal was then dried overnight in an oven at about 100 °C and weighed. This was then refluxed under milder ambient pressure conditions with the amount of the co-solvent (EDA) and the solvent (NMP) as shown in Table 6 for about 2 h. Following the same procedure as already mentioned, the extraction yield was calculated both on clean coal product basis and on loss in weight basis.

3 Results and discussion

3.1 Extraction time

In each run, the solvent (NMP–EDA) extraction time has been taken as 2 h. Another set of experiments with the variation in the experimental time and keeping the coal to EDA (1:1) (wt/vol) and coal to solvent (1:17) (wt/vol) ratio fixed were carried out. The graph obtained for the extraction yield (%) versus time (h) has been shown in Fig. 1.

About 26.2% of the extraction yield could be obtained in just first 2 h of extraction which is about 60% of the total

Table 6 15 Runs experimental set up obtained using response surface methodology-Box Behnken design

S. no.	Particle size	% Co-solvent (A) in solvent (A + B)	Coal/solvent (A + B) ratio	Experimental value–particle size (μm)	Experimental value–co-solvent (mL) (for 2 g coal)	Experimental value–solvent (mL) (for 2 g coal)
1	0	− 1	− 1	125–250	3.75	26.25
2	1	− 1	0	− 125	3	27
3	1	1	0	125–250	3	17
4	0	− 1	1	125–250	6	34
5	1	0	− 1	125–250	2	18
6	0	1	1	+ 250	5	35
7	− 1	0	1	− 125	4.5	25.5
8	0	0	0	+ 250	2.5	17.5
9	1	0	1	− 125	2.5	17.5
10	0	0	0	+ 250	4.5	25.5
11	0	1	− 1	125–250	4	36
12	− 1	− 1	0	125–250	3.75	26.25
13	− 1	0	− 1	+ 250	3	27
14	0	0	0	125–250	3.75	26.25
15	− 1	1	0	− 125	5	35

possible exhaustive extraction yield of 45% (obtained in 24 h). Therefore, keeping in mind, the added extra consumption of resources (electricity, equipment etc.) and avoiding much complexity in extraction conditions, the appropriate extraction time in each NMP–EDA run was taken as 2 h.

3.2 Screening and interpretation of data

Minitab(18) and JMP(11.0) were used to generate the runs and for the interpretation of the responses. The screening for the DSD for the optimization of five parameters gave relevant input that the significance of particle size, % co-solvent concentration and the coal/solvent concentration was much more than the other two parameters (acid pre-treatment time and coal to acid ratio) for the extraction. Their extent of influence was measured by their *p*-values

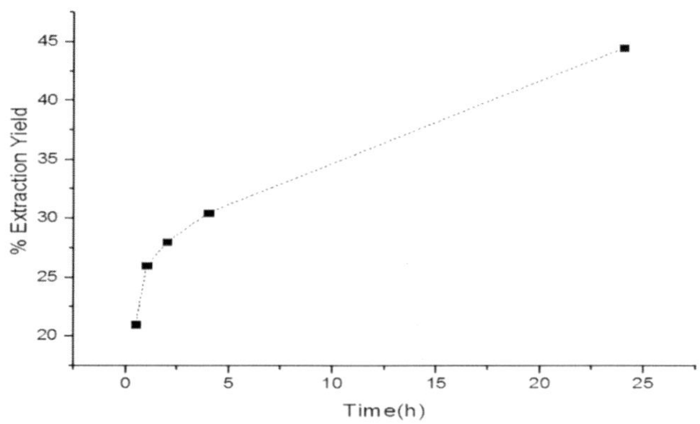

Fig. 1 A plot of extraction yield (%) versus time to optimize the extraction time of Kedla Coal

and *t*-ratios. The particle size shows a larger negative t-ratio value followed by the % co-solvent and then a positive t-ratio for the solvent concentration (Fig. 2). The effect of acetic acid pre-treatment and time of acid pre-treatment were not found to be as high as compared to the other parameters because acetic acid selectively acts on Ca and Mg cations in coal that are bridged through oxygen functional groups especially in low-rank coals (Masaki et al. 2005; Li et al. 2004). Its selectivity to act upon the certain cations in coal makes it relatively less significant to drastically influence the extraction yields. Moreover, NMP acts upon aromatic rings and breaks the strong cross-linkages in the coal macromolecular structure and EDA cleaves H-bonds in coals thereby promoting the action of the main solvent NMP (Pande and Sharma 2002). The particle size provides more surface area for the action of these solvents. Thus these three parameters—particle size, % co-solvent (EDA) concentration and the coal/solvent (NMP) concentration were used again for response surface methodology (RSM) in order to further maximize the extractability.

The responses (extraction yields) on clean coal product basis were found to be in good agreement with the loss in weight basis with a standard error of about 2%–3%. Using JMP (11.0), the effect of the three scaled down parameters-particle size, % co-solvent in solvent (A/A + B) and coal/solvent (Coal/A + B) ratio were studied (Table 7). The responses well fitted with the second order equation obtained with the RSM-Box Behnken method and the R^2 value (ratio of the sum of squares of the predicted responses to the sum of squares of the observed responses) was also found to be 0.966 (Table 8) which verified the good strength of model fitting. Final experimental

Term	Contrast	Plot of t-ratio	Length t-ratio	Individual p value	Value
Particle size	– 3.68627		– 1.93	0.0731	0.4614
%A in solvent	– 1.78657		– 0.94	0.3238	0.9950
Coal/(A + B)	1.34365		0.71	0.4707	1.0000
Coal/acid	0.75427		0.40	0.7251	1.0000
Time of pre-treatment	– 0.66393		– 0.35	0.7550	1.0000
Size * size	1.19698		0.63	0.5650	1.0000
Size * %A in solvent	1.89644		1.00	0.2976	0.9874
%A in solvent * %A in solvent	– 0.05121		– 0.03	0.9827	1.0000
Size * coal/(A + B)	– 1.16618		– 0.61	0.5773	1.0000
%A in solvent * coal/(A+B)	4.06476		2.13	0.0552	0.3600
Coal/(A + B) * coal/(A + B)	2.00977		1.05	0.2704	0.9762
Size * size * %A in solvent	– 0.18155		– 0.10	0.9335	1.0000

Fig. 2 Screening for five parameters of the definitive screening design (DSD). A: co-solvent (EDA) concentration, A + B: total solvent concentration (NMP + EDA)

Table 7 The responses obtained for the design of the experiments RSM

S. no.	Particle size (μm)	% Co-solvent (A) in solvent (A + B) (vol%)	Coal/solvent (A + B) ratio (wt/vol)	Response (Y) extraction yield (%) (on clean coal product basis)
1	0	– 1	– 1	25.6
2	1	– 1	0	13.3
3	1	1	0	11.2
4	0	– 1	1	22.1
5	1	0	– 1	15.4
6	0	1	1	26.5
7	– 1	0	1	37.3
8	0	0	0	19.4
9	1	0	1	18.5
10	0	0	0	18.1
11	0	1	– 1	24.8
12	– 1	– 1	0	36.2
13	– 1	0	– 1	34.0
14	0	0	0	18.2
15	– 1	1	0	36.0

validation was carried out to obtain the maximum extraction yield according the parameter values given by the prediction profiler (Table 9 and Fig. 3).

The most probable parameter values to obtain the maximum extraction yield were predicted by JMP (11.0) an Minitab(18) were : – 125 μm (particle size), 15% co-solvent (EDA) of total solvent (NMP) concentration and coal to solvent (NMP) ratio as 1:20 (wt/vol). The predicted extraction yield according to JMP(11.0) was 39.5% which was validated through experimentation that gave the

Table 8 Summary of fit

R^2	0.966148
R^2 Adj	0.763036
Root mean square error	4.186088
Mean of response	24.17333

extraction yields of 40.8% on clean coal basis and 39.9% on residual coal basis. This final run was repeated three times and the average extraction yields based on clean coal product basis and loss in weight basis have been presented in Table 9. Figure 4 shows the difference in the predicted values and the experimental values of extraction yields (Table 10).

The parameter values that are obtained through the profiler were found to be for factor X_1 (particle size), X_2 (co-solvent concentration) and X_3 (solvent concentration) are correspondingly of the levels − 1, 1, 1 respectively, i.e. minimizing the particle size and maximizing the co-solvent to solvent ratio and the solvent concentration gives the maximum output i.e. an extraction yield of about 39.5% (Table 9) which has been validated through experimentation as discussed earlier in the text. The values of the intercepts show the influence of each parameter individually and the interaction profilers show the influence of second order interactions among the parameters. The intercepts show that the influence of the particle size is the highest with a negative maximum intercept and the t-value depicting the increase in output with subsequent decrease in particle size. The impact of the co-solvent and the solvent is not as significant as compared to that of the particle size. The second order interactions show more impact on the extraction yield than the third order. The p-value for particle size factor was found to be closest to 1 further confirming how influential the factor is on the extraction yield and the next influential factor with the second order value is the solvent concentration that shows the relevance of the nature of the solvent for extraction. The measurement (least or highest) of the p-value is indicative of the significance of that factor. The p-value for the particle size ($\gg 0.05$) (Fig. 2) shows the much higher significance of this parameter on the overall extraction yield (significance of 95% confidence interval > 0.05). Using the above

Table 9 Final validation for maximum extraction yield

Final experimental validation	Particle size (μm)	% Co-solvent in solvent	Coal/solvent ratio (wt/vol)	Predicted value (%)	Experimental value (clean coal product basis) (%)	Experimental value (loss in weight basis) (%)
16	− 1 (− 125)	1 (15%)	1 (1:20)	39.5	40.87	39.98

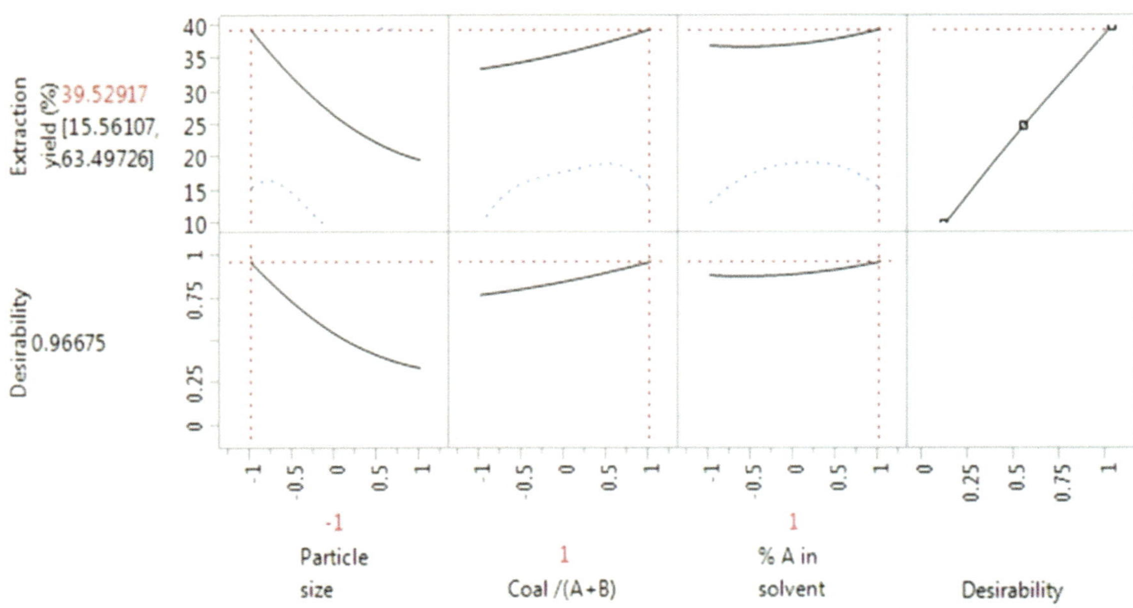

Fig. 3 Prediction profiler showing the variation of extraction yields with different parameters and the maximum desirability value

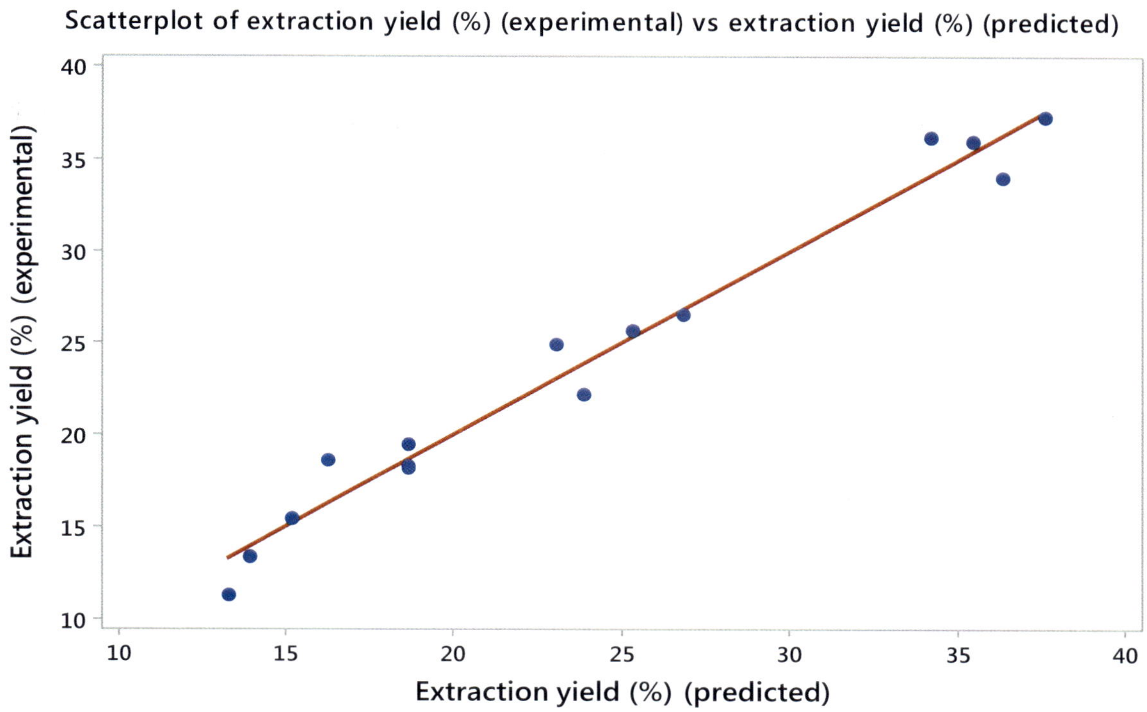

Fig. 4 Plot of difference showing the variation in the extraction yield obtained experimentally and the predicted values through the RSM technique

Table 10 Variation in experimental and the predicted values

Run. no.	Experimental extraction yield (%) (on clean coal product basis)	Predicted extraction yield (%)	Residual extraction yield (%)
1	19.4	18.56	0.84
2	15.4	15.13	0.27
3	36.2	34.17	2.03
4	18.1	18.56	− 0.46
5	26.5	26.78	0.28
6	18.2	18.56	− 0.36
7	37.3	37.56	− 0.26
8	25.6	25.31	0.30
9	34.0	36.31	− 2.31
10	18.5	16.18	2.32
11	24.8	23.03	1.77
12	36.0	35.45	0.55
13	13.3	13.85	− 0.55
14	11.2	13.22	− 2.02
15	22.1	23.86	− 1.76

values, the following equation for the output (extraction yield) based on intercept values (Table 11) has been predicted that helps to study the influence of each individual parameter on the extraction yield:

$$
\begin{aligned}
Y = {} & 20.5666 - 11.925 \times \text{particle size} - 0.45 \\
& \times \text{Coal}/(\text{A}+\text{B}) \\
& + \text{particle size} \times [\text{particle size} \times 2.5791] \\
& + \text{particle size} \times [\text{Coal}/(\text{A}+\text{B}) \times -0.05] \\
& + \text{Coal}/(\text{A}+\text{B})[\text{Coal}/(\text{A}+\text{B}) \times 3.4541] \\
& + \text{particle size} \times [\%\text{A in solvent} \times -0.475] \\
& + \text{Coal}/(\text{A}+\text{B}) \times [\%\text{A in solvent} \times 1.3] \\
& + \%\text{A in solvent} \times [\%\text{A in solvent} \times 1.0291] \\
& + \text{particle size} \times [\text{particle size} \times [\text{Coal}/(\text{A}+\text{B}) \times 2.05]] \\
& + \text{particle size} \times [\text{Coal}/(A+B) \\
& \times [\text{Coal}/(\text{A}+\text{B})2.575]] + \text{particle size} \\
& \times [\text{particle size} \times [\%\text{A in solvent} \times -1.475]]
\end{aligned}
$$

3.3 Contour plots

Figures 5, 6 and 7 show the contour plots and the response surface interactions between the different parameters. The response surface plots between the particle size and the % co-solvent in solvent (Fig. 7a) and particle size and coal to solvent ratio (Fig. 7b) are largely elliptical while the plot between the solvent and the co-solvent concentration (Fig. 7c) is parabolic in nature. It shows the major role played by the coal particle size on the extraction yield as it has already been discussed earlier by its intercept values. With the reduction in particle size, the interaction of the coal with the solvents increases. Giri and Sharma (2000) have reported mass transfer effects in the solvent extraction

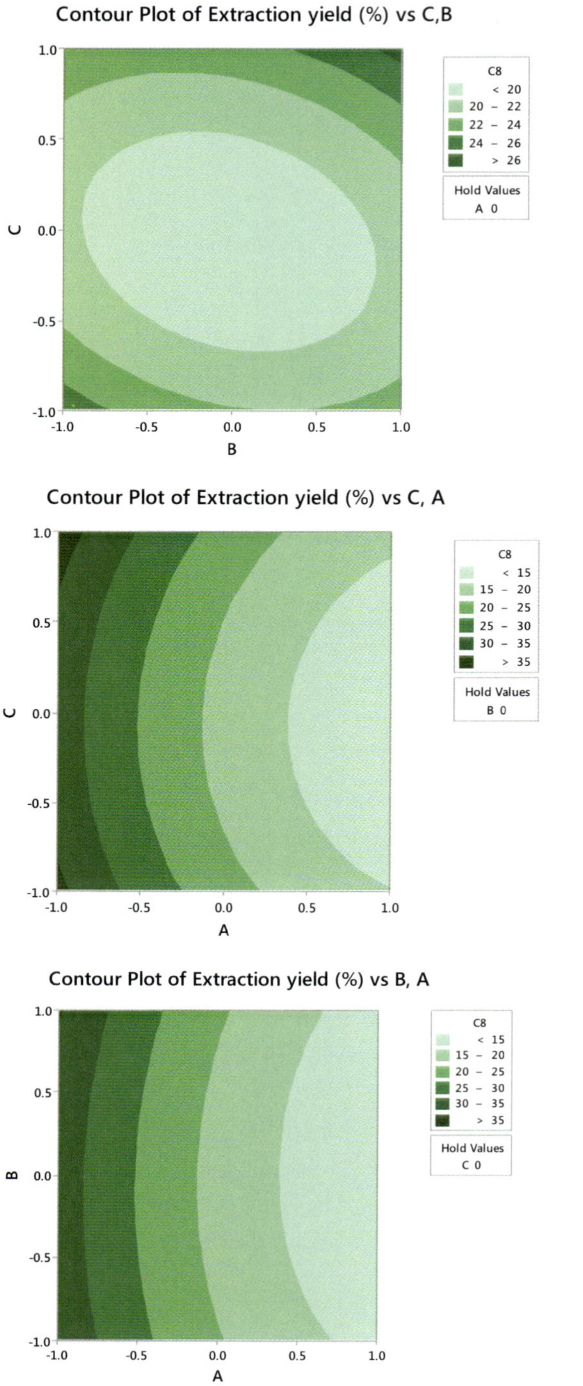

Fig. 5 Contour plots of % extraction yield versus different parameters (A: particle size, B: % co-solvent in solvent, C: total amount of co-solvent + solvent)

of coals. With more surface area available, the solvents NMP and EDA disrupt more coal–coal interactions by breaking H-bonds, π–π interactions, vanderwaal forces and charge transfer interactions. NMP then better interacts with the coal macromolecular structure especially aromatic rings thereby resulting in enhanced extraction (Pande and Sharma 2002; Iino et al. 1988).

Table 11 Screening for *Y* (% Extraction yield) by RSM Methodology

Term	Scaled estimate	Std. error	*t* ratio	Prob > t
Intercept	20.56	2.41	8.51	0.0135
Particle size	− 11.92	2.09	− 5.70	0.0295
Coal/(A + B)	− 0.45	2.09	− 0.21	0.8497
% A in solvent	0.9	2.09	0.43	0.7091
Particle size * particle size	2.57	2.17	1.18	0.3581
Particle size * coal/(A + B)	− 0.05	2.09	− 0.02	0.9831
Coal/(A + B) * coal/(A + B)	3.15	2.178	1.45	0.2846
Particle size * % A in solvent	− 0.47	2.09	− 0.23	0.8416
Coal/(A + B) * % A in solvent	1.3	2.09	0.62	0.5979
% A in solvent * % A in solvent	1.02	2.17	0.47	0.6832
Particle size * particle size * coal/(A + B)	2.05	2.96	0.69	0.5602
Particle size * coal/(A + B) * coal/(A + B)	2.57	2.96	0.87	0.4761
Particle size * particle size * % A in solvent	− 1.47	2.96	− 0.50	0.6677

3.4 Regression analysis

Minitab™ (version 18) was used for the regression analysis. The quality of the multiple regression model fitting was measured by calculating the coefficient of regression R^2. *F* test was used for testing its statistical significance. The analysis of variance test was used to calculate the significance of these linear parameters on the extraction yield. The results of which have been given in Table 12. The results again predict the much higher effect of the particle size on the extraction yield with the least *p* value and highest *F* value (Segurola et al. 1999). Thus the regression analysis was found to be in conformity with the design of experiments methodology.

3.5 Solvent recovery

Solvent recovery was also noted during the course of these extraction studies and the amounts of NMP and EDA recovered are shown in Table 13. The high recovery (greater than 90%) of NMP and EDA solvents in this process seems to be a step towards making the process economically attractive in future for the possibility of its use (Chandaliya et al. 2015).

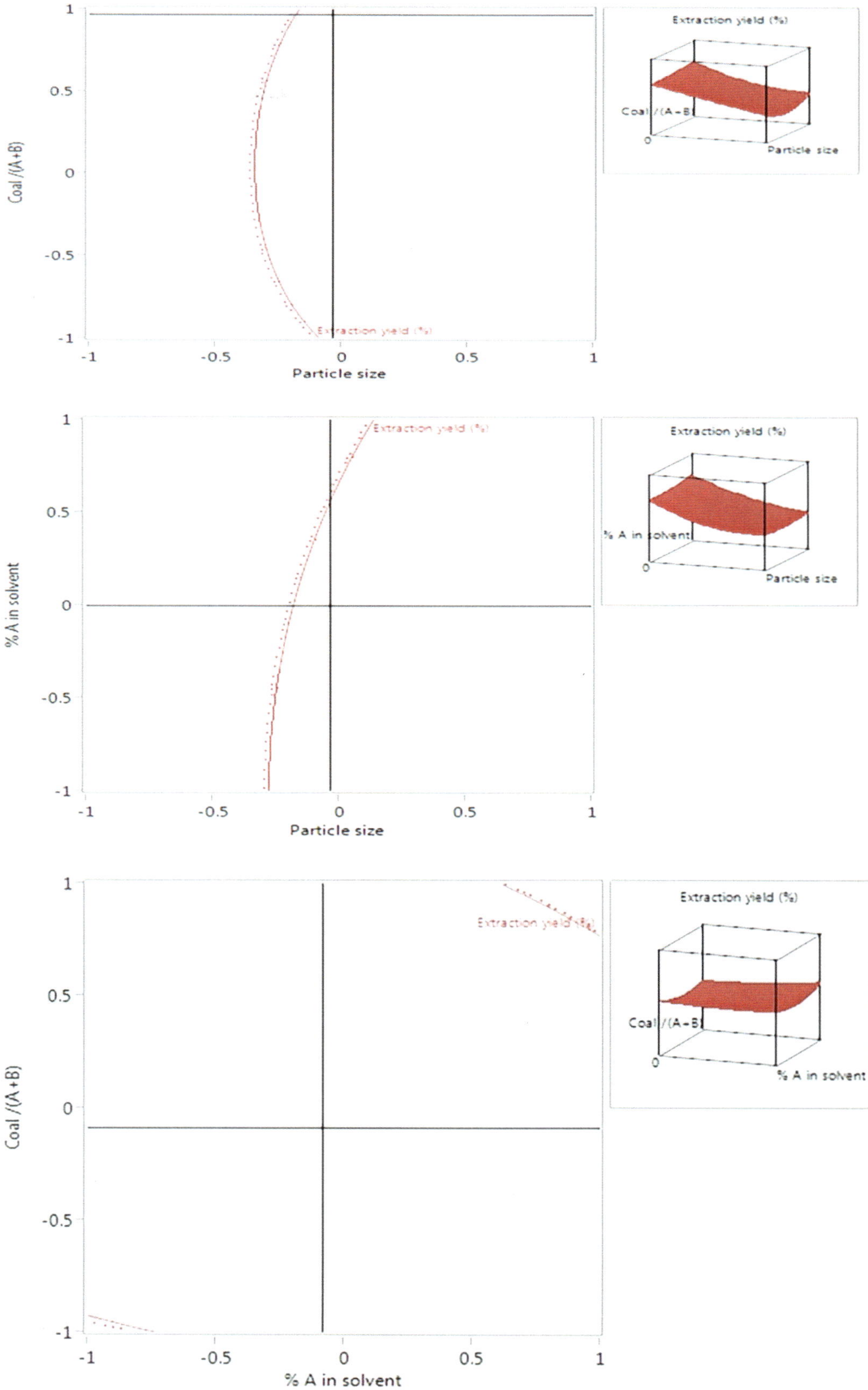

Fig. 6 Contour profilers for the parameters keeping one parameter constant and varying the other two [A: amount of co-solvent (EDA), A + B: total amount of co-solvent + solvent (EDA + NMP)]

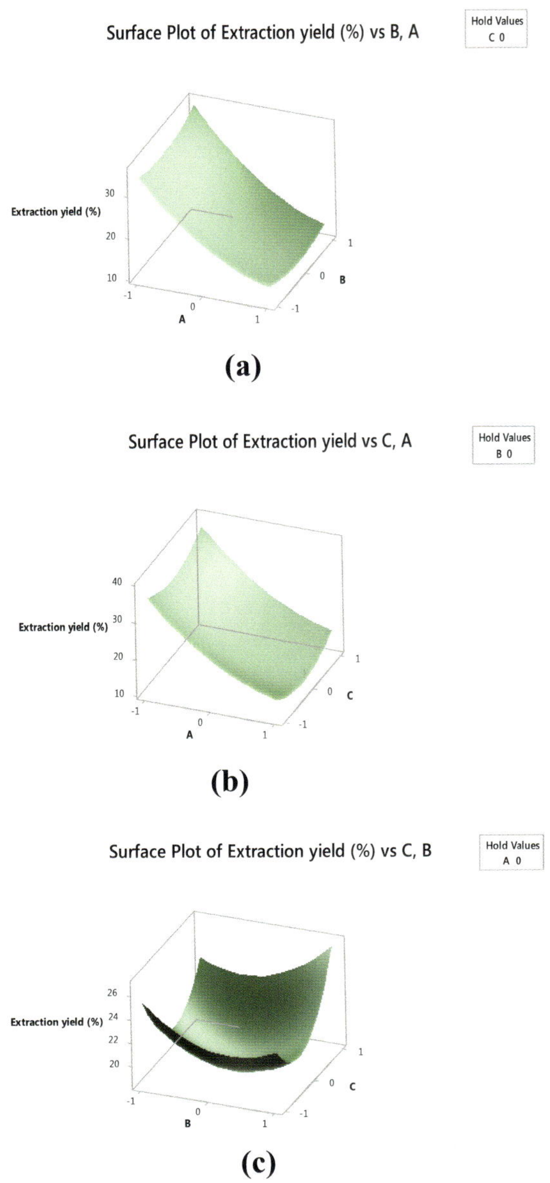

Fig. 7 Response surface plot showing the predicted values of the extraction yield with effect on particle size (A) and % co-solvent in solvent (B) (**a**), particle size with coal/total amount of co-solvent + solvent (C) (**b**) and % co-solvent in solvent (B) and coal/total amount of co-solvent + solvent (C) (**c**) [A: amount of co-solvent (EDA), A + B: total amount of co-solvent + solvent (EDA + NMP)]

Table 12 Analysis of variance model regression on extraction yield

Source	DF	Adj SS	Adj MS	*F* value	*p* value
Regression	3	908.11	302.702	26.18	0.000
Particle size	1	905.25	905.251	78.30	0.000
%A in solvent	1	0.21	0.211	0.02	0.895
Coal/A + B	1	2.64	2.645	0.23	0.642
Error	11	127.18	11.562		
Lack-of-fit	9	92.14	10.237	0.58	0.766
Pure error	2	35.05	17.523		
Total	14	1035.2			

R^2: 87.72%, R^2 (adj.): 84.36%, R^2 (pred.): 80.80%

DF, degree of freedom; SS, sum of squares; MS, mean of squares; F, *F* value; *p*, significance

Level of *P*-value (a significance level < 0.05)

Table 13 Amount of solvents recovered during the extraction process

Solvents used for extraction	NMP	EDA
Amount of solvents used (%)	100.0	100.0
Solvents recovered (%)	92.0	90.0
Unaccounted loss (%)	8.0	10.0

4 Conclusions

Solvent extraction of a high ash, low grade non-coking Indian coal, Kedla coal was performed using pre-treatment with acetic acid followed by extraction with NMP and EDA. Optimization of the extraction parameters was carried out. Using Definitive Screening Design for optimization of five parameters and RSM–Box Behnken experimental design for 3 parameters, maximum extraction yield under the given set of conditions was obtained. About 39.5% of the extraction yield could be achieved with maximum desirability output for these parameters obtained using the design of experiments (DSD and RSM) technique i.e. the particle size less than 125 micron, co-solvent to solvent concentration (vol%) of 15% of the total solvent concentration and the coal to solvent ratio as 1:20 (wt/vol). This technique also shows the linear and square interaction effects of the different parameters. The RSM model and the intercept values show the maximum impact of the coal particle size on the extraction yield. The regression analysis was found to be in good agreement with the RSM responses. This extraction process makes use of mild atmospheric pressure conditions unlike the high pressure and temperature conditions that make use of hydrogenating solvents where the clean coal product recovery is extremely difficult. Design of experiments reduces the number of experimental runs to be performed for optimization. The clean coal extraction yield is relatively high and this optimization consumes fewer amounts of the solvents. This technique results in leading to the conditions on obtaining enhanced amount of super clean coal without resorting to large number of experiments of process intensification studies on the organo-refining of power grade coal using the designer e, N solvent system. Organo-refining is an emerging clean coal technology and there is a need to make it economical. Present study is a step in this direction.

Acknowledgements The authors would like to thank NTPC Badarpur, New Delhi for arranging the coal sample for research work.

References

Aslan N, Cebeci Y (2007) Application of Box-Behnken design and response surface methodology for modelling of some Turkish coals. Fuel 86:90–97. https://doi.org/10.1016/j.fuel.2006.06.010

Azimi E, Karimipour S, Rahman M, Szymanski J, Gupta R (2013) Evaluation of the performance of air dense medium fluidized bed (ADMFB) for low-ash coal beneficiation. Part 1: effect of operating conditions. Energy Fuels 27:5595–5606. https://doi.org/10.1021/ef400456n

Balakrishnan S, Midhun Reddy V, Nagarajan R (2015) Ultrasonic coal washing to leach alkali elements from coals. Ultrason Sonochem 27:235–240

Chandaliya VK, Banerjee PK, Biswas P (2012) Optimization of solvent extraction process parameters of Indian coal. Min Process Extr Metall Rev 33:246–259

Chandaliya VK, Biswas P, Banerjee PK (2015) Process for production of low ash clean coal from high ash coal with total solvent recovery. US20150007494 A1

Chen C, Gao J, Yan Y (1998) Role of noncovalent bonding in swelling of coal. Energy Fuels 12(6):1328–1334

Dryden IGC (1948) Extraction of bright coals by anhydrous ethylene diamine. Nature 162:959–960

Elliott MA (1981) Chemistry of coal utilization, second supplementary volume. Wiley, New York

Giri CC, Sharma DK (2000) Mass-transfer studies of solvent extraction of coals in N-methyl-2-pyrrolidone. Fuel 79(5):577–585

Gunaraj V, Murugan N (1999) Application of response surface methodology for predicting weld bead quality in submerged arc welding of pipes. J Mater Process Technol 88:266–275. https://doi.org/10.1016/S0924-0136(98)00405-1

Haenel MW (1992) Recent progress in coal structure research. Fuel 71:1211–1223. https://doi.org/10.1016/0016-2361(92)90046-q

Hall PJ, Larsen JW (1993) Evidence for the formation of stable coal-N-methylpyrrolidinone and coal-pyridine gels. Energy Fuels 89:47–51

Hernández MR, Murcia CF, Gupta R, de Klerk A (2012) Solvent–coal–mineral Interaction during solvent extraction of coal. Energy Fuels 26:6834–6842

Hu YH, Painter PC, Coleman MM, Butera RJ (1998) Ramifications of chain connectivity in hydrogen-bonded polymer solutions. Macromolecules 31:3394–3396. https://doi.org/10.1021/Ma980248x

Iino M, Takanohashi T, Ohsuga H, Toda K (1988) Extraction of coals with CS_2-N-methyl-2-pyrrolidinone mixed solvent at room temperature: effect of coal rank and synergism of the mixed solvent. Fuel 67:1639–1647. https://doi.org/10.1016/0016-2361(88)90208-6

Jain N, Sharma DK (2004) Biohydrometallurgy for nonsulfidic minerals—a review. Geomicrobiol J 21:135–144

Kihong K, Heechan C, Sihyun L, Myoungwook M, Deayang L (2016) Coal and solvent properties and their correlation with extraction yield under mild conditions. Korean J Chem Eng 33(7):2142–2159

Kincl M, Turk S, Vrečer F (2005) Application of experimental design methodology in development and optimization of drug release method. Int J Pharm 291:39–49. https://doi.org/10.1016/j.ijpharm.2004.07.041

Kirk RE, Othmer DF (2006) Encyclopedia of chemical technology. Wiley, Hoboken. http://as.wiley.com/WileyCDA/WileyTitle/productCd-0471484970.html

Krevelen DW (1993) Coal—typology, physics, chemistry, constitution. Elsevier Science, Amsterdam

Kumar M, Patel SK (2008) Characteristics of Indian non-coking coals and iron ore reduction by their chars for directly reduced iron production. Miner Process Extr Metall Rev 29:258–273. https://doi.org/10.1080/08827500801997902

Kwak JS (2005) Application of Taguchi and response surface methodologies for geometric error in surface grinding process. Int J Mach Tools Manuf 45:327–334. https://doi.org/10.1016/j.ijmachtools.2004.08.007

Larsen JW, Mohammadi M (1990) Structural changes in coals due to pyridine extraction. Energy Fuels 4:107–110. https://doi.org/10.1021/ef00019a019

Larsen JW, Green TK, Kovac J (1985) The nature of the macromolecular network structure of bituminous coals. J Org Chem 50:4729–4735. https://doi.org/10.1021/jo00224a014

Li C, Takanohashi T, Yoshida T, Saito I, Aoki H, Mashimo K (2004) Effect of acid treatment on thermal extraction yield in ashless coal production. Fuel 83:727–732. https://doi.org/10.1016/j.fuel.2003.06.002

Masaki K, Kashimura N, Takanohashi T, Sato S, Matsumura A, Saito I (2005) Effect of pre-treatment with carbonic acid on "HyperCoal" (ash-free coal) production from low-rank coals. Energy Fuels 19:2021–2025. https://doi.org/10.1021/ef050030t

Mathews JP, Burgess-Clifford C, Painter P (2015) Interactions of Illinois No. 6 bituminous coal with solvents: a review of solvent swelling and extraction literature. Energy Fuels 29:1279–1294

Meshram P, Purohit BK, Sinha MK, Sahu SK, Pandey BD (2015) Demineralization of low grade coal—a review. Renew Sustain Energy Rev 41:745–761

Mishra A (2009) Assessment of coal quality of some Indian coals. Mining Engineering

Mishra S, Sharma DK (1990) Solvent extraction and extractive disintegration of coal in anthracene oil. Fuel 69(11):1377–1380

Nishioka M, Larsen JW (1990) Association of aromatic structures in coals. Energy Fuels 4:100–106

Nyathi MS, Clifford CB, Schobert HH (2013) Characterization of graphitic materials prepared from different rank Pennsylvania anthracites. Fuel 114:244–250

Otake Y, Suuberg EM (1997) Temperature dependence of solvent swelling and diffusion processes in coals. Energy Fuels 11:1155–1164

Painter PC, Park Y, Sobkowiak M, Coleman MM (1990) Coal solubility and swelling. 2. Effect of hydrogen bonding on calculations of molecular weight from swelling measurements. Energy Fuels 4(4):384–393

Pande S, Sharma DK (2002) Ethylenediamine-assisted solvent extraction of coal in N-methyl-2-pyrrolidone: synergistic effect of ethylenediamine on extraction of coal in N-methyl-2-pyrrolidone. Energy Fuels 16:194–204

Rahman M, Pudasainee D, Gupta R (2017) Review on chemical upgrading of coal: production processes, potential applications and recent developments. Fuel Process Technol 158:35–56

Renganathan K, Zondlo JV (1993) Non-destructive extraction mechanism of bituminous coals using K-IETBTL-2-PY1LR0-LID0NE. Fuel Sci Technol Int 11:677–695

Renganathan K, Zondlo JW, Mintz EA, Kneisl P, Stiller AH (1988)

Preparation of an ultra-low ash coal extract under mild conditions. Fuel Process Technol 18:273–278

Segurola J, Allen NS, Edge M, Mc Mahon A (1999) Design of eutectic photoinitiator blends for UV/visible curable acrylated printing inks and coatings. Prog Org Coat 37:23–37

Sharma DK, Giri CC (2016) CO_2 Gasification reactivity and kinetics studies of raw coal, super clean coal and Residual coals obtained after organo-refining (solvent extraction). J Power Technol 96(3):701-157–701-169

Shibaoka M, Stephens JF, Russell NJ (1979) Microscopic observations of the swelling of a high-volatile bituminous coal in response to organic solvent. Fuel 58:515–522

Song C, Schobert HH (1993) Opportunities for developing specialty chemicals and advanced materials from coals. Fuel Process Technol 34:157–1596

Song C, Schobert HH (1996) Non-fuel uses of coals and synthesis of chemicals and materials. Fuel 75:724–736

Suuberg EM, Otake Y, Langner MJ, Leung KT, Milosavljevic I (1994) Coal macromolecular network structure analysis: solvent swelling thermodynamics and its implications. Energy Fuels 8(6):1247–1262

Takanohashi T, Iino M (1995) Investigation of associated structure of upper freeport coal by solvent swelling. Fuel 9(5):88–93

Takanohashi T, Yanagida T, Iino M, Mainwaring DM (1996) Extraction and swelling of low-rank coals with various solvents at room temperature. Energy Fuels 10:1128–1132. https://doi.org/10.1021/ef960033t

12

Patterns and security technologies for co-extraction of coal and gas in deep mines without entry pillars

Nong Zhang[1] · Fei Xue[1] · Nianchao Zhang[1] · Xiaowei Feng[1]

Abstract Retaining gob-side entryways and the stability of gas drainage boreholes are two essential techniques in the co-extraction of coal and gas without entry pillars (CECGWEP). However, retained entryways located in deep coal mines are hard to maintain, especially for constructing boreholes in confined spaces, owing to major deformations. Consequently, it is difficult to drill boreholes and maintain their stability, which therefore cannot guarantee the effectiveness of gas drainage. This paper presents three measures for conducting CECGWEP in deep mines on the basis of effective space in retained entryways for gas drainage. They are combinations of retaining roadways and face-lagging inclined boreholes, retaining roadways and face-advancing inclined boreholes, and retaining roadways and high return airway inclined boreholes. Several essential techniques are suggested to improve the maintenance of retained entryways and the stabilization of boreholes. For the particular cases considered in this study, two field trials have verified the latter two measures from the results obtained from the faces 1111(1) and 11112(1) in the Zhuji Mine. The results indicate that these models can effectively solve the problems in deep mines. The maximum gas drainage flow for a single hole can reach 8.1 m^3/min and the effective drainage distance can be extended up to 150 m or more.

Keywords Retaining gob-side entryways · Stability of borehole · Gas pressure relief · Co-extraction of coal and gas without the entry pillar

1 Introduction

As an associated product of coal production, gas is an innovative clean energy and high-class chemical material, although it may cause major life threatening disasters in collieries and increase the greenhouse effect in the atmosphere (Zhang et al. 2013; Xue et al. 2014). The gas can be scientifically used and serve as one of the most important consequential energy resources in the 21st-century (Yuan 2008; Xie et al. 2014). Appropriate gas drainage in collieries can guarantee safe production, and decrease the emission of greenhouse gas (Bibler et al. 1998; Karacan et al. 2011; Keim et al. 2011; Xia et al. 2014; Zhou et al. 2014). The earliest gas drainage technique was applied in 1938, and two main approaches to drainage have been developed through the investigations of several generations (Yuan et al. 2013). They are known as surface and underground drainage. Multiple technical solutions for gas drainage have been developed based on related research and field trials. On account of different drainage situations, such as current coal seams, neighboring coal seams, and gob-areas, they can be divided into many categories. These include horizontal boreholes, intersection boreholes and grid-layout boreholes in current coal seams, boreholes through the roof or floor in neighboring coal seams, fracturing zone drainage, buried pipes drainage, and surface drainage aimed at gas in the gob area (Zheng et al. 2013; Pan et al. 2014). Research teams in China also developed deep-hole presplitting blasting, hydraulic cutting, hydraulic

✉ Nong Zhang
zhangnong_cumt@163.com

[1] School of Mines, Key Laboratory of Deep Coal Resource Mining, Ministry of Education, China University of Mining & Technology, Xuzhou 221116, China

drilling or reaming, to increase the permeability of coal or rock strata. These techniques have been widely applied in coal mines more than 1000 m deep. Nonetheless, in view of the poor permeability of coal seams, complex geological conditions, high gas content, potential explosions of coal and gas, and limited coal production in China, problems involving large investments in hole drilling and specially designed drainage entries will arise if we continue using these traditional measures.

Co-extraction of coal and gas without any entry pillars (CECGWEP) has been developed for gas extraction using pressure relief theory, which is illustrated in Fig. 1. Based on the conditions of coal seams, a technically minable coal seam is chosen as the first to be mined. This seam is called "the preferential coal seam" which is used to release the high initial gas pressure (Li et al. 2014; Yang et al. 2014). In CECGWEP, traditional gas drainage entries located in either the roof above or the floor below are replaced by gob-side retained entries in the preferential coal seam; a traditional U-type ventilation passageway is changed into a Y-type ventilation passageway. A series of potential problems, such as gas outbursts, coal mining problems, high temperatures controlling, relief of high underground pressure, and support of surrounding rock, are considered in the technology and it has already been widely applied by the Huainan Mining Group, Shanxi Coking Coal Group, Huaibei Mining Group, and Jincheng Anthracite Group.

However, although this technology has several merits, it is difficult to adopt in every coal mine, especially deep buried soft rock coal mines, where the entryway convergence is extremely severe (Zhang et al. 2012; Kong et al. 2014). It thus requires optimization on the basis of the complex geology. The purpose of this paper is to describe the implementation of CECGWEP in deep coal mines and present a systematic summary of the key techniques based on field trials in recent decades.

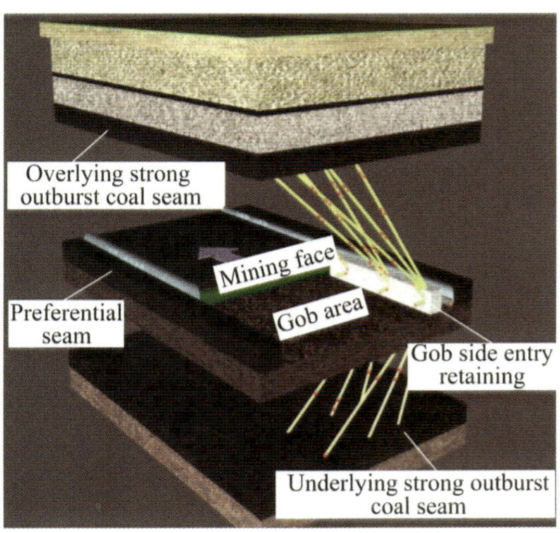

Fig. 1 Schematic diagram of CECGWEP

2 Technical challenges of CECGWEP in deep mines

2.1 Large deformations in rock mass lead to difficulties in drilling drainage boreholes in retained entryways

Gob-side entry retaining in deep coal mines is subject to high ground stress, high temperature, and severe mining pressure. The scope for breakage in rock masses is relatively large, and may cause a fracture line in the immediate roof which propagates into the coal seam. This will dramatically induce large subsidence and rotation moments in the roof, severe dilatation and buckling of coal sides, and uncontrollable floor heaving. These phenomena distinctly separate deep buried coal mines from shallow buried coal mines (Kang et al. 2010).

Figure 2 displays the relationship between the sectional area of the retained entryway and the correlated distance to the workface in the Zhuji and Panyidong Mines. It is apparent that the deformation of the surrounding rock in different faces shows different levels, which are directly related to their buried depth and the lithology of the roof and floor. All four original sectional areas are 12 m^2. The face 1252(1) is relatively shallow compared with the others, the sectional areas at 20, 40 and 60 m in front of the face are 11.8, 10.9 and 9.7 m^2, respectively, and the convergence rate of the stabilized segment in the retained entryway is 35 %. However, it is totally different for the face 1111(1), where it is buried relatively deeper than the other three, and the lithology consists of soft rocks. Its sectional areas at 20, 40 and 60 m in front of the face are 4.9, 4.3 and 4.2 m^2, respectively, and the convergence rate of the stabilized segment in the retained entryway is 66 %, which is almost double that of face 1252(1).

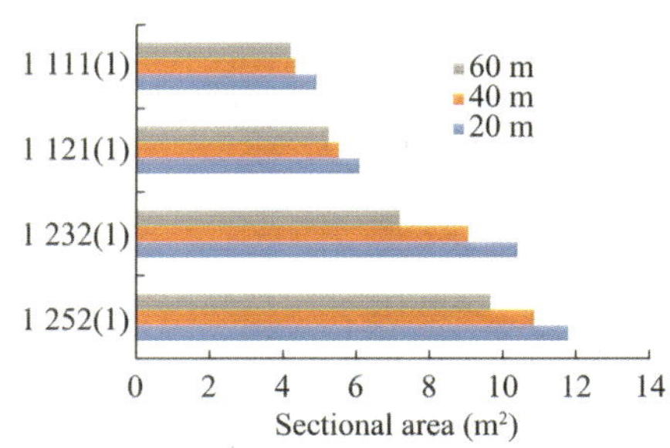

Fig. 2 Relationship between the sectional area of a retained entryway and the distance to the working face (All the original sectional areas were 12 m^2)

Additionally, to maintain total stability of the roof in the retained entryway, two or three rows of single hydraulic props are recommended to be suitably arranged and timber cribs are built along the backfilling wall to serve as auxiliary supports. It thus greatly restrains the space for drilling holes designed to drain gas.

2.2 Frequently occurring cutting and blocking of drainage holes and low efficiency in drainage

The drilled holes in an arrangement on the gob side are subjected to dramatic mining induced pressure because of the severe movement of strata in coal seam mining (Zhang et al. 2014). A frequent phenomenon is the cutting and blocking of drilling holes (Lu et al. 2014). A drainage pipe cannot be effectively guaranteed. This thus significantly restricts the high drainage rate of gas (Whittles et al. 2007). Field monitoring reveals that the average drainage rate of a single hole is less than 0.3 m^3/min. Hence, the drainage rate cannot be guaranteed. Concentration of gas around the working face always reaches the ultimate limit, which slows mining, and makes the co-excavation of coal and gas hard to implement.

2.3 Requirements for hole-creating and safety are hard to implement using lagging drilling behind the face

It has been established through field observations that the entryway rock from 20 to 60 m behind the workface is extremely fractured. Borehole drilling into this range is extremely difficult because the drill stem jams frequently, or boreholes collapse. Hence, the requirement for a standard borehole is hard to fulfil. At the same time, a large amount of gas can penetrate into the gob area through mining induced fissures. Gas escapes very easily into the entryway because of the borehole drilling when creating holes, and the local flooding of gas is extremely dangerous for the safety of miners.

Therefore, there are two main problems that need to be investigated when implementing CECGWEP in deep soft coal mines, namely, the technologies for supporting surrounding rock in the retained entryway, and for extracting gas safely. Different measures concerning CECGWEP need to be considered for the different geological conditions.

3 Implementation of CECGWEP

According to the degree of deformation of the surrounding rock, Table 1 and Fig. 3 divides CECGWEP into three measures based on the difficulties in controlling the surrounding rock.

Table 1 Implementation methods and applications of CECGWEP

Sectional convergence rate of entryway (%)	Implementation methods
<30	Roadway retention and face-lagging inclined borehole
30–50	Roadway retention and face-advancing inclined borehole
>50	Roadway retention and high return airway inclined borehole

3.1 Roadway retention and face-lagging inclined boreholes

This technique is schematically illustrated in Fig. 3a. It is capable of the convergence rate of entry under 30 % which can fulfill the requirements of space to let workers drill boreholes in retained entry. The geological conditions are simple and mining depth is small usually in this circumstance.

Since the stress has been redistributed after mining in the roof and floor of gob area, the stress concentration and rock mass rotation/subsidence of the face-lagging borehole are decreased, and then the stress environment of boreholes is also located in a relatively mitigated zone. Therefore, the borehole can be avoided sine potential breakage or cutting off and its stability can be maintained for a long time to guarantee the long-term drainage of gas. Total gas drainage and utilization rate of single hole also can be augmented.

3.2 Roadway retention and face-advancing inclined borehole

This technique is displayed in Fig. 3b. Compared with Fig. 3a, it can be seen that the boreholes are arranged in the advancing area of the face. The distance to the face can range from 60 to 80 m. To accommodate the in situ stress in the strata, it is recommended that high strength casing should be adopted.

The boreholes are then reamed to release the pressure and a flexible backfilling method is used to perform a comprehensive controlling technique. These measures are aimed at guaranteeing the stability of the boreholes during the whole procedure of coal mining, regardless of advancing or lagging the face. The merits of this measure are that it can intercept the relieved gas in advance and effectively drain it away before it enters the face area, eliminating the danger of gas flooding from its source. This can create "one borehole for the three stages", which can be used simultaneously before, during and after the mining which extends the drainage duration of the borehole. However, the drilling location in advance of the face is unaffected by the mining during the drilling of boreholes.

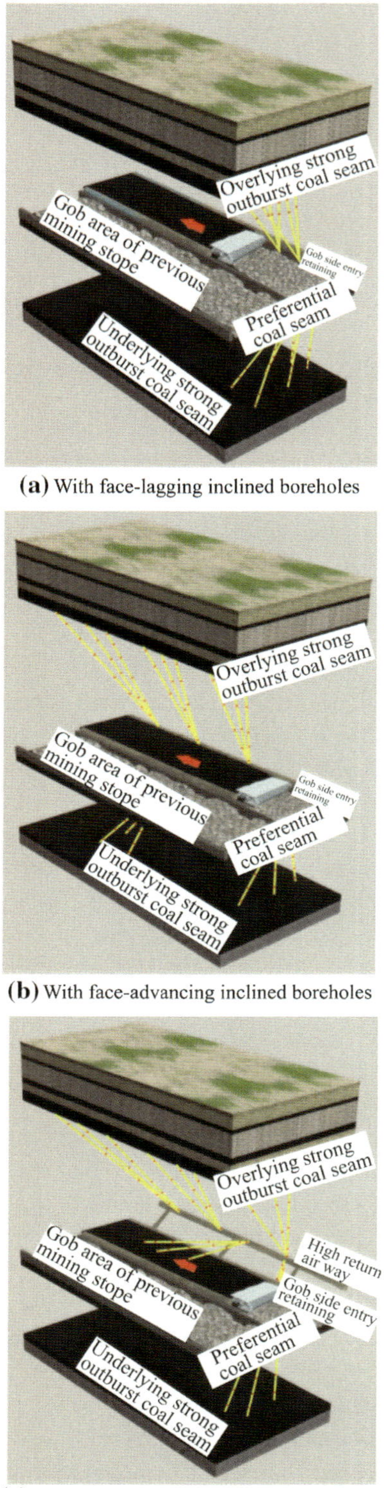

(a) With face-lagging inclined boreholes

(b) With face-advancing inclined boreholes

(c) With high return airway inclined boreholes

Fig. 3 Illustration of roadway retention of the three methods of implementing CECGWEP

Hence, drilling is easy to conduct. It decreases the requirements on the sectional area for drilling boreholes and plays an active role in controlling the surrounding rock mass. Because the entryway is little affected when the face is far enough from the drilling position, the rock mass is relatively intact, resulting in a high ratio of successfully drilled boreholes. Time can be saved for installing casing and for grouting or sealing the holes. Embedded gas will not be released in the areas in advance of the face, therefore making a large amount of gas emission during the drilling process impossible. The working conditions are safer than those illustrated in Fig. 3a.

3.3 Roadway retention and high return airway inclined boreholes

Retaining the gob-side entryway is characterized by continuous deformation. So short segment retention of the entryway is recommended when it is difficult to retain the whole entryway. This involves arranging an upper entryway above the original tailgate. These are connected by the return air crosscuts, as illustrated in Fig. 3c. Crosscuts are arranged at certain distances from each other in the direction of mining. The upper tailgate acts as drainage entry and a Y-type ventilation system can be created based on a combination of the upper drainage entryway, the original return air entryway, and the related crosscuts. The upper drainage entryway is barely influenced by the mining activity, which makes it possible to transfer the drilling work, initially assigned in the return air entryway, into the upper drainage entryway. With this approach, interference between the mining activity and drilling work can be avoided, thus providing adequate working space. In addition, boreholes drilled from the upper tailgate are far from the influence of mining activity taking place at the working face below. The maintenance cost of drainage boreholes can be significantly reduced compared with boreholes drilled ahead of the retained entryway or in the retained entryway (Liu et al. 2014). Additionally, boreholes beneath the current coal seam can also be drilled in the upper tailgate which avoids the additional drilling work such as arranging buried pipes in the gob area of the working coal seam or striking boreholes along the roof for gas drainage.

4 Key technologies of CECGWEP in deep mines

4.1 Pressure relief in strata using presplitting blasts

A presplitting blast is aimed at relieving the pressure in the rock mass that surrounds the retained entryway and is performed by blasting in deep drilled boreholes. This technique is appropriate for faces with thick and stiff roofs or where the immediate roof is too thin to accommodate the rotation and subsidence of the main roof. It consists mainly of two methods, the first is advanced deep-hole

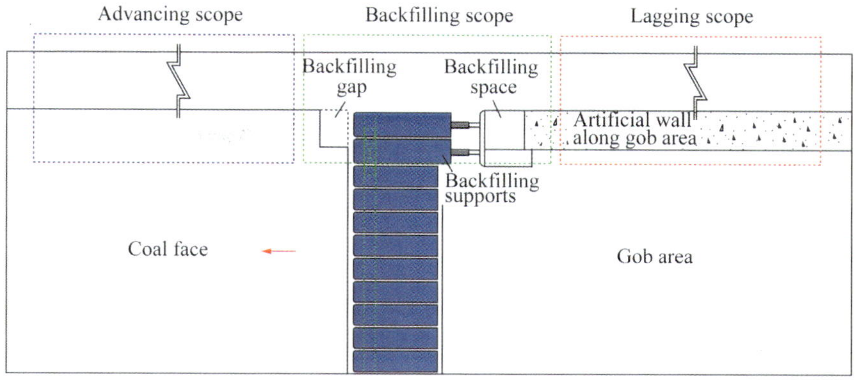

Fig. 4 Segments dividing gob-side entry retention

blasting for the caving roof along the inclination of the face, which can reduce the distance from the weight of the roof and reduce the concentration of the weight. The second advances the group of blasting holes along the strike direction of the face. They are horizontally distributed with small rotational angles from each other which eliminate the lateral roof cantilever located above the retained entryway. The regional structure of the roof can be adjusted accordingly and the stress field surrounding it can also be optimized. The layout form and parameters of blasting holes need to be optimized for the specific geological conditions.

4.2 Divisional control technology on surrounding rock in retained entryways

On the basis of the characteristics of the deformation and the distribution of the pressure on the abutment in the retained gob-side entryway, the procedure for retaining the entryway can be divided into advanced, backfilled, and lagging scopes, as schematically plotted in Fig. 4. Mining induced pressure will show different characteristics in different scopes, so their space–time conditions also vary with respect to each other. Therefore, different strategies need to be adopted. Maintenance measures in the advanced segment consist of secondary strengthening supports, advanced supports, and presplitting blasts to perform pressure relief. Maintenance measures in the backfilled segment consist of roof reinforcement in the scheduled backfilled area, and the construction of a backfilling wall along the gob side. Maintenance in the lagging segment contains auxiliary supports in the retained entryway, regional reinforcement, and related maintenance on the backfilling wall.

A certain amount of deformation should be designed to guarantee the normal sectional area of the retained entryway. However, the increased width of the entryway will increase the difficulties in controlling the roof and coal ribs. To solve this problem, the optimal ratio of width-height of the entryway is proposed to guarantee its safety, as demonstrated by Eq. (1). The width of the backfilling wall also influences the stability and cost of the whole system for retaining the entryway where the mining height is defined. To eliminate as far as possible the extra cost of retaining the entryway, and to ensure its technical feasibility, an optimal ratio of the width-height of the backfilling wall is suggested. Based on the appropriate value of this ratio, the backfilling wall can support the roof strata at the entrance, and also accommodate the deformation and rotation of upper strata in the future. The minimum value of the ratio that can withstand the roof and floor deformation is specifically expressed in Eq. (2)

$$\frac{1}{2} \le \frac{b}{h} \le \frac{1}{h}\left\{\frac{\left[R_{\mathrm{m}}\left(\frac{m_0}{E}+\frac{h_1}{E_1}+\frac{h_2}{E_2}\right)+\delta\right]l}{m_0-(K-1)h_1}-x_0\right\} \tag{1}$$

$$\frac{(b+x_0)\eta\gamma m}{h(\sigma_1-\eta\gamma m)} \le \frac{a}{h} \le \frac{\left[R_{\mathrm{b}}\left(\frac{m}{E}+\frac{h_1}{E_1}+\frac{h_2}{E_2}\right)+\delta\right]l}{h[m-(K-1)h_1]}-\frac{(x_0+b)}{h} \tag{2}$$

where b and h are the width and height of the entryway; R_m is the minimum strength of the roof and floor in the entryway; m the mining height; h_1, and h_2 the thickness of the immediate roof and floor; E, E_1, and E_2 the elasticity modulus of the backfilling wall, immediate roof, and immediate floor, respectively; δ the hanging roof behind the face; l the length of the key block; K the bulking coefficient of the immediate roof; x_0 the horizontal distance from the fracture point of main roof to the workface of the gob-side entryway; and η is a constant ranging from 4 to 8.

4.3 "The trinity" surrounding rock controlling techniques in retained entryways

The basis of supporting the retained entryway is to initially apply bolt support technology capable of adapting to the mining induced influences with the characteristics of "three highs" (high strength, rigidity, and pretension) to counteract shear failure. This can create high strength and high resistance stability in the surrounding rock bearing structure that can more effectively control the deformation of the surrounding rock of the roadway (Zheng et al. 2014). Auxiliary reinforcement uses auto shift roof supports, which mutually control the roof and floor. Accompanied by the high strength wall of roadside filling, this forms the total structure of "the trinity" supporting system, effectively controlling the deformation of the surroundings of the retained entryway.

4.4 Regional control technology for borehole maintenance

The deformation of the roof strata varies in different areas and the strata stress will redistribute itself. Thus, the form of damage to borehole casings in different areas is diverse. As shown in Fig. 5, the axial length, according to the broken line and the drilling strata after pressure relief, can be divided into three regions: the compression failure zone, the compression-shear transition zone, and the shear failure zone. The damage characteristics of different zones need different measures for borehole maintenance. In the compressive failure zone, the borehole casing should be thickened to resist the compressive stress, while in the shear failure zone, the key shear strata should be reamed to increase the diameter of the borehole.

The speed of building the backfilling wall should match the advancing speed of the fully mechanized working face. Therefore, a technique for fast building of the backfilling wall was developed. Bagged backfill material is transported to the workplace by a mine car; a vacuum chuck transports the material to an automatic unpacking machine; a through-the-pipe belt conveyor then sends the material to a blender and mixed with water; a backfill pump sends the material for backfilling to behind the formwork supporting the roof. The backfilling wall is ready after curing and removing the formwork. The speed of backfilling can reach 10 m per day. By optimizing the ratio of the materials, the final strength of the backfilling wall achieves at least 30 MPa.

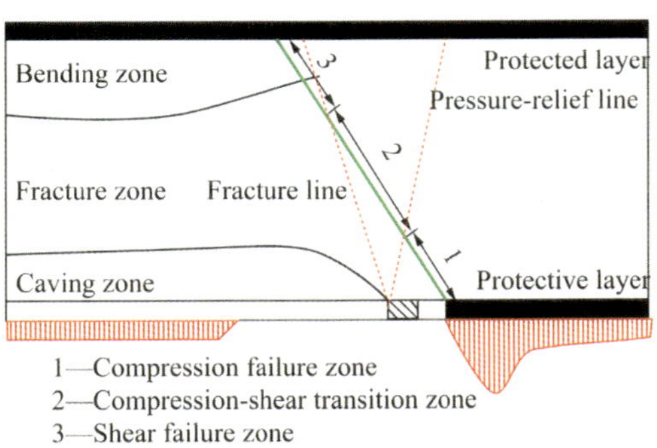

Fig. 5 Distribution of damage in the roof strata

This strength can satisfy the basic requirements for the retained entryways in thousand-meter-deep coal mines.

4.5 Rapid casing installation and borehole sealing technology

To ensure the stability of a borehole in the extraction process, the casing is installed after the borehole is created. However, metal casing is heavy and difficult to install manually. Therefore, in designing the casing, the casing connector should match the drill pipe. With the appropriate assistance of the drilling machine, the construction quality and efficiency of the installation of the casing can be improved.

The borehole sealing quality decides whether the air will escape through the borehole from the casing and fracture zones, which will reduce the gas concentration of extraction, and also influence the stability of the borehole. Therefore, the technology of injection sealing was developed. Both sides of the borehole (beginning and end segments) are sealed using the high polymer material, Marithan, which has an excellent adhesion, expansion and sealing performance, to achieve rapid sealing of the borehole. High strength and zero shrinkage cement material is injected into the sealed boreholes. This slurry is composed of cement and additives with a cement ratio as high as a 1/4. The compressive strength can reach 70 MPa and enhance the compressive properties of the materials compared with traditional cement. At the same time it can improve the sealing performance of drilled boreholes.

5 Engineering implementation

5.1 Roadway retention and face-lagging inclined borehole

Yuan (2008) introduced this technique in detail and it was proved to be effective and efficient in CECGWEP by field test. The study is not repeated here.

5.2 Roadway retention and face-advancing inclined borehole

5.2.1 Engineering and geological conditions

The trial area was located in panel 1111(1) of the Zhuji Coal Mine, operated by the Huainan Mining Group, China. The face elevation was −877.6 to −907.0 m. As shown in Fig. 6, the panel was 1608 m long and 220 m wide. It had an average thickness of 1.26 m and an average dip angle of 3°, ranging from 1° to 5°. The CH_4 content was 4.73–5.15 m^3/t. The face was mined using retreating long-

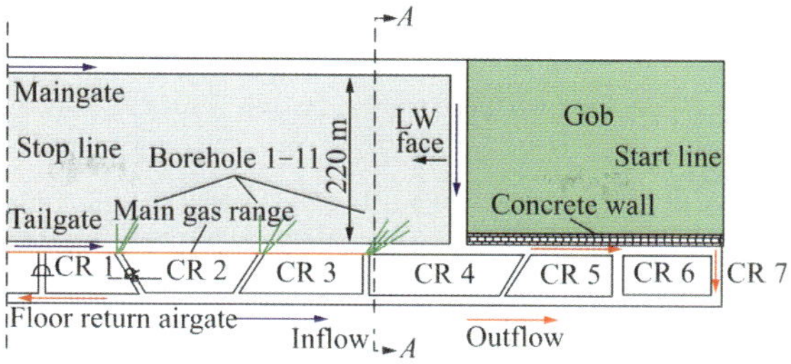

Fig. 6 Plan view of panel 1111(1)

wall mining. The working seam was overlain by Seam 13-1, which was outburst-prone coal with gas content of 6.98 m^3/t, at a distance of 65 m. There were seven connection roadways (CR) 1–7 from the starting line to the finishing line between the tailgate and the floor return airgate (Fig. 6).

5.2.2 Roadway retention surrounding rock control parameters

(a) Initial support: The roof bolt was designed as seven non-longitudinal-ribs rebar bolts at intervals of 750 mm × 800 mm. The diameter and length of the bolt were 22 and 2800 mm. The yield strength of the bolt was 400 MPa. The roof cable chosen was a cable-beam with four holes. The beam was made from No.20 channel steel and the cable was made from high strength and low relaxation steel fiber, with a diameter of 21.8 mm and yield strength of 1860 MPa. The interval between the cable-beams was 800 mm. The rib bolt used full screw and identity intensity bolts, with intervals of 650 mm × 800 mm. The bolt diameter was 20 mm and its length 2500 mm. There were five anchors on each rib wall. The roof and two sidewalls all adopted No.8 barbed wire and M-steel belts.

(b) Advanced reinforcement: The roof was strengthened using two cables, whose specifications were the same as the upper roof. The sidewall of the non-mining area was strengthened by two rows of horizontal cable-beams. The beam was made of No.11 joist steel and the cable length was 4300 mm.

(c) Auxiliary support: DZ35–42 single props, four meters No.11 joist steel and a 1.2 m long articulated top beam were used to provide auxiliary strength to the support up to 60 m from the advanced working face.

5.2.3 Test borehole layout parameters

As shown in Fig. 6, a total of 11 inclined cross-measure drainage boreholes in three successive groups were drilled at CR 4, CR 3, and CR 2. They were separated from the start line by 840, 1180, and 1460 m. To ensure the stability and reliability of the boreholes, a variety of casings were used to protect them. In the first group of boreholes, ϕ 89 × 6.45 mm geological drill rod was used as the protective casing. In the second group, a dual combined casing (ϕ 89 × 6.45 mm geological drill rod inner and ϕ 139.7 × 9.17 mm oil casing outer) was used. In the third group, ϕ 89 × 13 × 1500 mm and ϕ 177.8 × 19 × 1500 mm special thick-wall oil casing and the double-layer casing were used.

5.2.4 Effect of roadway retention and extraction

The length of successful roadway retention reached 1608 m. In the whole process of roadway excavation for its reuse, the deformation of the sidewall, roof, floor and backfill wall were 760, 235, 1215 and 50 mm, respectively. When the lagging length from the working face reached 80–100 m, the roadway achieved stability. The sectional area of the roadway was approximately 4–5 m^2. The actual appearance is shown in Fig. 7.

The methane flow rates for the face-advancing inclined boreholes of panel 1111(1) are shown in Fig. 8. The maximum methane flow of the extraction borehole reached 4.9 m^3/min and the average methane flow reached 1.9 m^3/min. The effective extraction length was more than 150 m. It met the needs of working face of gas extraction.

5.3 Roadway retention and the high return airway inclined borehole

5.3.1 Engineering and geological conditions

As highlighted in Fig. 9, the average mining depth of panel 1112 (1) in the Zhuji Mine was 960 m. The mining height was 1.8 m, the CH_4 content 4.73–5.15 m^3/t, the ground temperature 36.5–45 °C, the panel length 2059 m and the thickness of the working coal seam 1.37 m. The roof of the seam was made up of, from the bottom up, 4.8 m thick mudstone, a 0.2 m thick coal streak and 4.2 m thick fine sandstone. The floor of the seam was made up of, from the top down, 4.5 m thick mudstone, a 0.8 m thick coal seam and 3.2 m thick mudstone.

5.3.2 Roadway retention surrounding rock control parameters

(a) Initial support: The roadway cross section was 5 m × 3 m. The "three highs" bolt and cable-beam were used for the active support. The roof bolts were sinistral non-longitudinal-rib high strength bolts. The

Fig. 7 The appearance of panel 1111(1) roadway retention

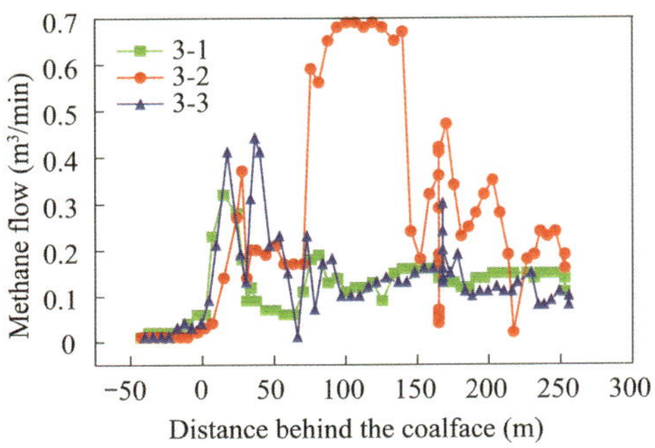

Fig. 8 Methane flow curve of the face-advancing inclined boreholes

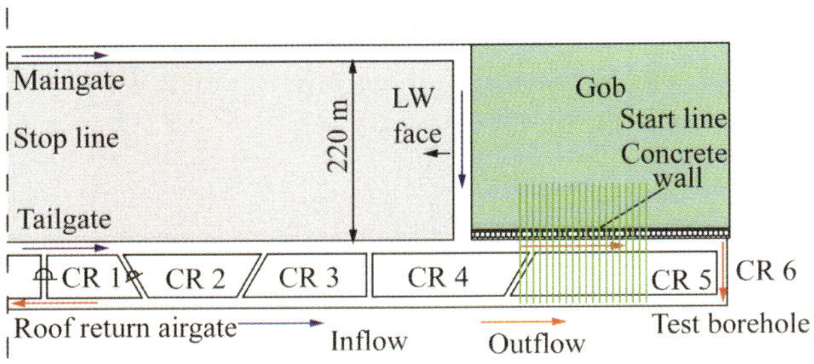

Fig. 9 Plan view of panel 1112(1)

sidewall bolts were dextrorotatory full-thread bolts at intervals of 700 mm × 800 mm.

(b) Advanced reinforcement: The çable-beam and grouting reinforcement had to be carried out ahead of the working face by at least 200 m. The method meant that the original roadway roof was reinforced by cable-beams, while two cable-beams were installed on the non-mining sidewall 650 mm from the roof and floor by the alternate application of single-beam hollow grouting cables and common cables. The installation angle of the upper cable was 30° upward, while it was downward for the lower cable. At the same time, at the advancing working face and the mining sidewall of the retained roadway, the coal section of 1.7 m × 3.5 m was broken manually for the bolt system reinforcing, and an articulated beam was installed as auxiliary support.

(c) Auxiliary support: DZ35–42 single props, 4 m No.11 joist steel and a 1.2 m long articulated top beam were used to provide auxiliary strength to the support up to 50 m from the advanced working face.

5.3.3 Test borehole layout parameters

The location of the installation of the boreholes was changed to a high return airway from the retained roadway along the advancing direction of the working face. The borehole spacing was 10 m with an inclination angle of 30° and a borehole length of 80 m. A ϕ 91 × 4.5 mm core barrel was installed in each borehole for protection.

5.3.4 Roadway retention and extraction effect

The successfully retained roadway reached 2059 m. In the whole process of reusing the roadway excavation material, the deformation of the sidewall, roof, floor and wall was 110–450, 257–450, 1311–2500 and 61–90 mm, respectively. Within the lagging length of 300 m from the working face, the width remained 4.4–4.9 m, and the height exceeded 1.5 m. The actual appearance is shown in Fig. 10. The sectional area of ventilation also exceeded 6.5 m^2 and satisfied the needs of the working face.

The methane flow rates of the high return airway inclined boreholes of panel 1112(1) are shown in Fig. 11. Because the borehole was far from the mining-induced area, the strength and thickness of the core barrel installed in the high return airway inclined boreholes were lower than that of the face-advancing inclined boreholes. The cost of borehole protection dropped sharply, and the extraction of gas was guaranteed. The maximum methane extraction flow of a single borehole reached 8.1 m^3/min,

Fig. 10 The appearance of the working face 1112(1) roadway retention

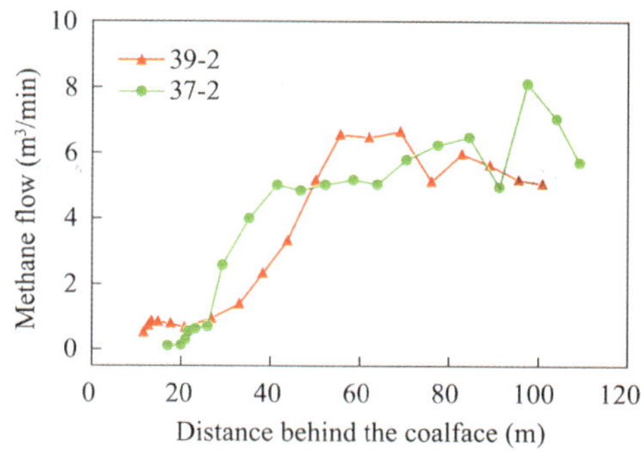

Fig. 11 Methane flow curve of the high return airway inclined boreholes

and the effective extraction reached more than 100 m. The required volumes of working face gas extraction were met.

6 Conclusions

(1) Three measures for the co-extraction of coal and gas without an entry pillar (CECGWEP) were introduced. They were roadway retention and face-lagging inclined boreholes, roadway retention and face-advancing inclined boreholes, and roadway retention and high return airway inclined boreholes.

(2) Several essential techniques were put forward to improve the maintenance effects for retained entryways and to stabilize the boreholes which provide pressure relief in the strata. These used presplitting blasts, divisional control technology on surrounding rock in the retained entryways, "the trinity" surrounding rock controlling technique in retained entryways, regional control technology of borehole maintenance and rapid casing installation and borehole sealing technology.

(3) The faces 1111(1) and 11112(1) in the Zhuji Mine were considered for the latter two measures. According to the results, the CECGWEP technology can effectively solve the problems in deep mines. The maximum gas drainage flow of a single hole can reach 8.1 m^3/min, and the effective drainage distance can be extended up to 150 m or more.

Acknowledgments The research was supported by Program for Changjiang Scholars and Innovative Research Team in University (IRT_14R55), and the National Natural Science Foundation of China under Grant No. NSFC-51274193.

References

Bibler CJ, Marshall JS, Pilcher RC (1998) Status of worldwide coal mine methane emissions and use. Int J Coal Geol 35:283–310

Kang H, Niu D, Zhang Z, Lin J, Li Z, Fan M (2010) Deformation characteristics of surrounding rock and supporting technology of gob-side entry retaining in deep coal mine. Chin J Rock Mech Eng 29:1977–1987

Karacan CO, Ruiz FA, Cote M, Phipps S (2011) Coal mine methane: a review of capture and utilization practices with benefits to mining safety and to greenhouse gas reduction. Int J Coal Geol 86:121–156

Keim SA, Luxbacher KD, Karmis M (2011) A numerical study on optimization of multilateral horizontal wellbore patterns for coalbed methane production in Southern Shanxi Province, China. Int J Coal Geol 86:306–317

Kong S, Cheng Y, Ren T, Liu H (2014) A sequential approach to control gas for the extraction of multi-gassy coal seams from traditional gas well drainage to mining-induced stress relief. Appl Energy 131:67–78

Li W, Cheng Y, Guo P, An F, Chen M (2014) The evolution of permeability and gas composition during remote protective longwall mining and stress-relief gas drainage: a case study of the underground Haishiwan Coal Mine. Geosci J 18:427–437

Liu C, Zhou F, Yang K, Xiao X, Liu Y (2014) Failure analysis of borehole liners in soft coal seam for gas drainage. Eng Fail Anal 42:274–283

Lu S, Cheng Y, Ma J, Zhang Y (2014) Application of in-seam directional drilling technology for gas drainage with benefits to gas outburst control and greenhouse gas reductions in Daning coal mine, China. Nat Hazards 73:1419–1437

Pan R, Cheng Y, Yuan L, Yu M, Dong J (2014) Effect of bedding structural diversity of coal on permeability evolution and gas disasters control with coal mining. Nat Hazards 73:531–546

Whittles DN, Lowndes IS, Kingman SW, Yates C, Jobling S (2007) The stability of methane capture boreholes around a long wall coal panel. Int J Coal Geol 71:313–328

Xia T, Zhou F, Liu J, Hu S, Liu Y (2014) A fully coupled coal deformation and compositional flow model for the control of the pre-mining coal seam gas extraction. Int J Rock Mech Min 72:138–148

Xie H, Zhou H, Xue D, Gao F (2014) Theory, technology and engineering of simultaneous exploitation of coal and gas in China. J China Coal Soc 39:1391–1397

Xue S, Yuan L, Wang Y, Xie J (2014) Numerical analyses of the major parameters affecting the initiation of outbursts of coal and gas. Rock Mech Rock Eng 47:1505–1510

Yang W, Lin B, Yan Q, Zhai C (2014) Stress redistribution of longwall mining stope and gas control of multi-layer coal seams. Int J Rock Mech Min 72:8–15

Yuan L (2008) The technique of coal mining and gas extraction by roadway retaining and borehole drilling. J China Coal Soc 33:898–902

Yuan L, Xue J, Zhang N, Lu P (2013) Development orientation and status of key technology for mine underground coal bed methane drainage as well as coal and gas simultaneous mining. Coal Sci Technol 9:6–11

Zhang N, Yuan L, Han C, Xue J, Kan J (2012) Stability and deformation of surrounding rock in pillarless gob-side entry retaining. Safety Sci 50:593–599

Zhang NC, Esterle J, Zhang N et al (2013) Analysis on Wedge-shaped roof mechanical behaviours under a thick sandstone layer with different longwall mining speed. In: Proceedings of 3rd international workshop on mine hazards prevention and control, Brisbane, pp 369–375

Zhang N, Zhang NC, Han C, Qian D, Xue F (2014) Borehole stress monitoring analysis on advanced abutment pressureinduced by Longwall Mining. Arabian J Geosci 7:457–463

Zheng X, Feng X, Zhang N, Liu J, Zhang L (2013) Fissures evolution law and gas drainage characteristics of remote decompressed mining. Disaster Adv 6:164–176

Zheng X, Feng X, Zhang N, Gong L, Hua J (2014) Serial decoupling of bolts in coal mine roadway supports. Arabian J Geosci: 1–14

Zhou H, Yang Q, Cheng Y, Ge C, Chen J (2014) Methane drainage and utilization in coal mines with strong coal and gas outburst dangers: a case study in Luling mine, China. J Natural Gas Sci Eng 20:357–365

13

Application of highwall mining system in weak geological condition

Takashi Sasaoka[1] · Tri Karian[1] · Akihiro Hamanaka[1] · Hideki Shimada[1] · Kikuo Matsui[1]

Abstract Almost all the coal is produced from open cut mines in Indonesia. As a consequence of open cut mine application, a great deal of coal is left out in the highwalls of the mined-out pits. Highwall mining systems can be used to recover this coal. The use of highwall mining systems has increasingly come into play in the US and Australia. However, it is not common in Indonesia. Moreover, Indonesia coal measure is categorized as weak geological condition. Some problems are likely to arise during the application of the highwall mining system for example instability of openings and highwalls due to the roof and pillar failures. Therefore, study of highwall mining system application in Indonesia is needed in order to increase the recovery rate of coal mining in Indonesia. This paper described the characteristics of the highwall mining system and discussed the appropriate highwall mining system application in weak geological condition, Indonesia. From the results of a series of laboratory tests and numerical analyses, it can be concluded that the stability of pillars and mine openings in auger mining systems is much higher than that in CHM and an auger mining system is suitable for such as very weak/poor strata conditions. Moreover, the application of backfilling system is very effective for improvement of the stability of pillar and openings.

Keywords Open cut mine · Left out coal · Highwall mining systems · Weak geological condition

1 Introduction

Indonesia produces over 400 Mt of clean coal and is the second largest coal exporter to Japan, accounting for about 32 Mt of coal annually. Over 99 % of the coal produced in Indonesia comes from surface mines. More open cut mines will be developed and more coal will be mined in order to fill the great demand both in Indonesia and the rest of the world.

In open cut mines in Indonesia, there are many sites where mining operations have developed long highwalls which have been abandoned due to the current economic condition. Mining operations is then transferred to lower stripping ratio blocks of coal or overlying seams. It is estimated that a great deal of coal beneath abandoned in the highwall. Considering these situations, it seems to be worth introducing highwall mining systems into Indonesian open cut mines.

Final highwalls of open cut mines can form the starting point for other mining methods, such as highwall or underground mining. In its basic application, highwall mining is a technique utilized after an open cut portion of a reserve has been mined, sometimes prior to the introduction of underground mining (Seib 1993). In this system, the coal seam is mined by remotely operated equipment, such as an auger machine or a modified continuous miner incorporated into a highwall mining system (Fig. 1). Major issues of highwall mining systems are less coal recovery due to leaving coal as the pillars and the instability of openings and highwalls due to the pillar and roof failures. These geotechnical and operational issues have been occurring in Indonesia when the conventional highwall

✉ Takashi Sasaoka
sasaoka@mine.kyushu-u.ac.jp

1 Department of Earth Resources Engineering, Kyushu University, Fukuoka 819-0395, Japan

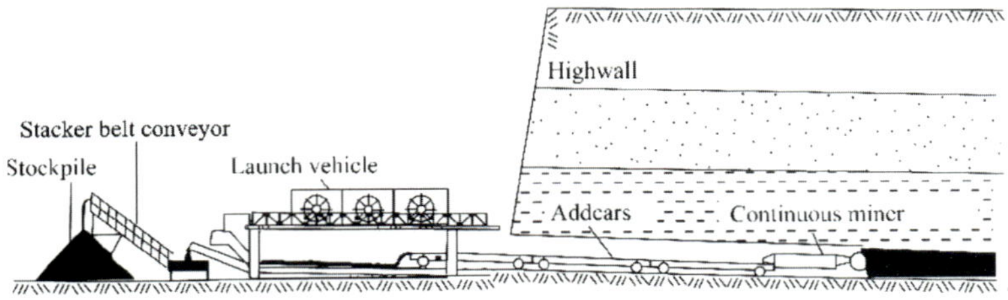

Fig. 1 Schematic of highwall mining system (Seib 1993)

mining system was introduced because of its weak geological conditions.

This paper describes the characteristics of the highwall mining system and discusses the appropriate highwall mining system in Indonesia including the application of the backfilling system to improve the stability of excavated highwall.

2 Highwall mining systems

Many coal seams that are presently uneconomical or technically unsuitable for conventional surface mining techniques can be recovered using highwall mining systems. Final highwalls of open cut coal mines can form the starting point for other mining methods, such as highwall or underground mining. In its basic application, highwall mining is a technique utilized after the open cut portion of a reserve has been mined, sometimes prior to the introduction of underground mining (Seib 1993).

In these mining systems, a continuous miner or an auger machine is primarily used to extract coal from the highwalls. Recently other systems such as Bucyrus Highwall Miners (formerly Superior Highwall Miners) and American Highwall Systems have been introduced in the US (The Virginia Center for Coal and Energy Research 2008).

The continuous miner system is called a continuous highwall mining (CHM) system and is categorized into two types: one is an Addcar system and the other is an Archveyor system. The former consists of a continuous miner, addcars (belt conveyor cars), a launch vehicle, a stacker conveyor and a loader as shown in Fig. 2. The latter consists of a continuous miner, archveyor chain conveyor that conveys coal and trams the machine system itself, and a loadout vehicle as shown in Fig. 3.

The CHM system can excavate rectangular holes over 350 m long and the size of the hole depends on the specifications of the continuous miner. Fully automated control of the system is achieved with the aid of advanced navigation technology, including a roof and floor passive gamma detector system, inclinometers, a ring laser gyroscope and programmable logic controller (PLC).

The auger system, as shown in Fig. 4, is very simple and can excavate holes over 100 m long and 0.5 m in diameter or greater into the coal seams in highwalls, depending on the application methods. This system is more maneuverable than the CHM system and can be applied when shorter lengths of highwall are presented. It is also suited to recovering open cut end walls and major pillars of coal remaining after the CHM operation (Matsui et al. 2008).

One of the first true "highwall mining systems" to be demonstrated, the Metec miner system, was introduced in 1981 and featured in a number of trade exhibitions during the 1980s. As a successor of the Metec miner, the Bucyrus (formerly superior) highwall mining and American highwall systems are now attracting significant attention in the US. This system consists of a continuous miner, double auger flights, a stacker conveyor and a front-end loader basically as shown in Fig. 5. This system can excavate rectangular holes of over 300 m long, has maneuverability along the bench which is important for the higher

Fig. 2 Addcar system

Fig. 3 Archveyor system

Fig. 4 Auger system

productivity, and the machine is equipped with four crawlers allowing for rapid positioning from hole to hole.

Highwall mining systems have improved mining operations due to their safety and productivity. However, the application is restricted by coal seam conditions such as the dip, thickness, roll, split of the seam, the existence of faulting and folding, etc. It is difficult or almost impossible to apply the system into steeper dipping coal seams, over 15°. Therefore, another type of system was developed, aimed at coal seams in the 16°–25° dip range. A high-powered auger machine, for use in seam dips of up to 23°, was developed and introduced at a coal mine in Australia (Matsui et al. 2003).

Highwall mining systems have been used extensively in the US and Australia due to their safe and economical mining method (Robertson et al. 1988; Michell 1999). However, in Indonesia, only a few highwall mining operations has been done (Furukawa et al. 2009).

3 Application of highwall mining system in Indonesia

Considering Indonesian coal seam conditions, only a few highwall mining operations had been conducted as shown in Fig. 6. According to their experience, dipping and rolling coal seam conditions always controlled its performance. Practical experience shows that a face of a 6 m-thick seam would be mined by two augering lifts using 1.9 m diameter auger according to the original mining plan. However, the actual mining was quite different from the plan due to the coal seam conditions. As there were many dips, rolls and thickness changes in the seam, a single lift augering had to be used in the center of the seam. Therefore, the coal recovery was much less than had been planned.

Indonesian coal measure rocks are generally very weak (Anwar et al. 1999). In Indonesian archipelago, coal deposits are mainly distributed in the Islands of Sumatra and Kalimantan. The sedimentary basins within which the coals were deposited are very young geologically. Based on the time of deposition, coal measures are divided on the basis of their age into Paleogene and Neogene coal measures. In general, the majority of the coal is Eocene and Miocene in age. Sedimentary rock becomes stronger with age and tectonic stress. Uniaxial Compressive Strength also increases with age in most sedimentary rocks due to increased lithification and reduced porosity (Waltham 1994). Therefore, due to its very young geological age, Indonesia coal measure is a weak geological condition.

Some coal measures rocks such as shale, mudstone, and siltstone show an excessive slaking behavior leading to a severe deterioration of their mechanical properties. The weak immediate roof tends to fall easily and the continuous miner also tends to sink into the softened floor or slip on the floor. Moreover, the shape of openings may affect the stability of openings/pillar remarkably in such as weak geological conditions. Therefore, in order to discuss a suitable highwall mining design in Indonesia, a series of

Fig. 5 Metec miner system

Fig. 6 Auger mining system in Indonesia

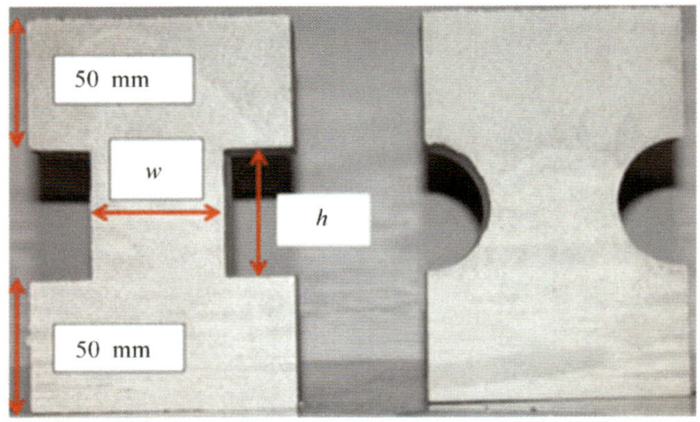

Fig. 7 Specimens for laboratory tests (*left, square openings; right, circular openings*)

laboratory tests and numerical analysis were conducted in this research.

4 Stability of openings/pillars in auger versus CMH mining systems

4.1 Laboratory test

The formula for evaluating pillar strength for CHM (rectangular shape openings) has already been developed (Mark et al. 1995). However, the ones for the auger mining system have not been done yet. Therefore, in order to evaluate pillar strength of the auger mining system (circular shape openings) and discuss the difference between the stability of openings/pillar in auger mining system and that in CMH mining system, the laboratory tests using a test model were conducted.

Figure 7 shows the specimen used in this research. The specimen dimension was set such that it can simulate various pillar width to opening width ratio. The dimension of this specimen was 200 mm in thickness, 200 mm long and 150 mm high. The specimen consists of gypsum, cement and water (weight ratio, gypsum: cement: water = 1:1:1). The shapes of openings were a circle with a diameter of 50 mm or a square 50 mm on a side. The pillar width was varied in accordance to opening width. Four different pillar widths (w), 25, 50, 75 and 100 mm, were evaluated. These specimens were stored under dry conditions for 3 days and then a loading test was conducted.

Figure 8 shows the relationship between the pillar strength index and ratio of pillar width/opening width of circular and square openings. Here, the pillar strength was defined as the following equation in order to eliminate the variation of material strength,

$$\text{Pillar strength index} = \frac{\text{Failure load}}{\text{Cross section of pillar} \times \text{UCS of material}} \tag{1}$$

It can be seen from Fig. 8 that the pillar strength index increases with the increasing pillar width for both shapes of openings. The pillar strength for circular shape openings is about 1.2–1.3 times stronger than that of the rectangular shape. In order to evaluate the shape effect quantitatively, more detailed study has to be conducted under different geological conditions, in-situ stress conditions, etc.

4.2 Numerical analysis

In order to discuss the stability of pillars and mine openings related to the shape of openings/pillars or the different mining systems, finite element modeling was also applied in this research. Non-linear analysis was also performed using the two-dimensional finite element method (FEM) cord 'Phase^2' (Rocscience Incorporation 2007). For simplicity, a two dimensional model was used in this analysis as shown in Fig. 9. The coal seam thickness is 3 m. The immediate roof and floor are siltstone being 4 m thick and 3 m thick, respectively. The initial vertical stress (σ_v) was assumed to be the depth pressure ($\sigma_v = \gamma h$, γ is the unit weight of overburden and h is the cover depth or highwall height) and the initial horizontal stress (σ_h) was variable. Roller boundary was set at the outer side of the model as a boundary condition. Five circular openings with a diameter of 3 m or square openings 3 m on a side were excavated one after the other in the coal seam. The pillar width (w) between the two openings was assumed to be 3 and 1.5 m wide. The following well-known Hoek and Brown criterion was employed as a failure criterion:

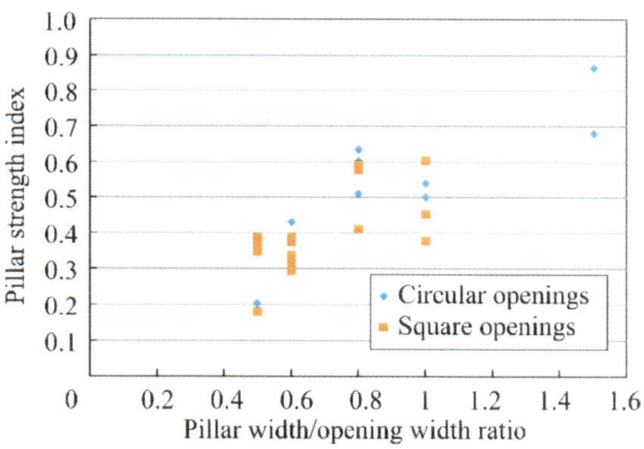

Fig. 8 Relationship between pillar strength and pillar width

$$\frac{\sigma_1}{\sigma_C} = \frac{\sigma_3}{\sigma_C} + \left(m\frac{\sigma_3}{\sigma_C} + s\right)^{\frac{1}{2}} \quad (2)$$

where, σ_1 is the major principal stress at peak strength, σ_3 is the minor principal stress, m and s are constants that depend on the properties of the rock and the extent to which it had been broken before being subjected to the failure stresses, and σ_C = uniaxial compressive strength of the intact rock material. The mechanical properties and parameters of the coal and rocks are listed in Table 1.

Figures 10, 11 and 12 show the results of the numerical analysis. From these figures, it can be seen clearly that the stability of openings/pillars in the auger mining system is higher than that in CHM. Figure 10 shows the failure development around the openings at different depths. With increasing the depth, the failure zones develop around the openings and in the pillars. Figure 11 also shows the failure development around the openings under the different initial stresses. The failure conditions depend on the stress conditions. Under high horizontal stress conditions ($\sigma_h/\sigma_z = 1.5$), the roof and floor of the openings tend to induce shear failure easily. Figure 12 shows the failure development around the openings with narrow pillars. Compared with Fig. 10, the narrow pillar in Fig. 12 shows less stability and the pillars fail much more severely in square openings. This situation will lead to unstable work conditions, and sometimes the cutting machine is caught in the opening being impossible to withdraw. On the other hand, in circular openings, no obvious and/or large pillar failure can be recognized even until a depth of 100 m.

It can be concluded that the stability of pillars and mine openings in auger mining systems is much higher than that in CHM and an auger mining system is suitable for such as very weak/poor strata conditions. However, this mining system has to be improved to cope with steeper coal seams and water problems in the face in order to increase the extraction ratio and maintain the highwall stability.

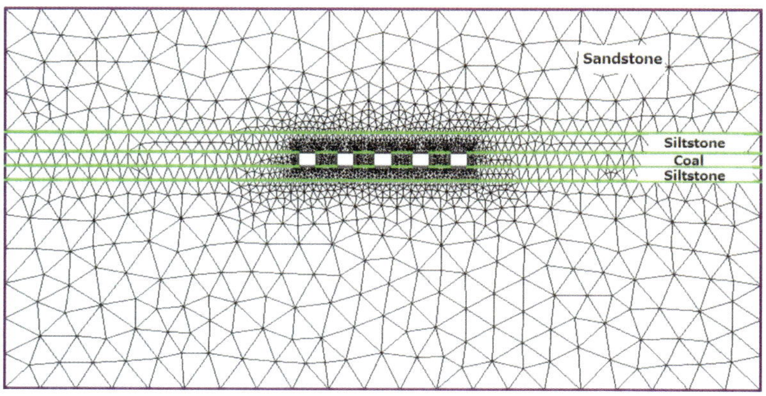

Fig. 9 Two-dimensional finite element model

Table 1 Mechanical properties of rocks used in numerical analysis

Rocks	Young's modulus (MPa)	Poisson's ratio	Uniaxial compressive strength (MPa)	*m*	*s*	Unit weight (MN/m^3)
Sandstone	5700	0.28	26.6	7.50	0.10	0.0243
Siltstone	4650	0.27	20.9	2.00	0.02	0.0247
Coal	1530	0.30	11.9	2.08	0.07	0.0134

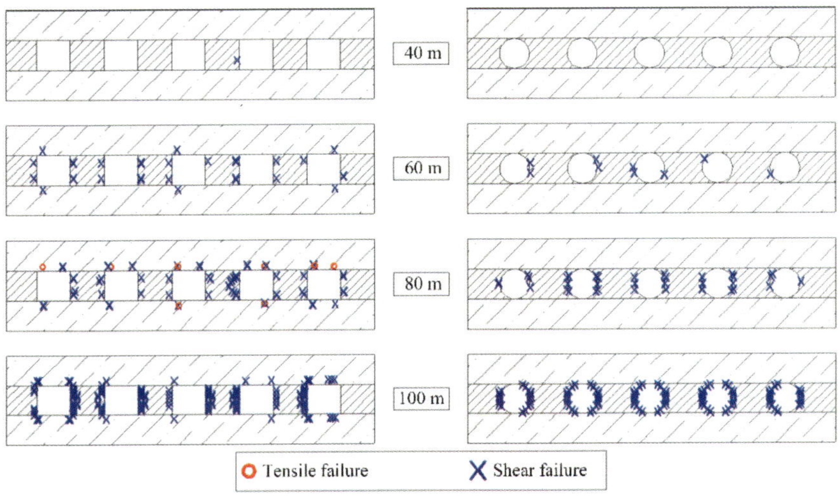

Fig. 10 Failure development around the openings at different depths (*left, square opening*; *right, circular opening*; $\sigma_h/\sigma_z = 0.75$; width of opening: width of pillar = 1:1)

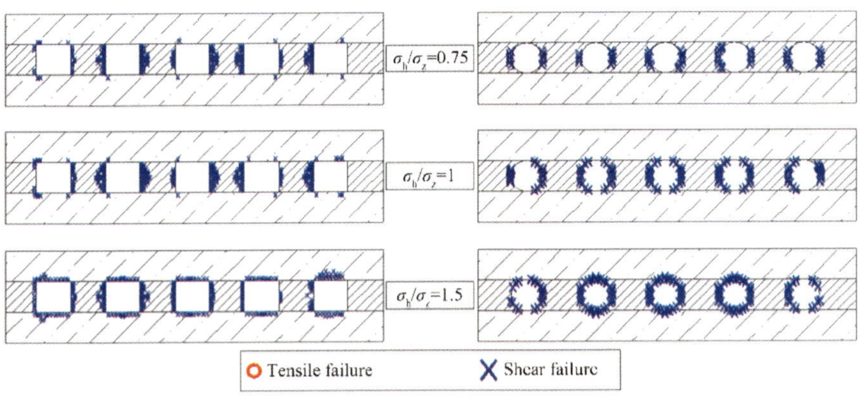

Fig. 11 Failure development around the openings under the different initial stress (*left, square opening*; *right, circular opening*; depth = 100 m; width of opening: width of pillar = 1:1)

5 Application of backfilling system

In order to improve the stability of openings and pillars and to increase the coal recovery, the application of backfilling system was also discussed by means of numerical analysis and laboratory tests.

5.1 Backfilling systems

Waste disposal in underground mines is not a revolutionary concept. Waste rocks have been used in the world for many years as backfilling materials to provide additional support to underground excavations in mines. Recently, for the lack

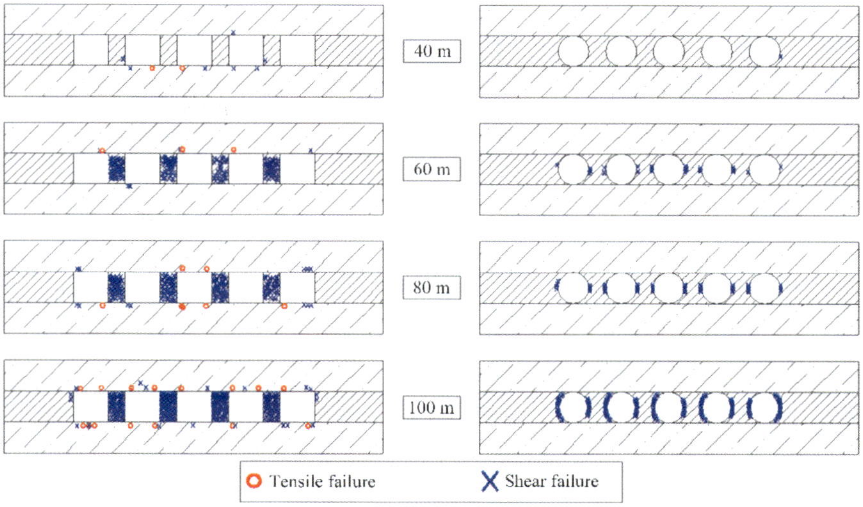

Fig. 12 Failure development around the openings with narrow pillars (*left, square opening*; *right, circular opening*; $\sigma_h/\sigma_z = 0.75$; width of opening: width of pillar = 2:1)

of a dumping site the use of backfill for regional and local support has been receiving increasing attention in the mining industry.

According to the previous research work (Afrouz 1994), backfilling presents the following merits to the underground coal mines:

(1) Fills the excavated areas, promoting better support and ground control.
(2) Provides a better environmental control on the waste rocks and coal preparation plant wastes.
(3) Increases coal recovery, especially in room and pillar mining and highwall mining systems.
(4) Reduces ventilation short-circuiting between adjacent mining sites.
(5) Reduces cost of waste transportation to the mine surface dumping sites and tailing ponds, and associated up-keeping and monitoring of these facilities.

In the USA, increased opposition from environmental groups is severely restricting the operation and planning of large-scale surface mines. Some projects of mountaintop removal mining have to be cancelled and downsized. In this case, as discussed in the previous section, the highwall mining system would be applicable and useful for the protection of the environment and reclamation. The backfilling would increase the coal extraction ratio, keep the pillars and the highwall stable and control the subsidence at the surface.

The required strength of the backfill depends on the strata and mining conditions: cover depth, rock types and properties, mining method, etc. At a shallow mine, the required strength is not critical compared to that at a deep mine. According to the research work in South Africa, backfilling shallow underground room and pillar mines with pulverized fly ash slurry could significantly improve the mine operating conditions by providing improved roof support and increased extraction ratio (Wagner and Galvin 1979). The research also suggests that the complete filling of the rooms, up to the roof level, is not necessary for improved roof control. Filling up to 70 % of the pillar height provided adequate confinement that constrained the lateral expansion of the pillar under concentrated compressive stresses. Moreover, by providing confinement to the pillars, thick coal seams have been successfully mined with a subsequent reduction of pillar height to width ratio and improved load bearing capacity of the pillars.

Backfilling materials that are of concern to the mining industry can be broadly classified into the following three categories:

(1) Waste originating from coal mines.
(2) Waste originating from coal burning power plants.
(3) Waste originating from other industries.

5.2 Finite element analysis about backfilling effects

Finite element modeling was used to clarify the effectiveness of the backfilling system on the stability of pillar, openings and highwall. For simplicity, the two dimensional model was used as shown in Fig. 9. The height of the highwall (h) was assumed to be 60–100 m. Five rectangular openings being of 4 m wide and 3 m high were excavated one after the other in the coal seam. After the excavation of each opening, the opening was backfilled. And then the next opening was excavated. The pillar width (w) between two openings was assumed to be 4 and 2 m wide. The mechanical properties and parameters of two different backfilling materials were simulated and listed in Table 2.

Table 2 Mechanical properties of backfilling materials used in numerical analysis

Materials	Young's modulus (MPa)	Poisson's ratio	Uniaxial compressive strength (MPa)	*m*	*s*	Unit weight (MN/m^3)
Backfill 1	2000	0.3	20	10.0	0.1	0.020
Backfill 2	500	0.3	10	7.5	0.1	0.017

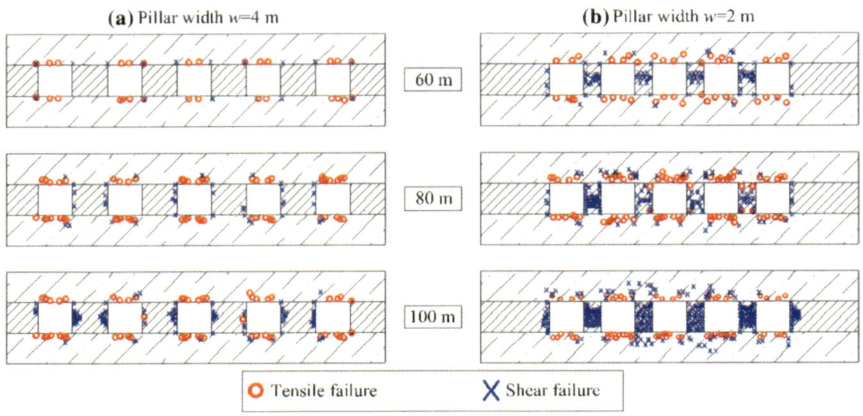

Fig. 13 Failure development around the openings with wide and narrow pillar ($\sigma_h/\sigma_z = 0.75$)

Figure 13 shows the failure development around the openings with wide and narrow pillars. With the narrow pillar, the openings show the less stability and the pillars fail much severer than the wide pillar case. This situation will lead to unstable work conditions, and sometimes the cutting machine would be caught in the opening being impossible to withdraw.

Figure 14 shows the failure development around the openings with backfilling. It is clear that the backfilling helps the openings keep stable compared to the non-backfilled cases as shown in Fig. 13. However, at greater depths, the effectiveness of backfilling cannot be obtained well. In such a case, we have to consider the use of wider pillar, narrow opening and stronger backfilling material.

Although it is clear from the numerical model that the backfilling improves the pillar stability shown by the reduction of plastic zone, the mechanism and reason is not well understood. Therefore, laboratory test must be carried out to unveil the role of backfilling in strengthening the pillar.

5.3 Laboratory tests about backfilling effects

Result in Fig. 14 suggests that the backfilling improve the pillar stability. This could be caused by confinement effect of backfilling and the infiltration of backfilling slurry content into the failure zone improves mechanical properties of the broken pillars. In order to examine the behavior of backfilled pillars, modeled pillar was set up and tested in the laboratory. Cylindrical cores were drilled out and prepared to simulate pillar models from *Ainoura* sandstone block sample.

5.3.1 Test procedures

The size of the prepared pillar models is around 25 mm in diameter and 60 mm in length. These samples were cured for 4 weeks under the constant environmental conditions: 100 % relative humidity and 25 °C.

After the above cure, the pillar models that were not backfilled were set in the stiff testing machine and tested under compressive loading to obtain the complete stress–strain curves. This test is the same as the standard compression test.

In order to investigate the backfilling effects, two tests were performed. One is for backfilled pillar models and the other is for the backfilled fractured pillar models. The backfilled model was made from the pillar model that was set in the polyvinyl chloride (PVC) tube and filled with the backfill mixture as shown in Fig. 15. Three backfill heights such as 25, 40 and 50 mm were considered in the tests. Five backfill mixtures consisting of Portland cement, fly ash and water were considered as listed in Table 3. These backfilled pillar models were cured under the same conditions as pillar models mentioned before.

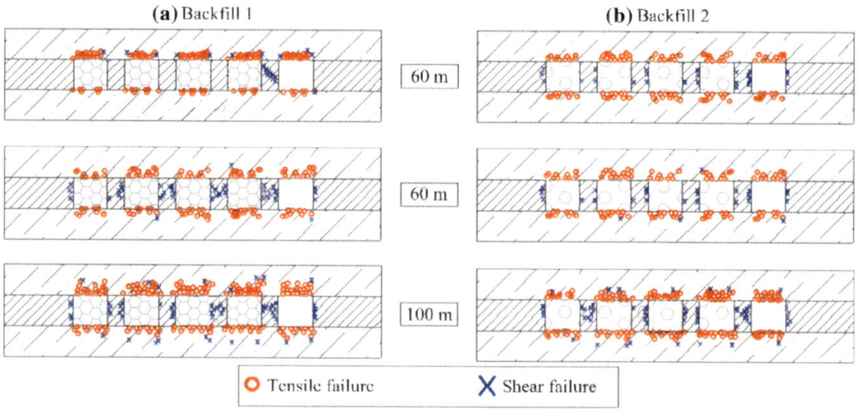

Fig. 14 Failure development around the openings with backfilling ($\sigma_h/\sigma_z = 0.75$, pillar width $w = 2$ m)

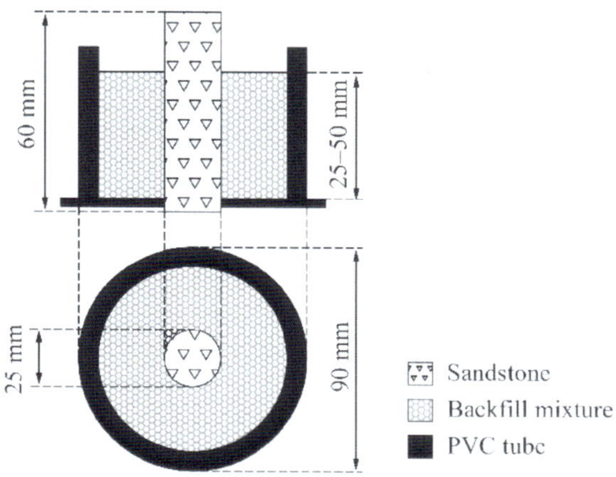

Fig. 15 Backfill model for standard compression test

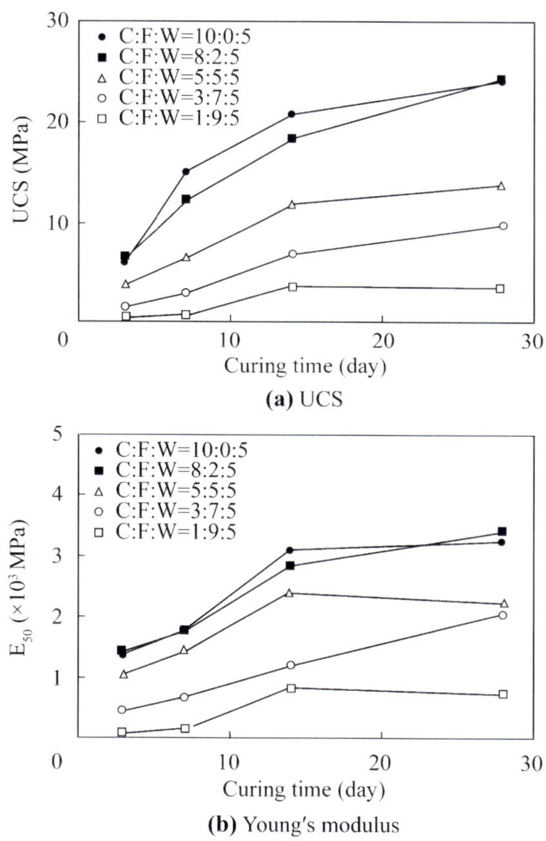

Fig. 16 Mechanical properties of backfill mixtures under different curing time

Table 3 Contents of each backfill mixture (weight ratio)

No.	Portland cement (C)	Fly ash (F)	Water (W)
1	10	0	5
2	8	2	5
3	5	5	5
4	3	7	5
5	1	9	5

The backfilled fractured pillar models were obtained from the above usual compression tests where the complete stress–strain curves were obtained. After the compression test, the fractured pillar was set in the PVC tube, and then filled up to the described height with the backfill mixture and cured as mentioned before. Three to five models were tested for each test condition.

5.3.2 Test results and discussion

Figure 16 shows the mechanical properties of backfill mixtures. Curing time shows definite effect on the mechanical properties such as uniaxial compressive strength (UCS) and tangent Young's modulus at 50 % of peak stress (E_{50}) in 2 weeks. Increasing the fly ash content decreases the mechanical properties.

Figure 17 shows the complete stress–strain curves of backfilled pillar models under the different backfilling heights. Backfilling improves the behavior of the pillar, increasing the Young's modulus, the peak stress and the residual stress.

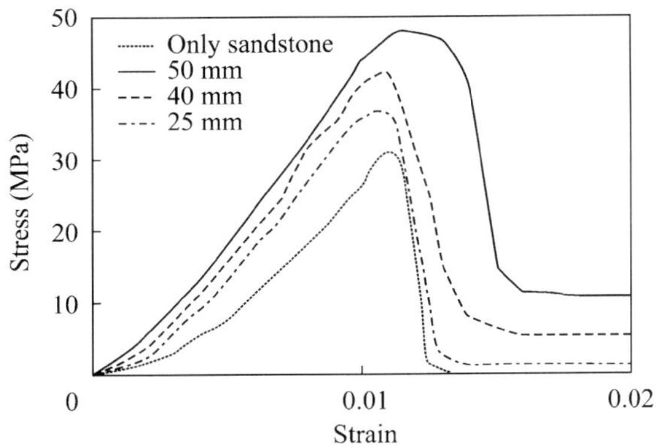

Fig. 17 Stress–strain curve of backfilled pillar models under different backfilling heights

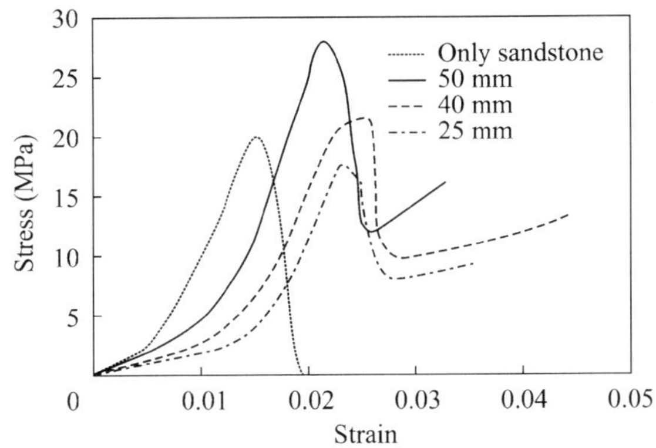

Fig. 19 Stress–strain curve of backfilled fractured pillar models under different backfilling heights

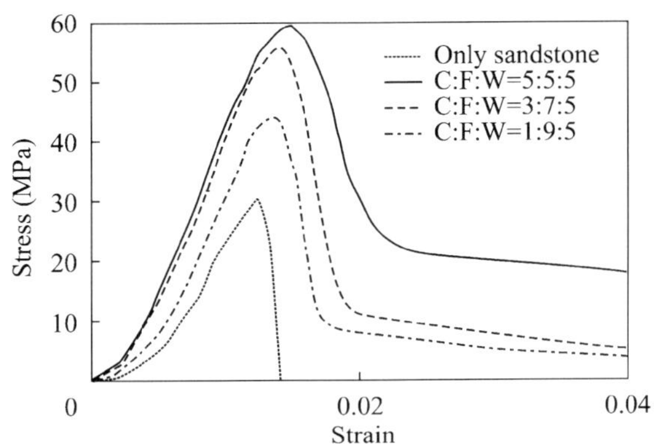

Fig. 18 Stress–strain curve of backfilled pillar models under different backfill mixtures contents

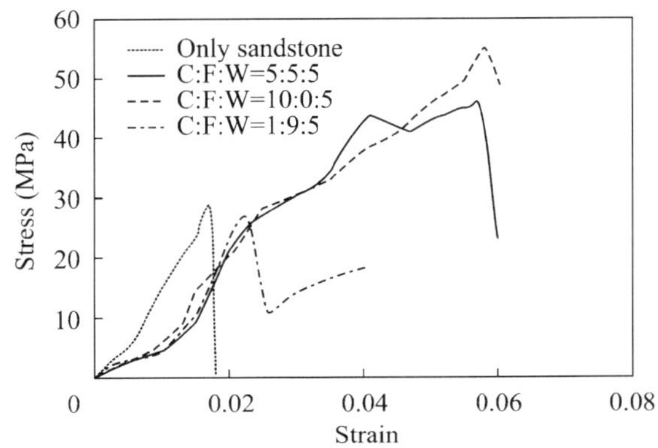

Fig. 20 Stress–strain curve of backfilled fractured pillar models under different backfill mixtures contents

Figure 18 shows the complete stress–strain curves of backfilled pillar models under the different mixture contents. Increasing Portland cement also improves the properties due to the high confinement of backfilling.

From these test results using the backfilled pillar models, the effectiveness of backfilling on the pillar behavior is shown especially in the peak stress and behavior in the post-peak region.

Figures 19 and 20 show the complete stress–strain curves of pillar models and backfilled fractured pillar models under the different backfilled heights and the different mixture contents, respectively. Although the pillar models show peak stress clearly, they show any residual stress after peak stress. It is difficult to mobilize the function as a pillar under poor strata and high stress conditions. However, backfilling plays an important role in increasing the peak stress and residual stress, though not increasing the Young's modulus.

From these test results using the backfilled fractured pillar models, fractured pillars restore their strength by backfilling. It means that backfilling improves the fractured pillar behavior very well.

6 Conclusions

Highwall mining systems can sometimes be the final mining method used in open cut mines, or it can serve as a means of transition, with low capital cost, from surface mining to underground mining. From the results of a series of laboratory tests and numerical analyses, stability of openings/pillars in the auger mining system is higher than that in CHM. The pillar strength for circular shape openings in the auger mining system is about 1.2–1.3 times stronger than that of the rectangular shape openings in CHM. The pillar strength index increases with the increasing pillar width for both shapes of openings. It is also found that backfilling improve the pillar stability. It improves the pillar stability by, increasing its Young's modulus, the peak stress and the residual stress. From test results using the backfilled fractured pillar models, fractured pillars restore their strength by backfilling. It means that backfilling improves the fractured pillar behaviour very well. However, at greater depths, the effectiveness of backfilling cannot be obtained well. Based on this result, the highwall mining operation can be applied even in weak geological conditions by application of the appropriate mining system and/or backfilling system.

References

Afrouz AA (1994) Placement of backfill. Min Eng 153:205–211

Anwar HZ, Shimada H, Ichinose M, Matsui K (1999) Slaking durability of the improvement of coal mine shales behavior. Proceeding of the international symposium on mining science and technology, Beijing, pp 343–346

Furukawa H, Matsui K, Sasaoka T, Shimada H (2009) Application of a punch mining system to Indonesian coal mines. Proceedings of the 2nd international symposium of novel carbon resources science, earth resource science and technology, Bandung: III1-9

Mark C, Chase FE, Campoli AA (1995) Analysis of retreat mining pillar stability. Proceedings of 14th conference on ground control in mining, pp 49–59

Matsui K, Yabuki A, Shimada H, Sasaoka T, Ueda T, Yuasa T (2003) A practical trial to increase the coal recovery in highwall auger mining in Australia. Proceedings of 12th international symposium on mine planning and equipment selection, pp 205–208

Matsui K, Sasaoka T, Shimada H, Ueda T, Kramadibrata S, Sulistianto B (2008) Highwall mining systems at surface coal mines in Indonesia. Coal Int 256(3):28–31

Michell G (1999) Benefits of longwall punch mining. Proceedings of the 2nd annual longwall mining summit, p 8

Robertson BN, O'Regan G, McKew M (1988) Longwall punch mining from open cut highwalls. Proceedings of the 21st century higher production coal mining systems-their implications, pp 215–220

Rocscience Incorporation. (2007) Application of the finite element method to slope stability. http://www.rocscience.com/library/pdf/SlopeStabilityUsingPhase2.pdf. Accessed 17 Feb 2007

Seib WT (1993) Strip mining. In: Hargraves AJ, Martin CH (eds) Australasian coal mining practice, 2nd edn. Australasian Institute of Mining Metallurgy, Carlton, pp 238–242

The Virginia Center for Coal and Energy Research (VCCER) (2008) Meeting projected coal production demands in the USA, chapter 3: mining technology and resource optimization. Blacksburg: The Virginia Center for Coal and Energy Research

Wagner H, Galvin JM (1979) Use of hydraulically placed PFA to improve stability in bord and pillar workings in South African collieries. Proceedings of symposium on the utilization of pulverized fuel ash, p 27

Waltham AC (1994) Foundations of engineering geology. The Alden Press, Oxford

14

Infrastructure-based localisation of automated coal mining equipment

Chad O. Hargrave[1] · **Craig A. James**[1] · **Jonathon C. Ralston**[1]

Abstract A novel radar-based system for longwall coal mine machine localisation is described. The system, based on a radar-ranging sensor and designed to localise mining equipment with respect to the mine tunnel gate road infrastructure, is developed and trialled in an underground coal mine. The challenges of reliable sensing in the mine environment are considered, and the use of a radar sensor for localisation is justified. The difficulties of achieving reliable positioning using only the radar sensor are examined. Several probabilistic data processing techniques are explored in order to estimate two key localisation parameters from a single radar signal, namely along-track position and across-track position, with respect to the gate road structures. For the case of across-track position, a conventional Kalman filter approach is sufficient to achieve a reliable estimate. However for along-track position estimation, specific infrastructure elements on the gate road rib-wall must be identified by a tracking algorithm. Due to complexities associated with this data processing problem, a novel visual analytics approach was explored in a 3D interactive display to facilitate identification of significant features for use in a classifier algorithm. Based on the classifier output, identified elements are used as location waypoints to provide a robust and accurate mining equipment localisation estimate.

Keywords Localisation · Waypoint navigation · Machine learning · Radar · Underground · Longwall mining · Automation

1 Introduction

1.1 Background

The inherent dangers of mining operations, particularly underground, are driving a growing demand for automation solutions that will remove personnel from hazardous and unhealthy environments. However, due to the relatively unstructured nature of the production environment and the attendant difficulties regarding accurate equipment monitoring, the mining domain represents a particularly challenging industry sector for the application of automation technology. Traditional closed-loop automation techniques drawn from the factory floor typically assume a high degree of certainty regarding key equipment parameters (such as position or status) that is often unavailable in the mining context. For this reason, probabilistic methods based on data analytics have shown more promise for managing and automating the mining process. This paper examines the use of such techniques in the context of a key mining automation challenge: the development of robust and accurate equipment localisation systems.

1.2 Problem description

While surface mining applications are now employing both RTK-GPS (Real Time Kinematic Global Positioning System—an enhanced version of standard GPS that employs a ground station to improve position accuracy) and extensive wireless communications systems to assist with accurate

✉ Chad O. Hargrave
chad.hargrave@csiro.au

[1] Mining and Processing Technologies Research Group, Coal Mining Research Program, CSIRO Energy, Brisbane, Australia

equipment localisation (Wei et al. 2014), underground operations such as longwall coal mining (Singh 1997) have no GPS access and typically very limited wireless access. A significant problem that must be solved in order to achieve continuous autonomous mining in longwall coal mines is the measurement of small changes in the position or orientation of specific sections of mining equipment over extended operational periods. For the case of longwall mining, two key parameters are longwall creep (cross-track position) and longwall retreat (along-track position) with respect to gate road travel. An outline of the longwall mining process is necessary in order to explain these problems in detail.

1.3 The longwall mining process

A coal seam is typically a contiguous wedge of coal, of varying height, sandwiched between layers of rock. The goal of the coal mining process is to extract as much of this seam as possible (or at least, the high-quality coal portions thereof) in an efficient and safe manner, with minimal impact on the surrounding environment. In a *longwall* coal mine two long, horizontal and permanent tunnels known as *gate roads* are cut into the coal seam to form the main boundaries for a large rectangular block of coal, known as a longwall *panel*. The gate roads are typically five to six metres in width. A typical panel is 200–300 m wide, and two to three kilometres in length (the length of the gate roads). At the end of the gate roads the major pieces of mining equipment are installed across the back of the panel, creating the *longwall face* (Fig. 1). At regular, surveyed intervals along the gate roads, cross-tunnels known as *cut-throughs* are established to provide access and assist with the management of mine ventilation (Singh 1997).

The key parameters to be measured and tracked are known as *longwall retreat* and *longwall creep*. Longwall retreat is simply the progression of the longwall face into the unmined coal panel as shown in Fig. 1. As the mining progresses, however, geological and equipment interactions creates variation in pressure across the longwall face, leading to incremental sideways slippage of the face equipment into the gate road, known as longwall creep (Fig. 2). This sideways movement is undesirable, and if left unchecked can lead to the longwall equipment eventually running up against the gate road rib, effectively bringing the face to a halt: a situation that is very difficult to resolve. Retreat and creep thus correspond to the along-track and across-track position of the mining equipment in the longwall gate road. Because this equipment is mechanically linked to the entire longwall face, accurate localisation of the gate road machinery enables improved management of the longwall process. Both relative localisation (i.e., measurement with respect to surrounding mine infrastructure such as tunnel walls) and absolute positioning within the overall mine region are critical.

1.4 Candidate sensors for position measurement

To achieve accurate localisation, a variety of sensing technologies have been considered in various mining contexts. For relative localisation (e.g., vehicle collision avoidance applications), active ranging technologies such as laser scanners and radar-based sensors are solutions that have been employed in similar non-mining applications. For absolute localisation, the absence in the underground of a pervasive positioning infrastructure such as GPS has led to the use of alternative technologies, in particular distributed sensor networks. For applications also requiring high accuracy, inertial navigation systems have been successfully applied by Australia's CSIRO to machine guidance of both longwall and continuous miner applications (Ralston et al. 2014). Nevertheless, even high-end inertial systems are prone to drift with regard to the position estimate accuracy over time, and an external reference sensor is therefore required to place a bound on this source of error by correcting the inertial estimate on a periodic basis.

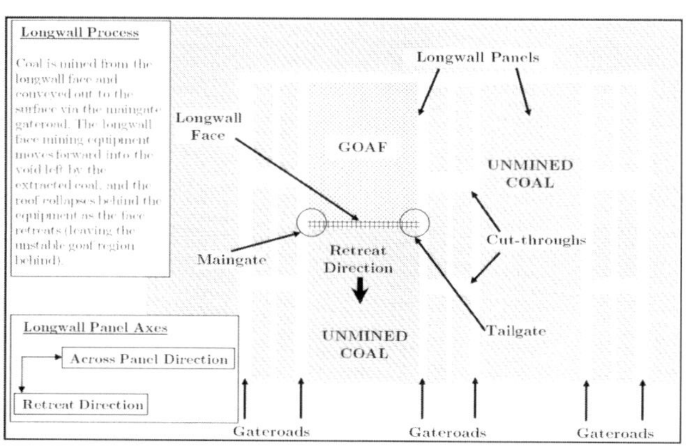

Fig. 1 Plan view of a longwall coal mine

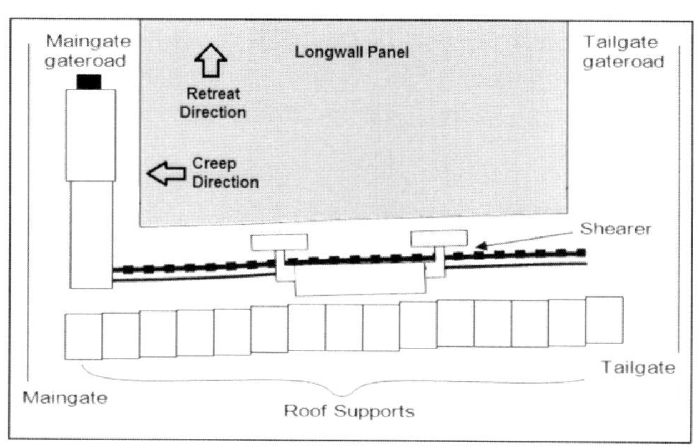

Fig. 2 Detail view of longwall face showing *creep* and *retreat*

A scanning laser sensor has previously been employed to perform this position correction task in the context of *longwall automation*, however this is an expensive solution that requires a large flameproof cabinet to house the scanning laser for safe operation in the explosive environment of an underground coal mine. More significantly, laser scanner ranging performance is easily degraded by the presence of airborne or surface dust, which is a major limitation in a mining context. When water is present, as is always the case in an underground coal mine, the surface can rapidly become coated with an opaque layer of mud that completely obscures the instrument's field of view (Fig. 3).

1.5 Radar-based measurement sensor

In response to these limitations, it has therefore been proposed that a radar-based sensor be trialled as a more suitable external reference measurement system, given the superior performance of radar in dusty environments. Previous published work has discussed the development of this radar sensor, and the initial results from a mine-site trial (Hargrave et al. 2014). The current paper examines the application of various probabilistic data analysis techniques to this raw radar data, to determine if sufficient information can be extracted to provide a useful monitor for longwall creep and retreat in a production context. Of the two measures, creep (the across-track distance to the gate road rib-wall) is both more straightforward (a direct application of radar ranging) and the more critical measurement, as it provides direct feedback of a potential collision between the equipment and the tunnel walls. The retreat measurement is more challenging, as it requires tracking of the along-track position of the longwall equipment based on indirect inference from the adjacent rib-wall, as there are no stable along-track targets in a typical gate road. The initial hypothesis for the study was that the ubiquitous roof and rib structural bolts, with their large plate washers, would make excellent candidates for radar tracking, as they present a strong radar target that would contrast well with the coal wall background (Fig. 4). A series of measurement campaigns, both synthetic in-house testing and actual underground trials, were therefore planned to test this hypothesis and attempt to resolve both the retreat and creep measurement challenges.

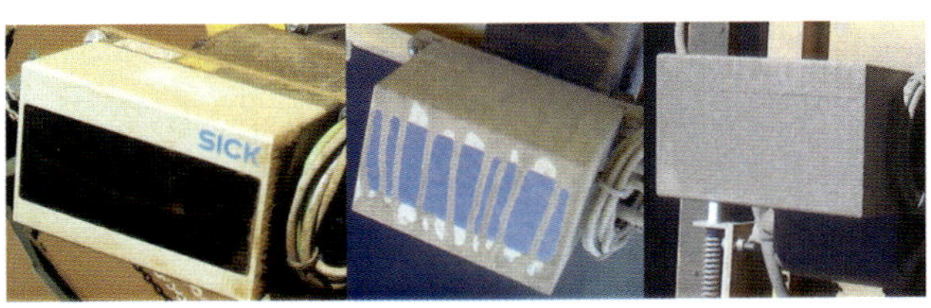

Fig. 3 Surface mud coating a laser scanner employed in a mining operation

Fig. 4 Rib wall with bolt-plates visible on the surface

The specific radar system developed for this application was adapted from an Endress-Hauser FMR 250 industrial radar level sensor (Hargrave et al. 2014). The primary consideration in selecting this sensor was its certification for use in explosive environments, which is a critical requirement for operation of electrical equipment in the methane-rich atmosphere of an underground coal mine. Further details regarding the radar system development can be found in (Hargrave et al. 2014).

2 System trials

2.1 Initial radar system testing

Initial radar testing was carried out at CSIRO Energy's premises in Brisbane. The radar unit was mounted on a moveable gantry above a box filled with coal to allow measurements at various points across the coal pit. Testing included basic ranging to the coal surface, and also detailed analysis of the changes in the radar response based on the presence of various targets on the coal surface that would typically be encountered in the mine environment, e.g., the plate-washers (Hargrave et al. 2007).

The first field test of the radar sensor was carried out in a disused section of an underground colliery in New South Wales, Australia, and targeted a very small section of a tunnel rib wall, scanning only three consecutive bolt plates (Hargrave et al. 2014). The rib-wall face was otherwise clear of any clutter, which provided ideal conditions for initial proof-of-concept testing. For the full field trial of the sensor under realistic conditions, a significant section of mining roadway was necessary, under more realistic conditions including the presence of piping, signage and other metallic structures found in a typical coal mine.

2.2 Description of underground mine test site

Permission was obtained to use a pre-production gate road at an operational longwall mine to conduct the field trial. This gate road was under development for future production, and hence was as close to an operational environment as was possible given that the test platform contains numerous uncertified items of electrical equipment. The mine provided support to carry out this field trial, including examination of all the non-certified equipment, the necessary dispensations to use these items in the proposed test location, and the provision of personnel to supervise the trial.

The test site was a 50 m length of gate road, close to the high wall of the mine. The testing took place between two side-roads branching off from the main roadway tunnel, known as cut-throughs. This was an ideal test location as cut-throughs are a common feature of all roadways in an underground coal mine. Cut-throughs are a very obvious feature for detection, and their locations are accurately surveyed, hence they can be used as correction points for any cumulative localisation errors. Roadway development for the test location was complete at the time of the trial, including all the typical mine site infrastructure such as airlines, hydraulics, signage, and electrical distribution systems.

2.3 Sensor platform system

The underground trial featured a comprehensive set of sensor hardware (Fig. 5), mounted on a test platform designed to set the instruments at a similar height to that of the gate road mine machinery. A series of customised software applications were developed to acquire and synchronise data from the various sensing instruments.

The sensors employed for the trial were:

(1) The primary radar unit, based on a 26 GHz Endress-Hauser instrument, as discussed in Sect. 1.
(2) Three SICK LMS 200 laser scanners, two mounted broadside to the direction of travel scanning both tunnel walls (to verify the creep measurements) and the third in the direction of travel (to provide a basis for comparison with the system retreat estimate).
(3) A video camera with low-lux capability for use in the underground.
(4) A data acquisition PC (standard laptop).

The real-time data acquisition system time-synchronised the data from the radar and the laser scanners, allowing rapid post-processing, comparative analysis and data fusion. The post-trial data processing was carried out in Matlab, however the data storage and retrieval protocols were developed for eventual integration into an existing real-time longwall automation software system.

2.4 Description of system testing

The field trial was carried out over two days, beginning with a series of measurements along a defined section of gate road at various offset distances from the rib wall. Targets were positioned at each end of the test course to provide a reliable along-track reference from the forward-looking laser scanner for the radar data. One feature of note with regard to the gate road was the presence of extensive mine services (water, power, and ventilation) located along one side of the gate road on the same side rib-wall as the cut-throughs (Fig. 6). The other (non-cut-through) rib of the tunnel was relatively clean (see Fig. 4). This contrast between the service-side and clean-side ribs is typical of underground mines, and provided a good opportunity to examine the radar sensor performance for both cases. Multiple separate test runs were carried out facing both sides and at different offset distances with the radar sensor facing perpendicular to each wall in turn. In total, 26 datasets were acquired over the two day test period.

For the majority of the testing, the radar sensor was positioned at 90° to the along-track progress of the test trolley with the sensor pointed directly at the rib-wall to provide a direct measurement of the creep distance. The rib wall in this section of the mine was unmeshed (see Fig. 4), unlike many mines where a steel meshing, covering the entire rib surface, is secured to the walls using the bolt

Fig. 5 Sensor suite on the test platform, radar enclosure on the *right*

Fig. 6 Cluttered rib wall and cut-through

plates. This mesh is used to prevent the collapse of smaller sections of roof or rib that could cause injury to personnel. Because the metal mesh presents an additional clutter source for the radar sensor, additional scans were also made of the roof (which was meshed, Fig. 7) along the full trial length of the trial roadway. These additional scans were performed to ensure that a truly representative dataset could be generated of the most common roof and rib conditions encountered in the vast majority of underground coal mines.

2.5 Data analysis plan

To estimate across-track movement (creep) and along-track movement (retreat), the data analysis plan first sought methods to isolate these two independent parameters from the raw radar signal data. The initial processing for display of the raw data is presented in Sect. 3, while the techniques employed to extract the two parameters of interest are discussed in Sect. 4.

3 Trial results

Both the laser and radar data were processed for initial display immediately following each trial run. The raw radar data consist of the "envelope curves" of the radar signal. Through the application of masking and background subtraction filtering, signal artefacts such as gain bias and cross-talk between transmit and receive circuitry were removed (Skolnik 2001). The individual radar traces were then stacked to develop a three-dimensional, time-varying display of the radar output. Figure 8 shows a typical result for a test run at a radar offset distance of four metres from the rib wall. Two notable features seen in this figure are: the clear gap associated with the side tunnel corresponding to a cut-through, and the frequent peaks in the main signal return corresponding to the highly reflective wall-mounted bolt-plates.

Fig. 7 Meshed roof in the trial roadway

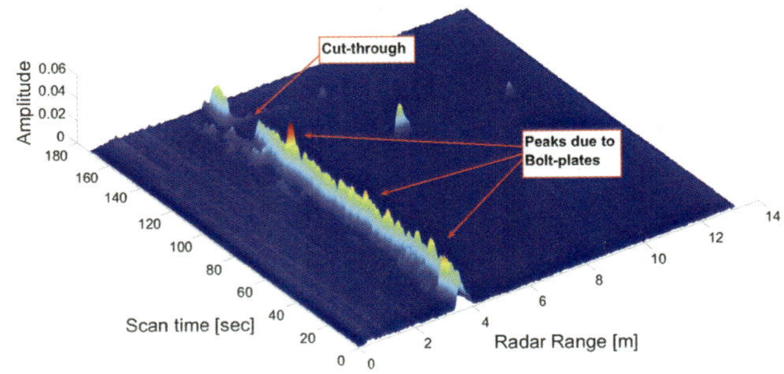

Fig. 8 Reconstructed radar data for a section of gate road with cut-though

Both of these features (the cut-throughs and the higher amplitude returns from the bolt-plates) are valuable aids for measuring the along-track distance.

The output from the laser scanners was used to provide ground-truth measurements for comparison with the radar data. By merging data from the forward-facing scanner, a reconstructed plan view of the section of gate road evaluated (Fig. 9). It shows both of the cut-throughs that bounded the region used for testing. Note that the figure's vertical axis is highly compressed in this reconstruction in order to represent the entire test site in a single image: the length of the test segment is approximately 50 m, while the width is only five metres. The deviation in the middle of the roadway section is therefore highly exaggerated in this figure, as the gate road appeared straight when viewed onsite. In addition, several tests were performed with the laser scanners oriented to capture the full surface of the gate road rib and roof (Fig. 10). These scans were performed in order to develop a more complete picture of the mine environment, and to provide an opportunity to trial some radar/laser data fusion techniques.

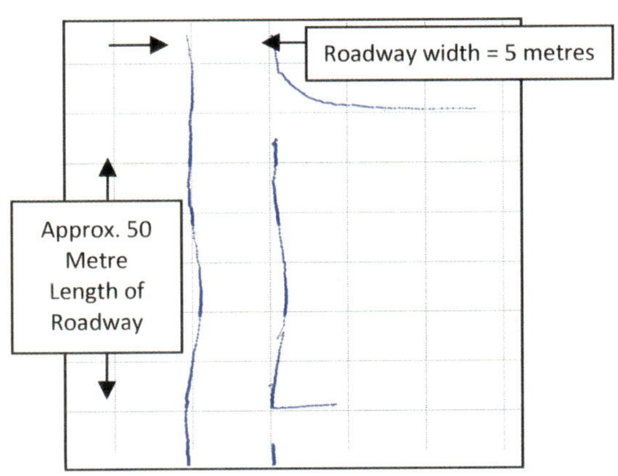

Fig. 9 Laser-reconstructed 2D view of the test site showing cut-throughs: note that the vertical scale is compressed in order to show detail of the horizontal features on a single image

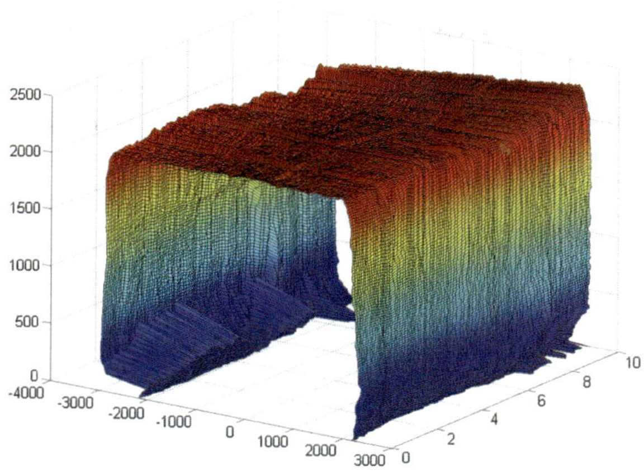

Fig. 10 Laser-reconstructed 3D view of a section of gate road ribs and roof

As noted in Section IID, scans of the roof with both radar and laser sensors were also acquired in order to test the efficacy of the technology in the presence of a meshed surface (Fig. 7). A detailed section of the roof reconstructed from the laser scanner data is shown in Fig. 11. In both this image and Fig. 10, the colour map variation corresponds to the laser range.

4 Analysis and discussion

4.1 Initial radar processing and creep measurement estimation

First-pass processing of the raw radar sensor data involved the isolation of both the amplitude and position of the maximum signal return for each radar trace. This feature ideally corresponds to the reflection from the rib-wall and thus may be used to provide a creep estimate. It should also vary depending on the presence or absence of the reflective bolt-plates at that point along the gate road and thus provide a waypoint reference for retreat measurement. The stacked raw radar trace data, the range to the maximum echo point and the maximum amplitude of the return for each trace are shown in Fig. 12.

As is clear from a cursory inspection of Fig. 12, the raw maximum-echo radar data is quite noisy, and thus filtering is required in order to obtain a reasonable position estimate for creep and to attempt to mask any peaks corresponding to the bolt-plate waypoints. For the creep estimate, both a Moving Average (MA) filter and a Kalman Filter (KF) were applied to the maximum echo radar range-to-wall estimate to smooth out discontinuities (Fig. 13).

As expected, a well-tuned KF provided a better estimate of the creep distance than the MA filter, reacting to rapid changes in the range more smoothly than the MA filter could do for a reasonable window size. The MA filter result could be made arbitrarily smooth by selecting a large enough window, but this introduces an unacceptable lag in the position estimate. The KF also provided a closer match to the creep estimate obtained from the laser scanner data when the two sensors were compared.

The creep position estimates obtained from the KF were compared to data from the laser scanner that was directed towards the same rib wall as the radar sensor. The laser scanner generated a 2D scan line of discrete points (rather than a single point measurement as is the case for the radar), so the comparison was performed using the mean value of the central 10° fan of laser data points. The laser data were otherwise unprocessed for the creep estimation task, and were synchronised with the radar traces based on the common system clock on the data acquisition PC.

For validation, a comparison of the typical system creep performance is shown in Fig. 14 for a 50 m run of the trolley at a nominal offset from the mine rib wall of 1 m. The radar sensor, when combined with a well-tuned filter,

Fig. 11 Laser-reconstructed view of a detailed roof section with meshing

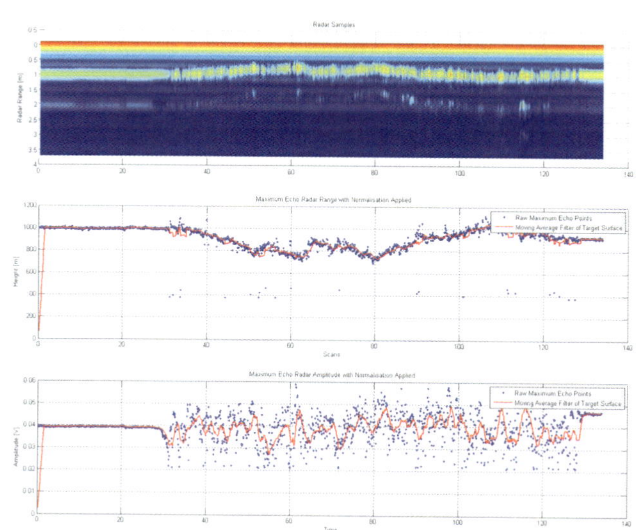

Fig. 12 Raw radar data with range estimate and maximum echo amplitude

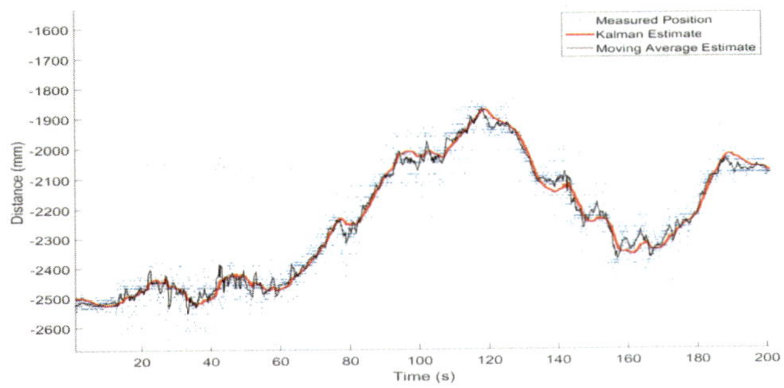

Fig. 13 Creep position estimate, comparison of moving average and Kalman filters based on the filtered versions of the raw radar signal

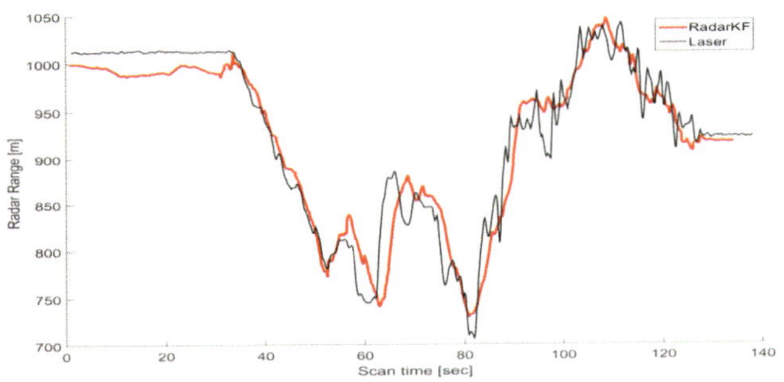

Fig. 14 Radar creep measurement compared with laser scanner data in close agreement

showed excellent correlation with the laser scanner data; its performance exceeded the accuracy and precision requirements for the basic creep measurement application. This result was found to be quite consistent across the many datasets generated in the field trial.

Despite the presence of mesh on the roof, similar results were also obtained for gate road height measurement as extracted from the raw roof scanning data (Fig. 11). A typical result for the radar versus laser tracking for the roof scenario is shown in Fig. 15, where the correlation between the two sensors is once again excellent.

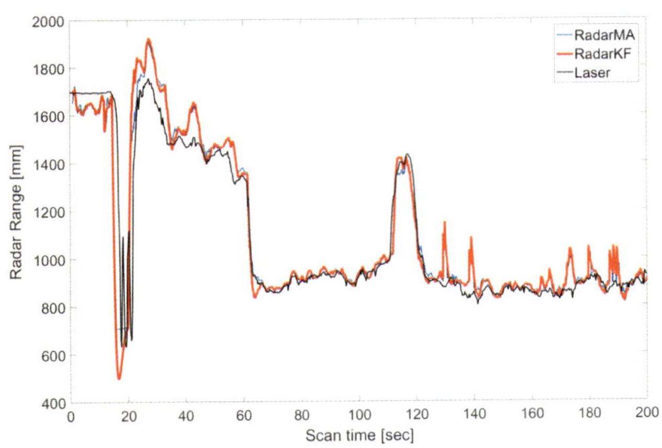

Fig. 15 Radar roof height measurement compared with laser scanner data

4.2 Extended trial of gate road height measurement

An application of the creep measurement technique was later applied on an actual longwall gate road (rather than a main roadway) at an underground mine in Queensland, with the goal of measuring the gate road height (i.e., sensing the range to the roof, rather than the rib). The length of gate road measured during this operational trial was over five hundred metres long (involving a trial period of almost two months), and both radar and laser-scanner data were acquired for validation purposes, as shown in Fig. 16 (results are shown for the first 10000 samples, or approximately 500 m). The correlation between the radar and laser data was quite strong for the majority of the roadway, with the main deviations being due to the effect of dust on the laser scanner system. The large vertical deviations in the blue laser data shown in Fig. 16 occurred when the window of the laser scanner flameproof became obscured by dust or mud, leading to inaccurate estimates for that sensor until the window could be cleaned. This result demonstrates the value of the radar-based sensing approach, as the radar was installed in an enclosure adjacent to the laser scanner yet remained unaffected by the same dust conditions that effectively blinded the laser. Smaller deviations occurred where the laser scanner result (which is obtained by averaging the height of all samples across an extended roof section) was affected by local pockets in the roof that led to an overestimate of the roof height, which again highlights the advantage of the radar for this application. This extended trial also demonstrated that the core radar technology is quite robust for use in practical mining environment.

4.3 Retreat measurement estimate

The working hypothesis for a retreat measurement system was based on discriminating the highly reflective rib-wall bolt-plates from the background coal, and using these

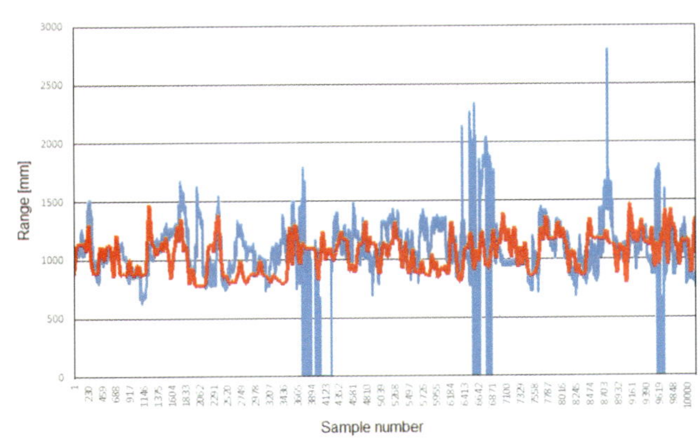

Fig. 16 Radar (*red*) versus laser (*blue*) data for a second, extended, longwall gate road height measurement trial

features as reference waypoints for a localisation solution. Bolting patterns are strictly controlled in underground mines, as they constitute the primary support preventing gate road collapse. The number of bolts installed in a segment of gate road (e.g., between two cut-throughs) is therefore a known environmental mining parameter, and by tracking the number of bolts encountered since the previous cut-through the along-track position of the mining machine can be accurately estimated.

Figure 17 shows the raw maximum echo output for the side-facing radar against the service-side rib-wall, together with the MA filter and KF results. The superiority of the KF was again notable, as the responsiveness of this filter tracks the rapid change between the bolt-plates and the background coal more closely, improving the discrimination result. In both filtered datasets, the consecutive peaks and troughs associated with many of the passing bolts are easily discerned.

Although an initial examination of the maximum radar return data in Fig. 17 appears encouraging, closer inspection reveals that the number of peaks do not correspond to the number of bolts present (40 bolts) in the tested segment of the gate road rib. While some bolt-plates provided a strong reflection and hence maximum signal amplitude, in other cases the contrast with the background coal was not sufficient to allow discrimination. This appears counter-intuitive given the strong scattering from a plate structure; the explanation lies in the fact that plates are highly specular reflectors, so that the strong reflection from the target will scatter in a narrowly focussed direction corresponding with Snell's law. If the plane of the plate is mounted at 90° to the incident wavefront it will provide a strong radar target, with most of the signal reflecting back to the radar receiver. However, if the plate is angled a few degrees off-axis then most of the signal energy will reflect away from the sensor and the overall return energy may be quite low. In addition, there are other significant scattering sources present (particularly on the service-side of the gate road) that will constitute unwanted radar "clutter" that further complicate the detection and measurement process.

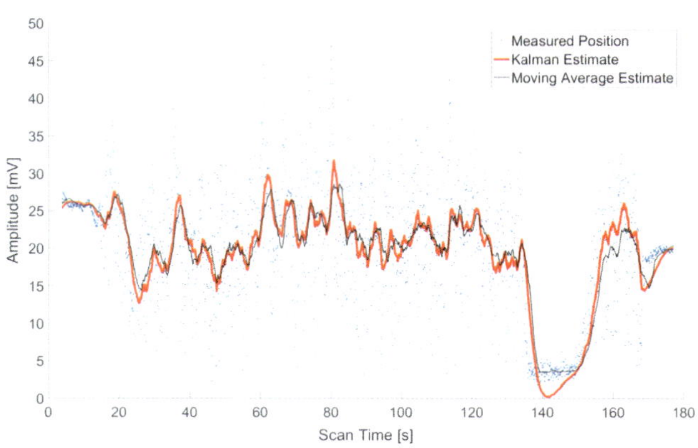

Fig. 17 Maximum amplitude echo points with KF and MA filters

Given the feature ambiguity associated with a technique based purely on maximum return amplitude, it was clear that a classification engine would require a richer feature set from the raw radar data in order to discriminate successfully between the bolt-plates and other structures, or even the coal wall itself. Identifying strongly correlating features for the bolt-plates therefore became an important task for the research.

4.4 Visual analytics for feature selection

Because of the complexity of the detection environment (the irregular surface of the rib walls, the presence of clutter sources, and the significant variation in the installed angle of the bolt-plates) recognising consistent features for the target was not a trivial task. In order to aid this process, a novel data fusion and visual analytics approach was employed based on combining the laser scanner and radar data in a 3D interactive display to facilitate examination of significant features (May et al. 2010). The goal of this work was to provide the researchers with a comparative tool whereby the raw radar data could be examined directly with the surfaces generated from the laser scanner data. By visual inspection of the data, it was hypothesised that different features in the radar data might be identified that could be useful for discrimination purposes.

Using the trial runs in which the laser units were oriented to scan in a vertical plane in order to capture the full surface of the gate road ribs and roof (as shown in Fig. 10), and exploiting the synchronisation between the laser and radar sensors, the radar return information could be overlaid against the high-resolution point-cloud structures from the laser data. The resulting 3D surfaces allowed the radar data to be examined in a highly detailed alignment with the physical surface structures as revealed by the laser scanner (shown conceptually in Fig. 18). By adjusting the section of the radar return overlaid on the projected rib wall structure, different aspects of the radar return could be

Fig. 18 Visual analytics concept showing mine structure based on laser point cloud overlaid with radar data

interrogated for changes and associated with specific features in the laser point cloud, in particular the bolt-plates.

This method of data fusion for visual inspection was applied to the combined laser and radar datasets from the mine site trial, allowing the radar return information to be projected onto the 3D surface generated by the laser scanner data. Figure 19 shows the resulting combined information for the section of roof shown in Fig. 10, with the laser-generated surface now colour-mapped to the corresponding radar return amplitude data. In Fig. 19 the strong reflections from various radar targets are clearly visible, and their correspondence to particular features of the laser data such as the bolt-plates and other infrastructure elements can be readily identified. The combination of radar and laser data provides a powerful mechanism for characterising the particular radar dataset features that correspond to the bolt-plate targets, which are the primary features of interest for establishing a way-point tracker for retreat measurement.

This visualisation process proved an effective way to reveal useful features beyond the initial strong echo. One important example was verification of the so-called "double bounce" effect which occurs when a secondary echo is observed for a given target at twice the true target range, due to a round-trip reflection between the radar sensor and the strongly reflective target (the phenomenon can be observed in the stacked raw radar traces shown in Fig. 12) (Skolnik 2001). Although not observed for every target, this phenomenon was clearly characteristic of the bolt-plates, and was not observed even in the strong scattering from clutter sources. Where present, this significant additional feature provided a useful supplementary source of information regarding the observation of a bolt-plate, and thus was incorporated into the classifier algorithm to improve discrimination. Other features such as characteristic variations in radar reflection intensity across the face of a bolt-plate are under investigation.

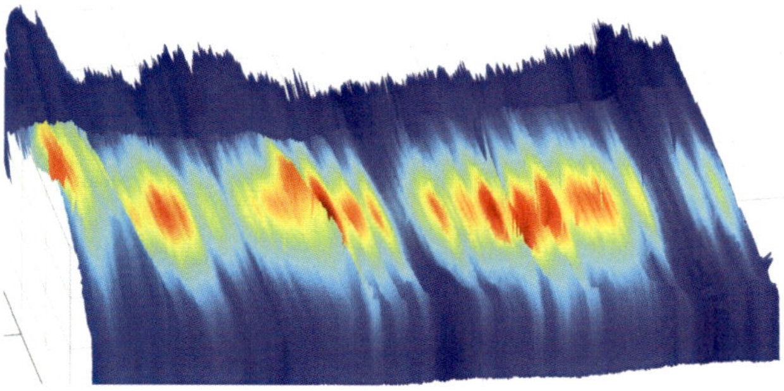

Fig. 19 3D roof surface with overlaid radar amplitude

4.5 Exploration of feature generation for visual analytics

Further improvements for estimating the position of the bolt plates could be realised by incorporating different feature sets into the visual analytics approach. For example, ambiguities associated with the class of target identification methods based on signal strength alone could be significantly mitigated through the use of resonant frequency target identification techniques (Hargrave et al. 2013), which have remarkable properties for aspect-independent imaging. The features extracted using this technique could be incorporated into the classifier algorithm to further refine the discrimination results.

A variety of classifiers have been trialled for use with the radar data. A decision tree structure, which allows a simple implementation of rules (e.g., if the maximum amplitude of the return is within a given range, classify the target as a bolt-plate), has shown promise as a suitable algorithm for the application (Fig. 20). This technique has the distinct advantage that any a priori information regarding the structural hierarchy of the gate road configuration and/or equipment can be meaningfully incorporated to improve the detection and estimation accuracy.

5 Conclusion and future work

We have described a novel radar-based system for underground mine machine localisation. The approach avoids the need to install additional reference infrastructure by exploiting mining structures already present in the underground mining environment, such as bolt plates and cut-throughs, to generate feature-rich radar signal returns. Different probabilistic data processing techniques were explored in order to provide estimates of the equipment's along-track and across-track position. These estimates were

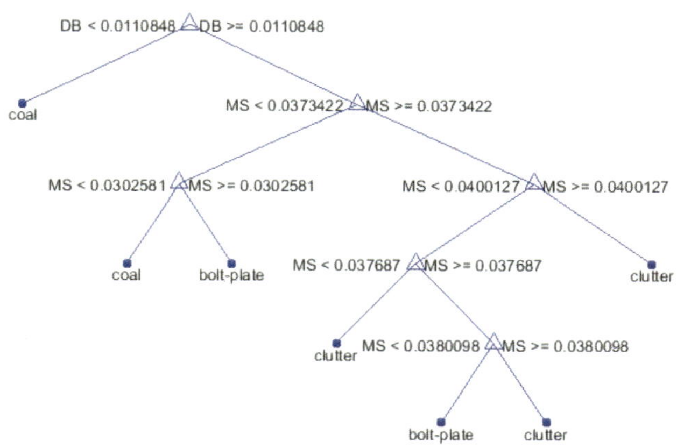

Fig. 20 Initial decision tree for classifying bolt-plates, coal and clutter

subsequently validated through the use of laser range scanners and found to be in close agreement. The success of these trials has spurred development of a modified system that is now ready for integration into an operational mine machine automation system. Work is ongoing to improve the classifier algorithm performance with the view to incorporating these results into an operational automation system.

References

Hargrave CO, Shuley NV, Ralston JC, Hainsworth DW (2007) Radar level sensor for longwall creep and retreat measurement. In: Industry applications conference, 2007. 42nd IAS annual meeting. Conference record of the 2007 IEEE, 2007, pp 2102–2109

Hargrave CO, Abbosh A, Shuley N, Clarkson V (2013) Radar target identification by histogram analysis of resonant frequencies. In: Presented at the 8th European conference on antennas and propagation (EUCAP), Gothenburg, Sweden, 2013

Hargrave C, Clarkson IVL, Hoi-Shun L (2014) Radar waypoint navigator for underground mining. In: Antennas and propagation (EuCAP), 2014 8th European conference on, 2014, pp 3587–3591

May R, Hanrahan P, Keim DA, Shneiderman B, Card S (2010) The state of visual analytics: views on what visual analytics is and where it is going. In: Presented at the visual analytics science and technology (VAST), 2010 IEEE symposium on, Salt Lake City, USA, 2010

Ralston J, Reid D, Hargrave C, Hainsworth D (2014) Sensing for advancing mining automation capability: a review of underground automation technology development. Int J Min Sci Technol 24:305–310

Singh RD (1997) Principles and practices of modern coal mining. New Age, New Delhi

Skolnik MI (2001) Introduction to radar systems, 3rd edn. McGraw Hill, New Delhi

Wei N, Collings IB, Ren Ping L, Zhuo C (2014) Relay-assisted wireless communication systems in mining vehicle safety applications. IEEE Trans Industr Inform 10:615–627

15

Surrounding rock control theory and longwall mining technology innovation

Guofa Wang[1] · Yihui Pang[1]

Abstract To accommodate surrounding rock structure stability control problem in underground mining, we study the coupling effect principle between hydraulic support and surrounding rock, and develop a series of longwall mining technology and equipment, which solves four common technical problems that significantly undermine coal mining safety, efficiency, and high recovery and extraction rates. Based on the coupling characteristic between mining-induced stress field and supporting stress field of hydraulic support, we identify six controllable factors in the application of hydraulic support to surrounding rock, and further reveal the relationship between hydraulic support and surrounding rock in terms of the strength, the stiffness, and the stability coupling. Our findings provide a plausible solution to the longwall mining technical problem with 6–8 m mining height. By analyzing the dynamic disequilibrium characteristics between hydraulic support and surrounding rock, we propose the intelligent top coal caving control method and the high-coal-recovery-rate technology for fully mechanized caving faces. With the invention of this technology, China is likely to lead the world in terms of the fully mechanized top coal caving mining technology. We are also the first to employ the intelligent coupling technology between hydraulic support and surrounding rock, and automated mining mode, and supporting system cooperative control with automatic organization. We develop the comprehensive multi-index intelligence adjusting height decision-making mechanism and three-dimensional navigation automatic adjusting straightness technology based on shearer cutting height memory association, cutting power parameters, vibration, and video information, leading to the first set of intelligent longwall mining technology and equipment for thin seam. Our innovation makes a solid contribution to the revolution of intelligence mining technology. With the innovative use of three-dimensional coupling control principle for surrounding rock, we successfully resolve the technological difficulties of longwall mining equipment and surrounding rock control for steep dipping seam, making a breakthrough of longwall mining technology with steep dipping seam.

Keywords Surrounding rock control · Hydraulic support · Thick coal seam · Thin coal seam · Steep dipping seam

1 Introduction

There have been four major stages of coal mining development in China, namely artificial mining, blasting mining, conventional mechanized mining, and fully mechanized mining. The mining practice suggests that the fully mechanized mining method was the basic solution for guaranteeing safety, efficiency, and high recovery ratio mining of underground coal mining. The development of longwall mining equipment technology epitomizes the mining method change. In this revolution, hydraulic support is the core equipment for surrounding control. The

Supported by the National Key Research and Development Program of China (2017YFC0603005).

✉ Yihui Pang
pangyihui@tdkcsj.com

[1] Coal Mining and Design Branch, China Coal Research Institute, Beijing 100013, China

improvement of hydraulic support model has become a milestone of longwall mining technology development and a decisive factor of mining technology innovation. The adaptation and reliability of hydraulic support and matching system determine the success of the longwall mining technology (Gong and Jin 2008; Wang 2010).

In the past 30 years, China's coal production has increased from 872 million tons in 1985 to 3685 million tons in 2015, and the death rate per million tons has dropped form 7.63 in 1985 to 0.157 in 2015, suggesting that the coal mining efficiency has increased more than 30 times, and the death rate per million has declined nearly 50 times, and the roof accident deaths has decreased by 99.4%. For some coal mines, the death rate per million tons is almost zero, which achieves safety, efficiency, and high-recovery-rate mining (Wang et al. 2006; Xu et al. 2015; Wang and Pang 2016). These achievements mainly benefit from China's development and progress in longwall mining technology and equipment.

At present, the key technical problems to achieve coal mining safety, efficiency, and high recovery ratio are the surrounding rock control in strong strata behaviors working face and the safety, efficiency mining technology, and equipment development in complicated mining seam. To accommodate the above-mentioned mining problems, this paper studies "hydraulic support and surrounding rock coupling control theory and technology" based on long-term technological research and innovation practice. The sets of longwall mining technology and equipment had developed for thick coal seams longwall mining with high cutting, super high seam full-mechanized caving mining with high cutting, thin coal seam automated and intelligent mining, and complicated steep dipping seam, which could solve four typical coal seams working faces mining problems with safety supporting, automated mining, high production, high efficiency, and high recovery ratio.

2 Hydraulic support and surrounding rock coupling principle

The surrounding rock coal mining brakes the equilibrium state of in situ stress filed. The surrounding rock experience deformation, destruction, caving and stability under the coupled action of mining stress field and supporting stress field (Gao and He 2010; Ju et al. 2012). Based on the independently researched and developed longwall mining working face supporting quality monitoring system, we identify the six controllable parameters of hydraulic support to surrounding rock through many field observations of hydraulic support and surrounding rock. These factors include the hydraulic support setting load and working resistance, the top beam action point of resultant force, the horizontal applied force from the top beam to roof rock, the face guarding moment, the distance from beam to rib, and working face mining rate. We optimize the six parameters to control the surrounding rock stability.

We improve the surrounding rock self-bearing capacity by studying the "multi-field coupling" between hydraulic support and surrounding rock. The essential of working face supporting is the coupling relationship in the strength, the stiffness, and the stability between hydraulic support and surrounding rock, as shown in Fig. 1.

In the dynamic equilibrium system of hydraulic support and surrounding rock, the surrounding rock deformation and instability induced by the mining stress field jointly determine the system stability. The hydraulic support could adapt and influence the process of surrounding rock motion by actively adjusting stress state to protect the safety working space in working face. In other words, the hydraulic support's supporting intensity should adapt the surrounding rock's strength to reduce the impact of dynamic load from the surrounding rock by actively protecting and reasonably yielding. The hydraulic support's supporting stiffness should adapt the surrounding rock's stiffness to change the dynamic load location of roof by optimizing the structural stiffness of hydraulic support, providing safe working space. The support system's stability should adapt the surrounding rock's stability to control the dynamic instability of surrounding rock based on the stability of hydraulic support itself.

Based on the coupling principle of hydraulic support and surrounding rock, we invent the hydraulic support three-dimensional design method and software system (Wang et al. 2011; Wang and Pang 2015), as shown in Fig. 2. The motion skeleton of hydraulic support is established by PROE 3D design software, and goes through dynamics analysis of hydraulic support state of motion to finalize the reasonable technical parameters of hydraulic support. We also analyze the assembly unit and whole hydraulic support using ANSYS to guarantee the strength design, the stiffness, and the stability of hydraulic support.

Based on the coupling relationship between hydraulic support and surrounding rock, we establish the adaptability evaluation model of hydraulic support and surrounding rock, which allows comprehensive evaluation of the adaptability of hydraulic support. We divide the hydraulic support and surrounding rock system into strength coupling subsystem, stiffness coupling subsystem, and stability coupling subsystem, and then build the evaluation index system based on the adaptive composite index of hydraulic support and surrounding rock, as shown in Fig. 3.

To meet the differential needs of the surrounding rock controlling effect of hydraulic support for different coal seams, we calculate the preference coefficient matrix of the different evaluation indicators. We use the weight as the

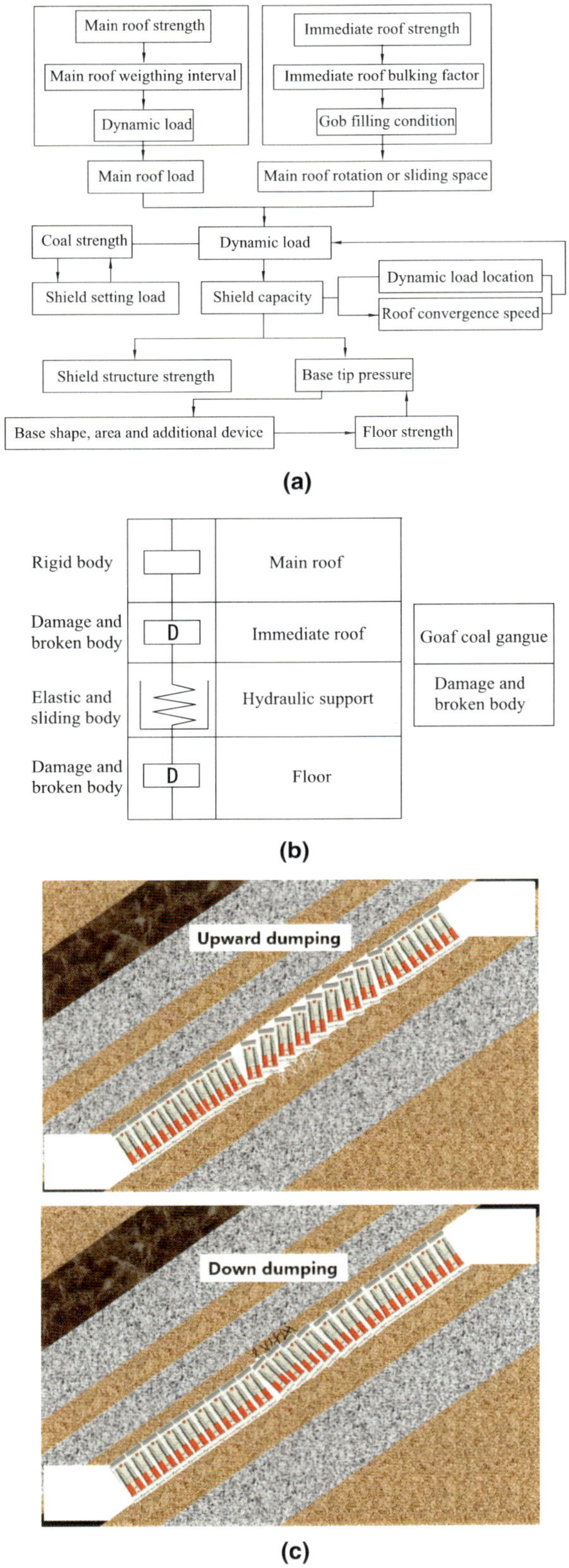

Fig. 1 Coupling relationship between hydraulic support and surrounding rock. **a** Strength coupling relationship, **b** Stiffness coupling relationship, **c** Stability coupling relationship

value of superiority degree corresponding to the evaluation index, and adopt the FCE method to build the adaptive evaluation matrix of hydraulic support and surrounding rock. The adaptive evaluation results of hydraulic support could guide the design to improve hydraulic support.

3 Key technology and equipment of longwall mining with super high cutting

The thick coal seam reserves account for 45% of total coal reserves, and many 6–8 m thick coal seams exist in the large-scale mining areas in Western China. The slicing method for this type of coal seam has problems of low efficiency, low profits, and great security rick, etc. (Sasaoka et al. 2016). The fully mechanized mining method with high or super high cutting suits this type of coal seam well, but increasing cutting height could lead to stronger mining pressure behavior and substantially raise the rib and roof control risk. Therefore, it is imperative to solve the problem of the multi-field coupling relationship of hydraulic support and surrounding rock as well as the system adaptability and reliability assurance technology for longwall mining with super high cutting.

Based on the coupling relationship between hydraulic support and surrounding rock, we propose the double factors control method to confirm the reasonable working resistance of hydraulic support. The first factor, the working resistance of hydraulic support should control the roof subsidence effectively by using the roof self-bearing capacity. The second factor, the working resistance and rib guard force of hydraulic support should prevent rib spalling and reduce the rib failure depth. The reasonable working resistance of hydraulic support should meet above two conditions at the same time.

Based on the surrounding rock deformation and instability feature of longwall mining working face with super cutting height, we establish the roof strata "cantilever + voussoir beam" structural mechanics model and rib "tension-slippage" coupling mechanical model (Pang and Wang 2017a, b), as shown in Fig. 4.

We develop the impact resistance column with diameter of 530 mm based on energy dissipation theory, the micro-clearance and quasi-rigid four bar mechanism, the three-level coordinating action rib guard appliance, and the large flow and high pressure intelligent supply system to solve the surrounding rock stability control problems caused by the massive space, the super strong mine ground pressure, super high rib, and strong disturb stratum movement in longwall mining working face with super cutting height. The life span of hydraulic support has increased from 5000 working cycles to 60000.

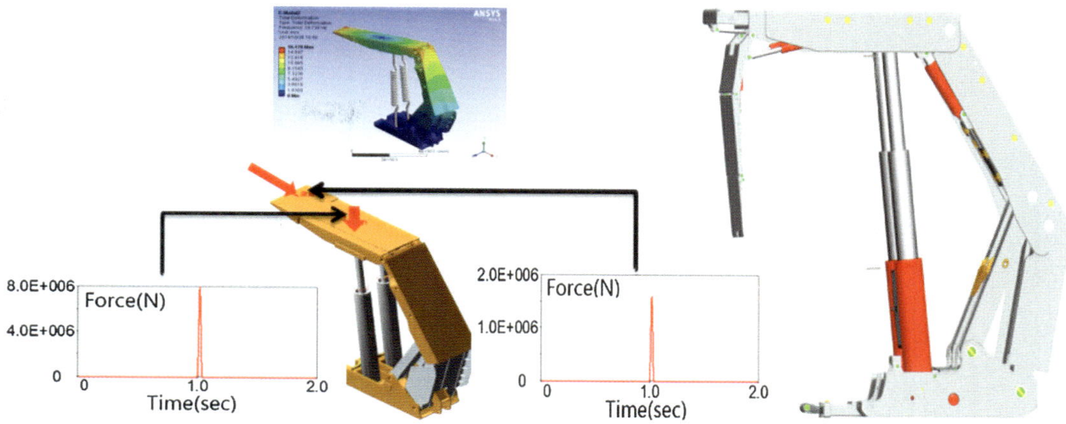

Fig. 2 Hydraulic support three-dimensional dynamic optimization design

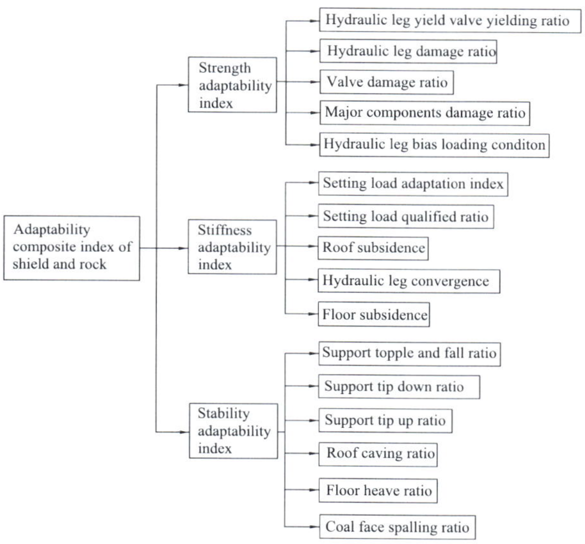

Fig. 3 Hydraulic support and surrounding rock adaptability evaluation indexes

Due to a height difference of 2–4 m between the working face and the roadway (the working face cutting height was 7–8 m, and the roadway height was 4–4.5 m) in longwall mining working face with super height cutting, the traditional step-by-step matching manner leads to substantial triangle coal loss. To solve this problem, we introduce the large gradient transition matching manner for 7 m cutting height or the large gradient + short gradual transition for 8 m cutting height and supporting equipment to achieve one-time transition from working face to roadway, which could recycle more than 0.4 million tons coal in one working face, as shown in Fig. 5.

The whole set of technology and equipment that we develop could exploit coal 6 million tons per year, which is the first time that imported products are completely replaced with the domestic high-end longwall mining technology and equipment. The first set of longwall mining technology and equipment for 7.2 m cutting height were successfully launched and were used at Hongliulin mines in 2011. The working face production reached 12 million tons per year. In 2014, to accommodate the 6–8 m hard and thick coal seam of Jinjitan coal mines, the hydraulic support (ZY21000/38/82D) and the set of longwall mining technology with the biggest cutting height and the greatest working resistance in the world were launched. The three-level coordinating action rib guard appliance for super height hydraulic support and the impact resistance column with diameter of 530 mm based on energy dissipation theory were introduced with a breakthrough of the hydraulic support structural, material and manufacturing technique, which solved the problem of stability and reliability due to the large-scale and high-dynamic-pressure sensitive structure of hydraulic support.

4 Fully mechanized caving mining technology and equipment with large mining height for super thick coal seam

The fully mechanized caving mining, marked by the employment of the top coal caving hydraulic support, tested and developed in the early 1980s went through three important stage of development: high-level caving, middle-level caving, and low-level caving. It has been widely shown that the top coal caving hydraulic support is the decisive factor that makes the fully mechanized caving mining the safe, efficient, and high-recovery mining technology for thick and super thick coal seam.

Through a number of observations and high-volume in-depth analyses for top coal deformation rule and mining pressure revealing rule, we find the dynamic unbalanced coupling characteristics between hydraulic support and top coal in fully mechanized caving face. The load of hydraulic support is time-varying load, which changes along with the

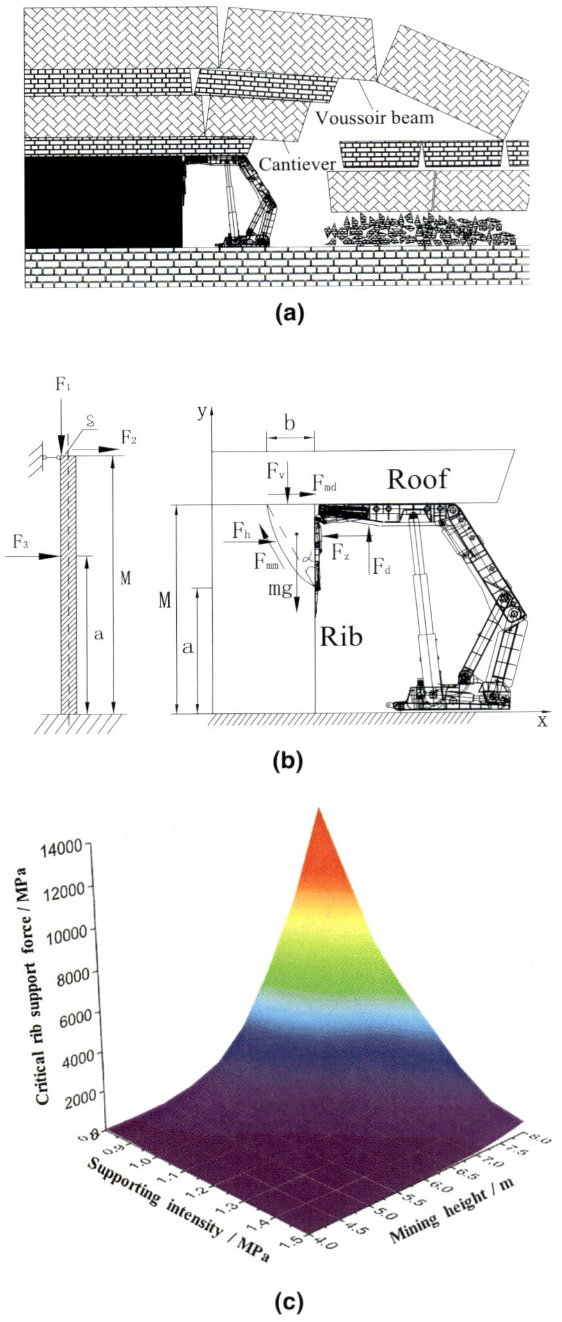

Fig. 4 Coupling mechanical model between hydraulic support and surrounding rock. **a** "Cantilever + voussoir beam" structural mechanics model, **b** "Tension-slippage" mechanical model, **c** relationship of supporting intensity and critical rib support force in different mining height

top coal caving (Chen et al. 2007; Wang et al. 2012; Liu et al. 2015). To solve the safety, efficiency and high-recovery-rate mining problem, we develop the fully mechanized caving mining technology and equipment with large mining height for super thick coal seam. We build the "combined cantilever" mechanical model for top coal and roof stratum, and further identify and solve the contradiction between rib spalling and top coal caving and drawing characteristics. This is the first study to quantitatively analyze the top coal failure depth and lumpiness from the hydraulic support to top coal, as shown in Fig. 6. This study solves the problems of super mining space overlying strata moving law, super high rib stability control, super thick top coal caving law due to 15–20 m super coal seam mining, and also improves top coal recovery ratio technology in Datongtashan coal mine, which is the world's first set of fully mechanized caving mining technology and equipment with large mining height for super thick coal seam, increasing the top coal caving efficiency by 40% and setting a world record of 1.3 million tons per month with no accidents.

To accommodate the hard and heavy coal seam condition in the mines of Western China, we introduce the intelligent top coal caving control method and set of technology. Based on the coupling relationship between hydraulic support and surrounding rock, we develop the intelligent coupling control system of hydraulic support and surrounding rock, which allows real-time monitoring for the posture and stress state of hydraulic support. Based on the hydraulic support intelligent coupling control appliance and top coal caving automatized control appliance, we study the sequential control automatized caving logical relationship based on the section multi-window multi-wheel coal technology, and propose the top coal caving time control method based on the fruit fly optimization algorithm and RBF mixed-forecast, which could adjust caving time automatically and record the manual intervention condition at the same time. The top coal caving experiment results show that the average response time of the top coal caving mechanism is 1.43 s.

To increase the top coal caving ratio of hard and heavy coal seam, we invent the strong disturb three-level high efficiency caving device to fit with the hard-top coal condition, as shown in Fig. 7. Compared to the conventional caving device, the caving space in the back of the hydraulic support increases by 23%, significantly decreasing the arching probability and improving the top coal recovery ratio.

To solve the problem that the transition hydraulic support cannot cave top coal when the rear scraper conveyer, which occupies the back-caving space with the drive unite, adopts the matching manner of end capacity discharge, we design the cross side-discharge scraper matching manner, as shown in Fig. 8, which raises the top coal recovery ratio by more than 1.5%. This design solves the problem of automatic control for transition hydraulic support of supporting system and increases the non-failure operation time by more than 10%.

The fully mechanized top coal caving automated mining has been used for super thick coal seam in Aosida coal

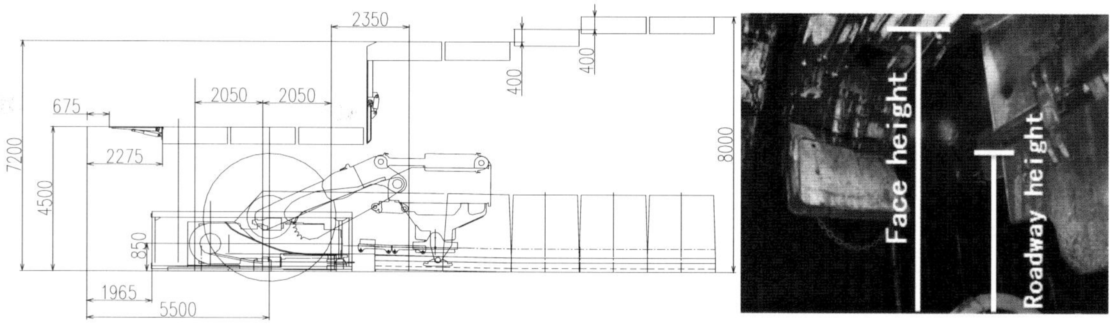

Fig. 5 Equipment matching manner for super cutting height working face

mine of Australia, which improved the recovery ratio by 12%, improved the labor efficiency by 41%, decreased the number of workers by 72%, and has been extended to regions such as Shandong, Shanxi, Inner Mognolia, etc.

5 Intelligent fully mechanized mining technology and equipment for thin coal seam

More than 85% of the mines in China have thin coal seam < 1.3 m, which is mainly mined as protective layer for the lower coal seam. Due to the unsteady thickness of the thin coal seam, the mining space is narrow and small. There has been a contradiction between the capacity and the size of equipment. The imported advanced fully mechanized coal mining equipment can only be applied to the coal seam more than 1.5 m and cannot be used for safety and high-efficiency mining for the coal seam < 1.3 m and has become the technology bottleneck for the coal sustainable development (Wang 2013, 2014; Pang et al. 2015).

To solve the problem of safety and high-efficiency supporting for thin coal seam with unsteady thickness, we introduce intelligent coupling technology between hydraulic support and surrounding rock, automatic mining patter and supporting system cooperative control with automatic organization (Hargrave et al. 2017). We develop the comprehensive multi-index intelligence adjusting height decision-making mechanism and three-dimensional navigation automatic adjusting straightness technology based on shearer cutting height memory association, cutting power parameters, vibration and video information, which solve the problem of the hydraulic support followed the coal cutter moving by automatic control and the coal cutter cleaning the float coal after beveling cutting the tringle coal automatically. We also invent the super large-scale magnification ratio hydraulic support, which adopts the new structure design of panel top beam, the double linkage and double equilibrium jack fold the position, and the double-telescopic prop with single liquid into the mouth to break through the minimum mining height limit. The super large-scale magnification ratio hydraulic support, which fits in with the 0.5–1.4 m coal seam, is built to tackle the technical challenges of small size equipment, superpower, and automatic control.

The automatic mining equipment and technology, which is developed for thin coal seam, solves the problem of coal seam unstable thickness, thick and hard dirt band, complicated condition for 0.6–1.3 m thin coal seam in Jizhongnengyuan, and realizes automatic mining with supervisor and automatic operation.

To accommodate the thin coal seam (1.7–2.5 m) intelligent mining problem in Huangling No.1 coalmine, as shown in Fig. 9, we address the following problems: automatic continuous mining in complicated coal seam, automatic supporting in the working face end and forward section, intelligent perception and steady transmission, mining height intelligent decision control and system reliability, etc. We carry out the top-level design of intelligent mining system and system integration matching to accomplish the remote monitoring normalization and the intelligent mining with unmanned operation in working face.

Authorized by the State Administration of Work Safety, the Innovation Center for Intelligent Mining of Coal Mines has been established, focusing on intelligent mining key innovation, promotion and technical standards system research, such as intelligent working face system integration matching design standard research, intelligent working face safety evaluation and security technology standard, intelligent working face quality management technique standard, intelligent working face remote control operation technique standard, intelligent working face manual work technique standard, intelligent working face equipment health management standard, etc.

6 Longwall mining technology and equipment for steep dipping seam

The steep dipping seam refers to the coal seam with the dip angle of 35°–55°. Since the dip angle of coal seam is relatively large, the working face is vulnerable to the instability

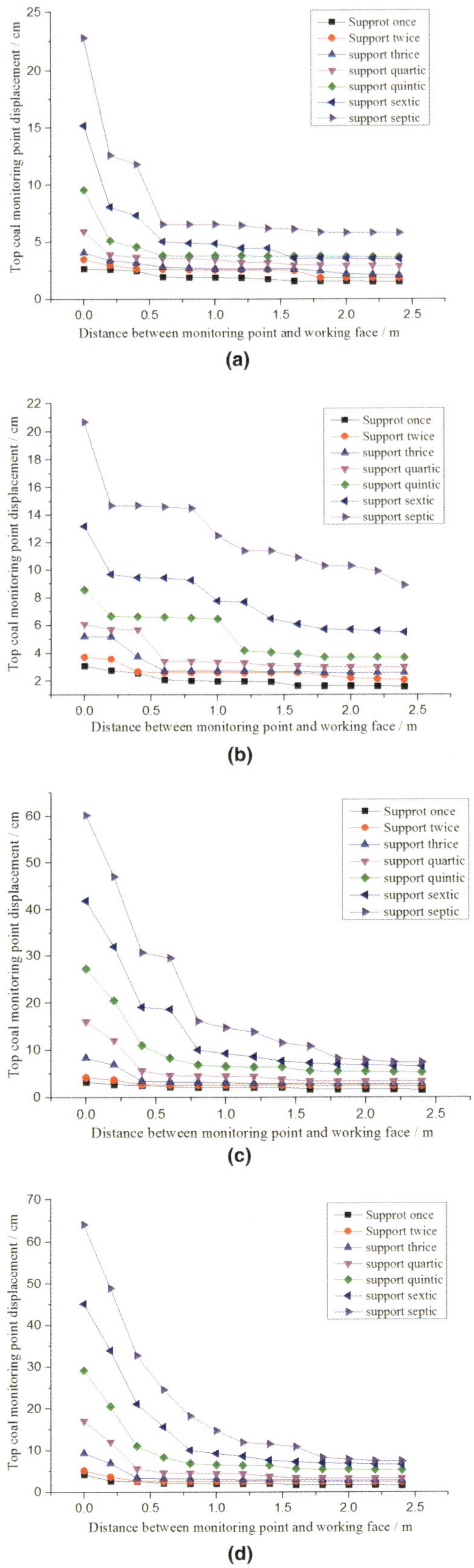

◀**Fig. 6** Top-coal time-displacement curve due to hydraulic support supporting again and again to top-coal. **a** Supporting stress 0.3 MPa, **b** Supporting stress 0.6 MPa, **c** Supporting stress 0.9 MPa, **d** supporting stress 1.2 MPa

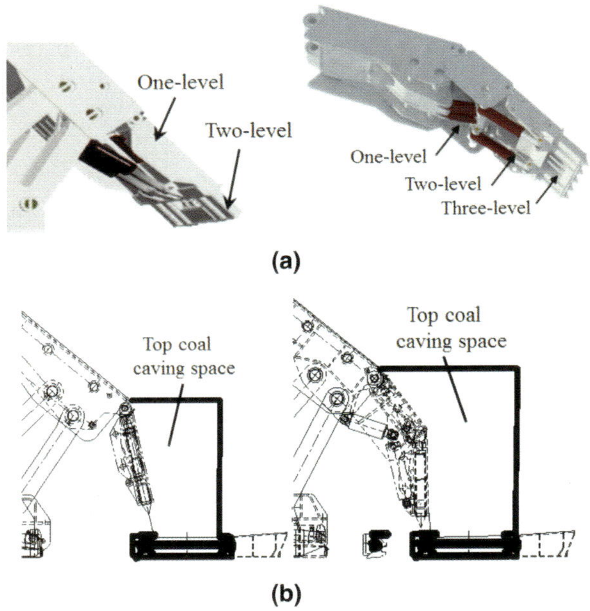

Fig. 7 Two types of top coal caving. **a** Top coal caving device contrast, **b** Top coal caving space contrast

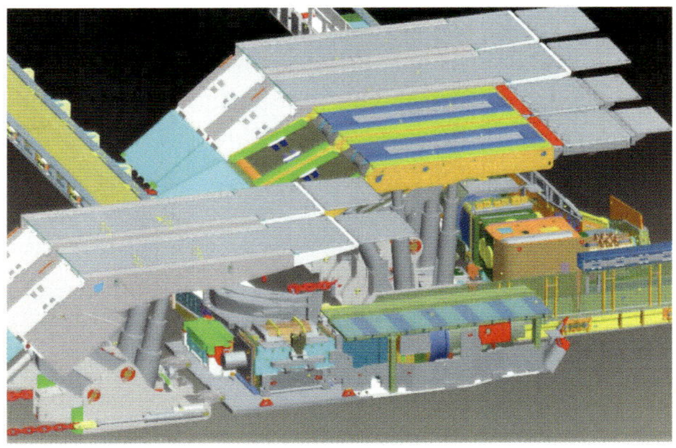

Fig. 8 Cross side-discharge scraper matching manner

such as the bottom heave, slippage, etc. The drawback would induce slanting, topple and fall, gliding of hydraulic supports, etc. In this respect, it is challenging to perform coupling stability control of hydraulic support and surrounding rock. The longwall mining system instability and coal gangue splashing menace the mining safety. The steep dipping seam mining has long been a challenge in global coal mining industry. In China, the steep dipping seam is evenly distributed in each major mining site. The reported reserves and production account for 10%–20% of the total coal reserves and 5%–8% of the total coal output in the country. In addition, more than 50% of the steep dipping

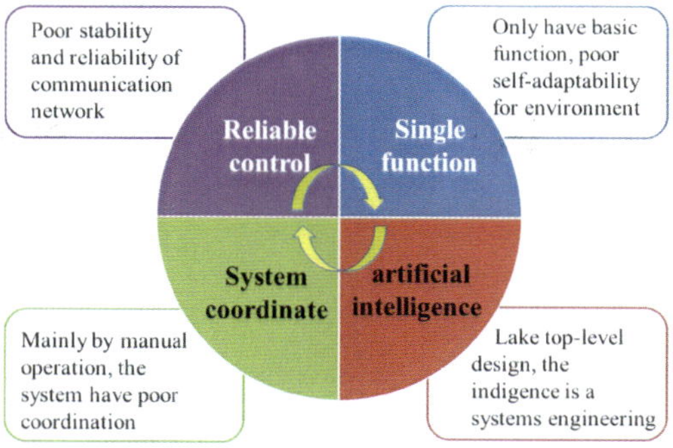

Fig. 9 Intelligent longwall mining key technical problems

seam mines are high-quality coking coal and anthracite coal. How to realize the stability coupling control of hydraulic support and surrounding rock in steep dipping working face restricts the safe and high-efficiency mining for steep dipping seam (Liu and Wang 2005; Luo 2015).

The final form of hydraulic support and surrounding rock instability in steep dipping face is the supporting failure and surrounding rock instability, which would induce surrounding rock accidents. Based on the coupling relationship between hydraulic support and surrounding rock, we analyze the mining pressure field distribution characteristics and the instability law of hydraulic support and surrounding rock, as shown in Fig. 10. We also examine the difference of supporting demand between steep dipping face and near horizontal or gentle dip face, and the steep dipping face needed 3D supporting for roof, floor and rib. The supporting device with self-supporting, next to pull, bottom to push and top beam squeeze is designed for steep dipping seam. In addition, we develop the hydraulic support with 3D preventing the spoil from splashing, which solves the instability problem of hydraulic support and surrounding rock.

To accommodate the steep dipping seam in Songzao coalmine with an average angle of 64° the working face adopts bow pseudo-inclined layout, and the rhombus hydraulic support is introduced to keep the constant roof distance, which realizes the double supporting function for sideway and roof surrounding rock.

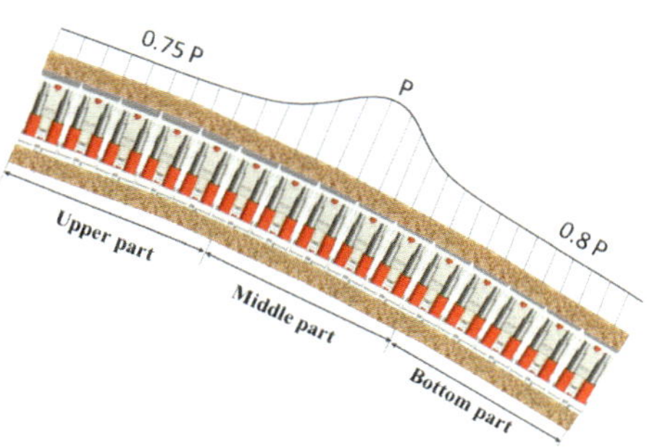

Fig. 10 Mining ground pressure distribution characteristics in steep dipping face

The largest mining angle is 55°, and the average coal seam thickness is 5.0 m of Fucheng coalmine in Inner Mongolia. Using longwall mining technology and equipment with high cutting and big angle, the working face output is 0.3 million tons per month, and the recovery ratio increases more than three times.

7 Conclusions

(1) The coupling relationship between hydraulic support and surrounding rock in terms of strength, stiffness and stability is the theoretical basis for surrounding rock stability control. Under the coupling action of mining pressure field and supporting stress field, the surrounding rock suffers dynamic instability. Although the hydraulic support does not change the final motion state of surrounding rock, it could influence the motion process in a way to protect the mining workers' working face.

(2) Based on the coupling principle between hydraulic support and surrounding rock, we develop a series of longwall mining technology and equipment, such as longwall mining technology and equipment with super cutting height for 6–8 m hard and heavy thick coal seam, fully mechanized caving mining technology and equipment with large mining height for 9–20 m hard and heavy super thick coal seam, which are used to build the system of longwall mining technology and equipment, and help fulfil safe, high-efficiency, and high-recovery mining for thick and super thick coal seams.

(3) To accommodate the unsteady 0.6–1.3 m thin coal seam, we invent the super large-scale magnification ratio strength hydraulic support and automatic mining pattern, which allow automatic and intelligent mining with supervisor and unmanned operation for thin coal seam.

(4) We develop the anti-skid and anti-collapse technology of self-supporting, next to pull, bottom to push and top beam squeeze for steep dipping hydraulic support and make a solid contribution to solving the stability control key technology of hydraulic support and surrounding rock for steep dipping working face.

References

Chen ZH, Feng JJ, Xiao CC, Li RH (2007) Fracture mechanical model of key roof for fully mechanized top-coal caving in shallow thick coal seam. J China Coal Soc 32(5):449–452

Gao J, He HT (2010) Application of fully mechanized full seam one passing mining technology to thick seam in Shendong mining area. J China Coal Soc 35(11):1888–1892

Gong PL, Jin ZM (2008) Mechanical model study on roof control for fully mechanized coal face with large mining height. Chin J Rock Mech Eng 27(1):193–198

Hargrave CO, James CA, Ralston JC (2017) Infrastructure-based localization of automated coal mining equipment. Int J Coal Sci Technol 4(3):252–261

Ju JF, Xu JL, Zhu WB, Wang QX, Hao XZ (2012) Strata behavior of fully mechanized face with 7.0 m height support. J Min Saf Eng 29(3):344–350

Liu JF, Wang GF (2005) Analysis on suitability of mini type hydraulic power caving support for deep inclined special thick seam mining. Coal Sci Technol 33(6):28–31

Liu CY, Yang JX, Yu B, Wu FF (2015) Support resistance determination of fully mechanized top-coal caving face in extra thick seam under multi-layered hard strata. J Min Saf Eng 32(1):7–13

Luo Y (2015) An improved influence function method for predicting subsidence caused by longwall mining operations in inclined coal seams. Int J Coal Sci Technol 2(3):163–169

Pang YH, Wang GF (2017a) Hydraulic support protecting board analysis based on rib spalling "tensile cracking-sliding" mechanical model. J China Coal Soc 42(8):1941–1950

Pang YH, Wang GF (2017b) Top-coal caving structure and technology for increasing recovery rate at extra-thick hard coal seam. J China Coal Soc 42(4):817–824

Pang YH, Liu XH, Ma L (2015) Key technologies of fully mechanized caving intelligent mining equipment in ten million tons of mines group. Coal Sci Technol 43(8):97–101

Sasaoka T, Karian T, H A, Shimada H, Matsui K (2016) Application of highwall mining system in weak geological condition. Int J Coal Sci Technol 3(3):311–321

Wang GF (2010) Innovation and development of completed set equipment and technology for high efficient coal mining face in underground mine. Coal Sci Technol 38(1):63–68

Wang GF (2013) Development of fully mechanized coal mining technology and equipment. Coal Sci Technol 41(9):44–48

Wang GF (2014) Theory system of working face support system and hydraulic roof support technology. J China Coal Soc 39(8):1539–1601

Wang GF, Pang YH (2015) Relationship between hydraulic support and surrounding rock coupling and its application. J China Coal Soc 40(1):30–34

Wang GF, Pang YH (2016) Shield-roof adaptability evaluation method based on coupling of parameters between shield and roof strata. J China Coal Soc 41(6):1348–1353

Wang YH, Guo DZ, Zhang HR, Shen BG (2006) Spatial distribution and application of coal resource potential in China. J Nat Resour 21(2):225–230

Wang GF, Liu JF, Ren HW (2011) Design and optimization of high seam-caving coal hydraulic support based on model of support and wall rock coupling. J China Coal Soc 36(1):145–151

Wang GF, Pang YH, Liu JF (2012) Determination and influence of cutting height of coal by top coal caving method with great mining height in extra thick coal seam. J China Coal Soc 37(11):1777–1782

Xu QY, Zhao YJ, Li YM (2015) Statistical analysis and precautions of coal mine accidents in China. Coal Eng 47(3):80–82

Permissions

All chapters in this book were first published in IJCS&T, by Springer; hereby published with permission under the Creative Commons Attribution License or equivalent. Every chapter published in this book has been scrutinized by our experts. Their significance has been extensively debated. The topics covered herein carry significant findings which will fuel the growth of the discipline. They may even be implemented as practical applications or may be referred to as a beginning point for another development.

The contributors of this book come from diverse backgrounds, making this book a truly international effort. This book will bring forth new frontiers with its revolutionizing research information and detailed analysis of the nascent developments around the world.

We would like to thank all the contributing authors for lending their expertise to make the book truly unique. They have played a crucial role in the development of this book. Without their invaluable contributions this book wouldn't have been possible. They have made vital efforts to compile up to date information on the varied aspects of this subject to make this book a valuable addition to the collection of many professionals and students.

This book was conceptualized with the vision of imparting up-to-date information and advanced data in this field. To ensure the same, a matchless editorial board was set up. Every individual on the board went through rigorous rounds of assessment to prove their worth. After which they invested a large part of their time researching and compiling the most relevant data for our readers.

The editorial board has been involved in producing this book since its inception. They have spent rigorous hours researching and exploring the diverse topics which have resulted in the successful publishing of this book. They have passed on their knowledge of decades through this book. To expedite this challenging task, the publisher supported the team at every step. A small team of assistant editors was also appointed to further simplify the editing procedure and attain best results for the readers.

Apart from the editorial board, the designing team has also invested a significant amount of their time in understanding the subject and creating the most relevant covers. They scrutinized every image to scout for the most suitable representation of the subject and create an appropriate cover for the book.

The publishing team has been an ardent support to the editorial, designing and production team. Their endless efforts to recruit the best for this project, has resulted in the accomplishment of this book. They are a veteran in the field of academics and their pool of knowledge is as vast as their experience in printing. Their expertise and guidance has proved useful at every step. Their uncompromising quality standards have made this book an exceptional effort. Their encouragement from time to time has been an inspiration for everyone.

The publisher and the editorial board hope that this book will prove to be a valuable piece of knowledge for researchers, students, practitioners and scholars across the globe.

List of Contributors

Mayuri Chabukdhara
Department of Environmental Biology and Wildlife Sciences, Cotton College State University, Guwahati, Assam 781001, India

O. P. Singh
Department of Environmental Studies, North Eastern Hill University, Shillong, Meghalaya 793022, India

Peter Wirth, Ralf-Uwe Syrbe and Wolfgang Wende
Leibniz Institute of Ecological Urban and Regional Development (IOER), Weberplatz 1, 01217 Dresden, Germany

Jiang Chang and Tinghao Hu
China University of Mining and Technology (CUMT), Daxue Road 1, Xuzhou 221116, China

Zhenqi Hu, Chao Chen, Wu Xiao and Mingjie Gao
Institute of Land Reclamation and Ecological Restoration, China University of Mining and Technology (Beijing), Beijing, China

Xinjing Wang
College of Resource and Environment, North China University of Water and Electric Power, Zhengzhou, China

Brenda K. Schladweiler
BKS Environmental Associates, Inc. and Accord Resource Solutions, LLC, Gillette, WY 82717, USA

Jinhua Wang, Hongpu Kang, Guofa Wang, Debing Mao and Pengfei Jiang
China Coal Technology & Engineering Group, Beijing 100013, China

Bin Yu
Datong Coal Mine Group Co., Ltd., Datong 037003, China

Yuntao Liang
CCTEG Shenyang Research Institute, Shenyang 110016, China

Limin Fan
Key Laboratory of Mine Geological Hazard Mechanism and Control, Ministry of Land and Resources, Xi'an 710054, China
Shaanxi Institute of Geo-Environment Monitoring, Xi'an 710054, China

Xiongde Ma
College of Environmental Science and Engineering, Chang'an University, Xi'an 710054, China

Liang Yuan
State Key Laboratory of Deep Coal Mining & Environment Protection, Huainan Mining Group Co., Ltd., Huainan 232001, China

Jiachen Wang, Renlun Wu and Peng Zhang
College of Resources and Safety Engineering, China University of Mining & Technology (Beijing), Beijing 100083, China
State Key Laboratory of Coal Resources and Safe Mining, China University of Mining & Technology (Beijing), Beijing 100083, China

Cansın Yüksel
Resource Engineering Section, Department of Geoscience and Engineering, Delft University of Technology, Stevinweg 1, 2628 CN Delft, The Netherlands

Jörg Benndorf
Institute for Mine Surveying and Geodesy, Faculty of Geosciences, Geoengineering and Mining, University of Technology Bergakademie Freiberg, Reiche Zeche, Fuchsmühlenweg 9, 09599 Freiberg, Germany

Matthias Lindig and Oliver Lohsträter
Mitteldeutsche Braunkohlengesellschaft mbH, Glück-Auf-Straße 1, 06711 Zeitz, Germany

Yi Luo
Department of Mining Engineering, West Virginia University, Morgantown, WV, USA

Heena Dhawan and D. K. Sharma
Centre for Energy Studies, Indian Institute of Technology, New Delhi 110016, India

Sreedevi Upadhyayula
Department of Chemical Engineering, Indian Institute of Technology, New Delhi 110016, India

Nong Zhang, Fei Xue, Nianchao Zhang and Xiaowei Feng
School of Mines, Key Laboratory of Deep Coal Resource Mining, Ministry of Education, China University of Mining & Technology, Xuzhou 221116, China

Takashi Sasaoka, Tri Karian, Akihiro Hamanaka, Hideki Shimada and Kikuo Matsui
Department of Earth Resources Engineering, Kyushu University, Fukuoka 819-0395, Japan

Chad O. Hargrave, Craig A. James and Jonathon C. Ralston
Mining and Processing Technologies Research Group, Coal Mining Research Program, CSIRO Energy, Brisbane, Australia

Guofa Wang and Yihui Pang
Coal Mining and Design Branch, China Coal Research Institute, Beijing 100013, China

Index